KU-429-202

ALLE ZEIT WACH
1842

A. F. Williams

A Theoretical Approach to
Inorganic Chemistry

With 144 Figures

Springer-Verlag Berlin Heidelberg New York 1979

Dr. A.F. Williams
Chef de Travaux, Département de Chimie Minérale, Analytique et Appliquée, Université de Geneve, 30 quai Ernest-Ansermet, CH - 1211 Genève

ISBN 3-540-09073-8 Springer-Verlag Berlin Heidelberg NewYork
ISBN 0-387-09073-8 Springer-Verlag NewYork Heidelberg Berlin

Library of Congress Cataloging in Publication Data.
Williams, Alan Francis, 1950-. A theoretical approach to inorganic chemistry.
Includes bibliographies and index. 1. Chemistry, Inorganic. 2. Molecular orbitals.
3. Quantum chemistry. I. Title. OD152.3.W54 546 78-26415

Offsetprinting and Binding: Konrad Triltsch, Würzburg
2152/3020-543210

Foreword

Dr. Alan Williams has acquired a considerable experience in work with transition metal complexes at the Universities of Cambridge and Geneva. In this book he has tried to avoid the variety of ephemeral and often contradictory rationalisations encountered in this field, and has made a careful comparison of modern opinions about chemical bonding. In my opinion this effort is fruitful for all students and active scientists in the field of inorganic chemistry. The distant relations to group theory, atomic spectroscopy and epistemology are brought into daylight when Dr. Williams critically and pedagogically compares quantum chemical models such as molecular orbital theory, the more specific L.C.A.O. description and related "ligand field" theory, the valence bond treatment (which has conserved great utility in antiferromagnetic systems with long internuclear distances), and discusses interesting, but not too well-defined concepts such as electronegativity (also derived from electron transfer spectra), hybridisation, and oxidation numbers.

The interdisciplinary approach of the book shows up in the careful consideration given to many experimental techniques such as vibrational (infra-red and Raman), electronic (visible and ultraviolet), Mössbauer, magnetic resonance, and photoelectron spectra, with data for gaseous and solid samples as well as selected facts about solution chemistry. The book could not have been written a few years ago, and is likely to remain a highly informative survey of modern inorganic chemistry and chemical physics.

Geneva, January 1979

C.K. Jørgensen

Introduction

"It is in the nature of an hypothesis, when once a man has conceived it, that it assimilates everything to itself, as proper nourishment, and from the first moment of your begetting it, it generally grows the stronger by everything you see, hear, read or understand. This is of great use."

Laurence Sterne
'The life and opinions of Tristram Shandy'

It has become almost traditional to begin inorganic chemistry books with a remark on the growth of the subject in the past 25 years. For the student of chemistry, this has resulted in a great increase in the amount of material he has to learn, much of which is separated into apparently unrelated topics such as ligand field theory, electron deficient compounds, the ionic model, etc. This book is intended to show that an approach that takes as its starting point the elementary molecular orbital model of the chemical bond may be developed painlessly to cover the whole subject, and to throw into relief the particularities, differences and similarities of the various sub-divisions of inorganic chemistry. I hope that this will not only aid the understanding and the memorisation of inorganic chemistry, but will also help the reader develop a certain chemical intuition. The book is intended to illustrate the use of theoretical models in studying chemistry, and *not* the use of chemistry in supplying facts for the delectation of theories.

The theoretical approach I advocate is essentially qualitative, and is intended to provide a basis for the subject as a whole, rather than an accurate method of calculation or prediction in a narrow field. The theories of chemical bonding themselves are not particularly complicated, but their application requires a critical understanding, and a good deal of common sense. I have attempted to follow a logical course in presenting the various subjects, but it should be borne in mind that the division into chapters is artificial: the reactivity of a molecule is directly related to its electronic structure, even if the subjects are discussed in different chapters. Certain sections treating more advanced or more mathematical topics are marked with an asterisk, and may be skipped at a first reading.

The quantum mechanics and group theory used in the book are introduced in the first chapter. I have used a certain amount of mathematics wherever it simplifies the discussion, but have also included non-mathematical summaries at the end of each section as it is important that the reader has a sound understanding of the physical principles to follow the remainder of the book. It is interesting to note that the 'complicated' mathematical part of the subject dates mostly from the 1920's, whilst the more qualitative application of quantum mechanics is much more recent.

Chapter 2 introduces L.C.A.O. molecular orbital theory for simple molecules, and endeavours to emphasise the physical principles of the method; in Chap. 3 this approach is used to describe the electronic structures of a wide variety of inorganic compounds. Chapter 4 treats the electronic spectra and magnetic properties of inorganic compounds, and is concerned mainly with d and f block elements. Chapter 5 discusses some other theories of chemical bonding, the use of thermodynamic data in inorganic chemistry, and some of the general concepts often used in descriptive chemistry. The relationship between electronic structure and reactivity is discussed in Chap. 6. Chapter 7 is an illustrative chapter showing the application of the approach introducted in previous chapters to the hard facts of descriptive chemistry. The final chapter gives a résumé of the spectroscopic methods referred to in the text, and discusses their chemical usefulness.

The book presents an approach to the subject, and not a complete treatment of inorganic chemistry, an impossible task in a book of this length. I have given references to more detailed treatments of the topics discussed in the bibliographies at the end of each chapter. At the end of each chapter there are also a few problems which further illustrate points discussed in the text, and indicate other applications. Most problems require only a few moments of reflection, and I hope that the reader will look at them as they are intended to encourage the use of his own critical faculties and common sense. Those nervous about quantum mechanics may find the first two chapters the most difficult, but, if they can understand the physical principles introduced therein, the rest of the book may be followed with little difficulty; the word 'theoretical' is not included in the title as a euphemism for complicated. I hope that the book will give a wide view of the subject, and will serve as a useful complement to more detailed descriptive studies of inorganic chemistry.

I should like to thank the Master and Fellows of Emmanuel College, Cambridge whose award of a Research Fellowship enabled me to take the opportunity of writing this book, and Professor *W. Haerdi*, Director of the Département de Chimie Minérale, Analytique et Appliquée of the University of Geneva where the book was finished. I am particularly grateful to Professor *C.K. Jørgensen* of Geneva, who kindly wrote the Foreword, and Dr. *A.G. Maddock* of Cambridge, both of whom not only read the whole manuscript and made many helpful comments and corrections, but also, by their enthusiasm for the subject and their willingness to discuss it, have made a substantial contribution to such knowledge of inorganic chemistry as the author may possess. I also thank Drs. *L. Balsenc, U. Burger*, Professor *M. Marcantonatos*, and Dr. *V. Parthasarathy* for having read and commented on parts of the book, and Dr. *N. Thalmann-Magnenat* for the figures reproduced from her thesis. Such errors and blemishes as remain are due to the author's intransigence or ignorance.

Two of the less agreeable aspects of writing a book are the effects in the author's temper and the volume of typing produced: both difficulties were faced with great tolerance by my wife, and this book is accordingly dedicated to her with apologies.

Geneva, January 1979

A.F. Williams

Contents

Sections marked with an asterisk may be omitted at the first reading.

Sections marked with an asterisk may be omitted at the first reading.

1. Quantum Mechanics and Atomic Theory

Inorganic chemistry is concerned with the chemistry of over an hundred elements, forming compounds whose stability ranges from that of mountains and minerals to species with lifetimes of less than a millisecond. Clearly, if the subject is to be more than a vast catalogue of apparently unrelated facts, we must seek a theoretical foundation which will enable us to rationalise and relate as many observations as possible. From this point of view, the development of ideas of chemical periodicity (by Mendeleyev and others) during the latter half of the nineteenth century stands as the starting point of theoretical inorganic chemistry. The first periodic table was drawn up on the basis experimental observations; soon after the introduction of quantum theories of the atom, it was shown by Rutherford and Bohr that the same table could be derived from the electronic structure of the individual elements. Following this demonstration, all theories of chemistry have been based more or less rigorously on the quantum theory of matter.

Exact calculations in quantum chemistry rapidly become very complicated, but the remarkable success of quantum mechanics in explaining spectra, and the reasonable success of approximate calculations give a sound justification for the use of quantum mechanics as a starting point. The quality of quantum chemical calculations is improving steadily, and for compounds of the first row of the Periodic Table, the accuracy and predictive power of some methods are now chemically useful. Nonetheless, for most inorganic compounds, calculations are only of value when a large number of effects can either be ignored, satisfactorily approximated, or replaced by experimentally determined values.

The approach in this book will be mainly qualitative, and we shall be more interested in physical principles than in mathematical details; we will try to justify the assumptions and approximations made by recourse to experimental evidence, most frequently that obtained from spectroscopic measurements. This chapter is concerned with establishing the elements of quantum mechanics that we shall need, the use of symmetry to simplify our calculations, and the approximate quantum treatment of atomic structure. Those who find quantum mechanics frightening should note that there is a simplified summary at the end of each section and that sections marked with an asterisk are rather more mathematical and may be omitted at a first reading. This is not a quantum chemistry book, and the treatment given is very brief; those completely unfamiliar with quantum mechanics may wish to consult one of the many introductions to quantum chemistry (see Bibliography, page 37).

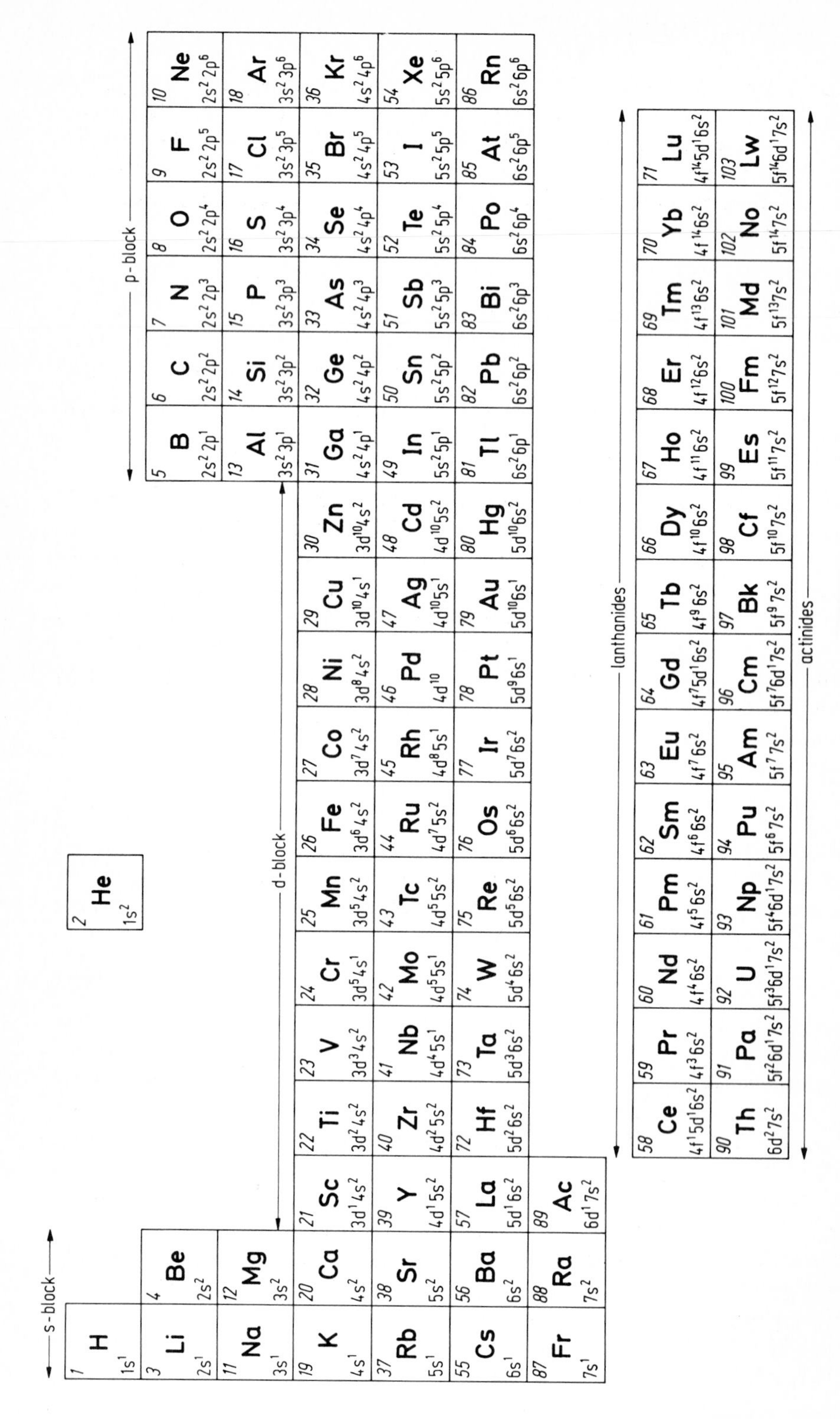

s-block		d-block										p-block					
1 H $1s^1$																	2 He $1s^2$
3 Li $2s^1$	4 Be $2s^2$											5 B $2s^2 2p^1$	6 C $2s^2 2p^2$	7 N $2s^2 2p^3$	8 O $2s^2 2p^4$	9 F $2s^2 2p^5$	10 Ne $2s^2 2p^6$
11 Na $3s^1$	12 Mg $3s^2$											13 Al $3s^2 3p^1$	14 Si $3s^2 3p^2$	15 P $3s^2 3p^3$	16 S $3s^2 3p^4$	17 Cl $3s^2 3p^5$	18 Ar $3s^2 3p^6$
19 K $4s^1$	20 Ca $4s^2$	21 Sc $3d^1 4s^2$	22 Ti $3d^2 4s^2$	23 V $3d^3 4s^2$	24 Cr $3d^5 4s^1$	25 Mn $3d^5 4s^2$	26 Fe $3d^6 4s^2$	27 Co $3d^7 4s^2$	28 Ni $3d^8 4s^2$	29 Cu $3d^{10} 4s^1$	30 Zn $3d^{10} 4s^2$	31 Ga $4s^2 4p^1$	32 Ge $4s^2 4p^2$	33 As $4s^2 4p^3$	34 Se $4s^2 4p^4$	35 Br $4s^2 4p^5$	36 Kr $4s^2 4p^6$
37 Rb $5s^1$	38 Sr $5s^2$	39 Y $4d^1 5s^2$	40 Zr $4d^2 5s^2$	41 Nb $4d^4 5s^1$	42 Mo $4d^5 5s^1$	43 Tc $4d^5 5s^2$	44 Ru $4d^7 5s^2$	45 Rh $4d^8 5s^1$	46 Pd $4d^{10}$	47 Ag $4d^{10} 5s^1$	48 Cd $4d^{10} 5s^2$	49 In $5s^2 5p^1$	50 Sn $5s^2 5p^2$	51 Sb $5s^2 5p^3$	52 Te $5s^2 5p^4$	53 I $5s^2 5p^5$	54 Xe $5s^2 5p^6$
55 Cs $6s^1$	56 Ba $6s^2$	57 La $5d^1 6s^2$	72 Hf $5d^2 6s^2$	73 Ta $5d^3 6s^2$	74 W $5d^4 6s^2$	75 Re $5d^5 6s^2$	76 Os $5d^6 6s^2$	77 Ir $5d^7 6s^2$	78 Pt $5d^9 6s^1$	79 Au $5d^{10} 6s^1$	80 Hg $5d^{10} 6s^2$	81 Tl $6s^2 6p^1$	82 Pb $6s^2 6p^2$	83 Bi $6s^2 6p^3$	84 Po $6s^2 6p^4$	85 At $6s^2 6p^5$	86 Rn $6s^2 6p^6$
87 Fr $7s^1$	88 Ra $7s^2$	89 Ac $6d^1 7s^2$															

lanthanides													
58 Ce $4f^1 5d^1 6s^2$	59 Pr $4f^3 6s^2$	60 Nd $4f^4 6s^2$	61 Pm $4f^5 6s^2$	62 Sm $4f^6 6s^2$	63 Eu $4f^7 6s^2$	64 Gd $4f^7 5d^1 6s^2$	65 Tb $4f^9 6s^2$	66 Dy $4f^{10} 6s^2$	67 Ho $4f^{11} 6s^2$	68 Er $4f^{12} 6s^2$	69 Tm $4f^{13} 6s^2$	70 Yb $4f^{14} 6s^2$	71 Lu $4f^{14} 5d^1 6s^2$
90 Th $6d^2 7s^2$	91 Pa $5f^2 6d^1 7s^2$	92 U $5f^3 6d^1 7s^2$	93 Np $5f^4 6d^1 7s^2$	94 Pu $5f^6 7s^2$	95 Am $5f^7 7s^2$	96 Cm $5f^7 6d^1 7s^2$	97 Bk $5f^9 7s^2$	98 Cf $5f^{10} 7s^2$	99 Es $5f^{11} 7s^2$	100 Fm $5f^{12} 7s^2$	101 Md $5f^{13} 7s^2$	102 No $5f^{14} 7s^2$	103 Lw $5f^{14} 6d^1 7s^2$
actinides													

Table 1.1. The periodic Table. The atomic numbers and valence shell electron configurations of the elements

A. Elements of Quantum Mechanics

Quantum mechanics postulates that any system may be completely described by a wave function ψ which is a function of all the variables of the system. For an isolated hydrogen atom, if the nucleus is taken as the origin of the coordinate system, the wave function will be a function only of the coordinates of the electron. We require for all wave functions discussed in this book that they be solutions of the time independent Schrödinger equation:

$$\mathscr{H}\psi = E\psi \tag{1.1}$$

where ψ is the wave function of the system we are discussing, E is the energy of the system described by ψ, and $\mathscr{H}$ is a well-defined mathematical operator called the Hamiltonian. An *operator* is a mathematical entity which, acting on a function, changes it (for example, in the expression $\frac{d}{dx}(f(x))$, $\frac{d}{dx}$ is an operator, since the result is a new function; similarly, $1/x$ may be regarded as an operator, since the product $\frac{1}{x} \cdot f(x)$ gives a new function). Operators are generally denoted by script letters. The Hamiltonian operator, which is related to the energy of the system, is constructed according to a set of quantum mechanical rules and consists of a sum of terms representing every contribution to the energy of the system. Thus, for the hydrogen atom, the operator will have components from the kinetic energy of the electron, and the potential energy of the electron nucleus interaction. A hydride ion (H^-) with two electrons will have kinetic energy terms (one for each electron), and potential energy terms due to electron-nuclear and electron-electron interactions.

The Schrödinger equation is a specific example of a general mathematical equation known as an *eigenvalue* equation, which has the form:

$$(\text{operator}) \cdot (\text{function}) = (\text{number}) \cdot (\text{the same function}) \tag{1.2}$$

In general, only a certain set of functions will obey Eq. (1.2), and when the product of an operator and a function gives the same function multiplied by a simple number, the function is said to be an *eigenfunction* of the operator, and the simple number is known as the *eigenvalue.* Thus:

$$(\text{operator}) \cdot (\text{eigenfunction}) = (\text{eigenvalue}) \cdot (\text{eigenfunction}) \tag{1.3}$$

The Schrödinger equation is thus nothing more than the eigenvalue equation of the energy operator, the Hamiltonian; furthermore, since the equation will hold only for specific values of E and ψ, we see that we have arrived at a quantisation of energy levels. The Eq. (1.3) is frequently found in quantum mechanics, and each experimentally observable quantity is the eigenvalue of a corresponding operator. If a given wave function is not an eigenfunction of a particular operator, then the experimental observable associated with the operator is not well defined; thus, if a wave function is *not* an eigenfunction of the Hamiltonian, then the energy of the system is not well defined, and is *indeterminate,* i.e. an exact value cannot be measured. Our requirement that the Schrödinger equation be obeyed is thus a requirement that our system has a well defined energy which does not change with time.

A wave function which is a solution to the Schrödinger equation (and henceforth, we shall use the term wave function only for such solutions) may also be an eigenfunction for other operators. In the case of the hydrogen atom, the wave functions are also eigenfunctions of the orbital angular momentum operators $\mathscr{L}^2$ and $\mathscr{L}_z$ (the square of the orbital angular momentum, and its component along one, arbitrarily chosen axis), and consequently the wave functions have well defined and measurable values of these quantities. The condition for a function to be an eigenfunction of two operators O_1, O_2 is that O_1 followed by O_2 has the same result as O_2 followed by O_1, or that O_1 and O_2 *commute.* The commutativity of operators, and consequent measurability of their observables is closely related to Heisenberg's uncertainty principle.

If every physically observable quantity is the eigenvalue of its corresponding operator, what is the physical interpretation of the wave function itself? Born suggested that the value of the square of the wave function ψ^2 (for complex wave functions, the square of the modulus $\psi^*\psi$ or $|\psi|^2$) of a particle at a point **r** is the probality that the particle is at that point. Turning again to the hydrogen atom, the value of $|\psi|^2$ at a given point is the probability of finding the electron there. However, Heisenberg's uncertainty principle warns us against regarding the electron as localised at a particular point; we should do better to regard the electron as having a certain probality density within a certain volume element which is equal to the integral of $|\psi|^2$ within that volume. The concept of electron density is extremely useful in discussing electronic wave functions, and we adopt Born's interpretation without further question.

If the electron exists at all, the sum of its probability density over all space must equal 1. This may be expressed by the integral

$$\int_{\text{all space}} \psi^* \cdot \psi \, \mathrm{d}\tau = 1 \tag{1.4}$$

The function ψ is said to be *normalised.* Since $\mathscr{H}\psi = E\,\psi$, multiplying both sides of Eq. (1.1) by ψ^* gives:

$$\psi^* \cdot \mathscr{H}\psi = \psi^* E\,\psi$$

We may now integrate both sides of this equation over all space, noting that E is a number and may therefore be taken outside the integral to give:

$$\int_{\text{all space}} \psi^* \mathscr{H} \psi \, \mathrm{d}\tau = E \int_{\text{all space}} \psi^* \psi \, \mathrm{d}\tau = E \tag{1.5}$$

We have thus obtained an explicit expression for the energy of the system.

The wave function ψ must satisfy certain conditions: it must be single valued, since at any point there can only be one value of the probability; it must be continuous, as must its first derivative (i.e. $\frac{\mathrm{d}\,\psi}{\mathrm{d}x}$); it must obey the boundary conditions of the system. Thus, for the hydrogen atom, the wave function must fall to zero at infinite distance from the nucleus for the electron clearly has an infinitely small probability density at this distance. The imposition of boundary conditions in Eq. (1.1) severely restricts the number of solutions, and is of vital importance in establishing quantisation of energy.

a) The Hydrogen Atom

This is the only system of chemical significance for which the Schrödinger equation can be solved exactly, apart from the trivial extensions to He^+, Li^{2+}, Be^{3+} etc., all having only a nucleus and one electron. The exact calculation is found in many texts[1], so let us look only at the qualitative features. First, we must construct the Hamiltonian operator – this contains a term for the kinetic energy of the electronic motion, and also for the electron nucleus attraction. This second term ($-e^2/4\pi\epsilon_0 \mathbf{r}$) falls off as $1/\mathbf{r}$, the distance of the electron from the nucleus; since this function ($1/\mathbf{r}$) has spherical symmetry about the nucleus, it will clearly be a good idea to adopt a coordinate system which reflects this. We therefore place the nucleus at the origin, and solve the equation using spherical polar coordinates, r, θ, ϕ for the electron's position. The equation is thus:

$$(\mathcal{H}_{\text{kin}} + \mathcal{H}_{\text{electron-nucleus}})\,\psi(\mathrm{r}, \theta, \phi) = E\,\psi(\mathrm{r}, \theta, \phi) \tag{1.6}$$

The solutions are given by:

$$\psi(\mathrm{r}, \theta, \phi) = R(n, l, \mathrm{r})\; Y(l, m, \theta, \phi) \tag{1.7}$$

where for Eq. (1.6) to hold, n, l, and m have integral values, and specify the forms of the two functions R and Y; it is more usual to specify these functions by writing n, l, and m as subscripts:

$$\psi_{nlm}(\mathrm{r}, \theta, \phi) = R_{nl}(\mathrm{r}) \cdot Y_{lm}(\theta, \phi) \tag{1.8}$$

n, l, and m are quantum numbers which specify the wave function. It is found that the energy E_{nlm} corresponding to the wave function ψ_{nlm} is

$$E_{nlm} = \mathrm{K}/n^2 \tag{1.9}$$

where K is a product of various fundamental quantities. Since E does not depend on l and m, the functions ψ_{nlm} correspond to the same energy of the system for all l and m if n is fixed. However, Eq. (1.8) is only valid if l and m have integral values obeying the following rules:

$$0 \leqslant l < n \;\; ; -l \leqslant \mathrm{m} \leqslant +l$$

thus for fixed values of n and l there are $(2l+1)$ solutions of the Schrödinger equation. For fixed n there are $\sum_{0}^{n-1}(2l+1) = n^2$ separate solutions all with the same energy. Separate solutions of the Schrödinger equation with the same energy are said to be degenerate. Thus we may say that solutions of the Schrödinger equation for the isolated hydrogen atom are of the form $\psi_{nlm}(\mathrm{r}, \theta, \phi) = \mathrm{R}_{nl}(\mathrm{r})\,\mathrm{Y}_{lm}(\theta, \phi)$, are degenerate for all l, m, given fixed n, and that $0 \leqslant l < n$ and $-l \leqslant m \leqslant l$, and that the energy of these solutions is given by:

$$E_{\mathrm{nlm}} = \int_{\text{all space}} \psi^*_{nlm}(\mathrm{r}, \theta, \phi)\,\mathcal{H}\,\psi_{nlm}(\mathrm{r}, \theta, \phi) = \mathrm{K/n^2} \tag{1.10}$$

Dirac introduced a very elegant system of notation which avoids the continual use of subscripts and integral signs:

(i) ψ_a is denoted by $|\,a\,\rangle$

(ii) ψ^*_a is denoted by $\langle\, a\,|$

(iii) $\int_{\text{all space}} \psi^*_a\,\psi_b$ is denoted by $\langle\, a \,|\, b\,\rangle$

Thus we may rewrite Eq. (1.10) as:

$$E_{nlm} = \langle\, nlm \,|\, \mathcal{H} \,|\, nlm\,\rangle = \mathrm{K}/n^2 \tag{1.10a}$$

[1] For example, see Murrell, Kettle, and Tedder (see Bibliography, page 37).

This 'bracket' notation is extremely convenient and we shall use it extensively.

Following Eq. (1.4), it is found that $< nlm \mid nlm > = 1$; this is the normalisation condition. If a wave function ψ_a is not normalised, so that $< a \mid a > = \alpha$ (for example), then the wave function $\psi_b = (1/\sqrt{\alpha})\ \psi_a$ will be normalised ($< b \mid b > = 1$). This is clearly a trivial alteration, and we shall henceforth assume all wave functions to be normalised.

Another property of the wave functions ψ_{nlm} is that the integral

$$\int_{\text{all space}} \psi^*_{nlm}\ \psi_{n'l'm'} = < nlm \mid n'l'm' > = 0$$

unless $n = n'$, $l = l'$, $m = m'$. This is a fundamental property of eigenfunctions, known as *orthogonality*. Any eigenfunction corresponding to a given eigenvalue will be *orthogonal* to all others corresponding to a different eigenvalue. For degenerate eigenfunctions, the problem is a little more complicated if two eigenfunctions ψ_a and ψ_b correspond to the same energy E, then, even if $< a \mid b > = 0$, the wave function $\psi_a \cos \alpha + \psi_b \sin \alpha$ will be an eigenfunction of the Hamiltonian for all α, and will not be orthogonal to ψ_a or ψ_b. This might appear to imply an infinite number of eigenfunctions corresponding to an infinite number of values for α. In fact, this is not the case and we require of all eigenfunctions that they are orthogonal to all other eigenfunctions, and we shall assume that all eigenfunctions have been orthogonalised.

A set of eigenfunctions which are orthogonal to each other, and are normalised are said to be *orthonormal.* This relationship may be summarised as

$$< \alpha\beta\gamma \ldots\ldots \mid abc \ldots\ldots > = \delta_{\alpha a}\ \delta_{\beta b}\ \delta_{\gamma c} \ldots\ldots \qquad (1.11)$$

where δ_{ij} is the Kronecker delta symbol such that $\delta_{ij} = 0$ unless $i = j$, when $\delta_{ij} = \delta_{ji} = 1$. All wave functions we discuss will be assumed to be orthonormal.

We have now discussed the general properties of our wave functions – their orthonormality and degeneracy, and such remarks apply to all wave functions, although the form of the solutions will be very different, and the degeneracy is usually much lower than that of the hydrogen atom. We now turn to the actual solutions ψ (r, θ, ϕ). It will be recalled that the wave function could be separated into the product of a radial function of r only, and an angular function of θ and ϕ only. We will discuss these two parts separately. Good discussions and diagrams are given in many physical and theoretical chemistry texts[2] – we discuss here only the fundamentals.

(i) The radial function R (r). We may notice immediately that, since the functions $Y_{lm}(\theta, \phi)$, $Y_{l'm'}(\theta, \phi)$ are orthogonal only if $l \neq l'$, $m \neq m'$, then for orthogonality, we require that the functions R_{nl} (r), $R_{n'l'}$ (r) should be orthogonal for $n \neq n'$. The general form of R_{nl}(r) for a series of n, l is shown in Fig. 1.1, together with the function R^2_{nl} (r) $\cdot$ 4 π r^2, the total electron density at a distance r from the nucleus. It will be seen that there are points where the wave function changes sign; these points are known as nodes, and, for the radial function R (r) form spheres.[3] A function R_{nl} (r) will have $(n-l-1)$ nodes. We label radial wave functions by the convenient (if not logical) notation in which the l value is represented by the letters s, p, d, f, g, ... as $l = 0, 1, 2, 3, 4, \ldots$ The radial function for $n = 4, l = 3$ is thus labelled 4f, for $n = 2, l = 0$ it is labelled 2s. There are three general features to note about the radial function:

---

2 E.g. Atkins; Coulson; Murrell, Kettle, and Tedder (see Bibliography, page 37).

3 Note that the electron has zero probability density at a node.

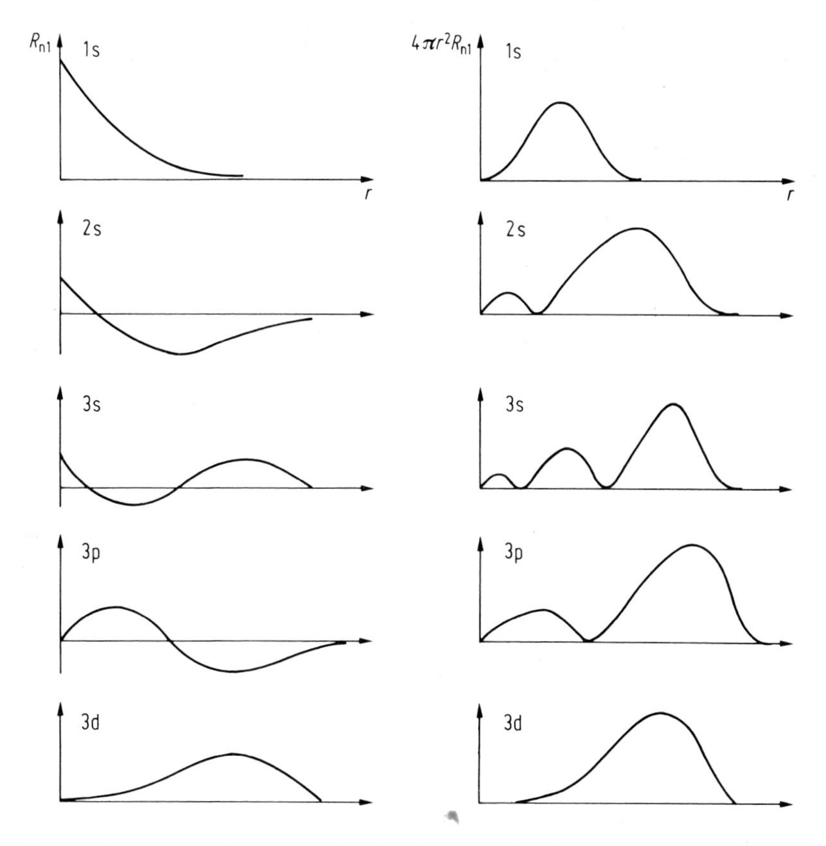

Fig. 1.1. The forms of the radial function R_{nl} (r) and radial distribution function 4π $r^2 R_{nl}$ (r) for various nl values of the hydrogen atom

a) As the value of n increases, the electron density moves away from the nucleus. This corresponds to a higher potential energy, and consequently a higher total energy (to be expected from the variation of E with n).

b) As the value of n increases, the wave function becomes more spread out for a given value of l. Thus the 6p radial function will have 6–1–1 = 4 nodes, five regions of non-zero electron density, the smallest closest to the nucleus, the largest furthest away from the nucleus.

c) All the radial functions have a vanishing electron density at the nucleus, with the exception of s-type functions (l = 0).

(ii) The angular function Y_{lm} (θ, ϕ). The angular functions Y_{lm} (θ, ϕ) are a class of functions well known in mathematics: the spherical harmonics. They have the particular property that the wave function f (r) · Y_{lm} (θ, ϕ) will be an eigenfunction of certain angular momentum operators. The two angular momentum operators $\mathscr{L}^2$ (giving the square of the total electronic angular momentum) and $\mathscr{L}_z$ (giving the value of the component of the angular momentum along one axis, arbitrarily denoted as the z axis) both commute with the Hamiltonian operator. This means that the eigenfunctions of $\mathscr{H}$ are also eigenfunctions of $\mathscr{L}^2$ and $\mathscr{L}_z$ however, unlike the Hamiltonian operator,

the eigenvalues associated with $\mathscr{L}^2$ and $\mathscr{L}_z$ are *not* the same for all *l, m* values. The eigenvalue equations are:

$$\mathscr{L}^2 \psi_{nlm} = \mathscr{L}^2 \mid nlm > = l(l+1)\hbar^2 \mid nlm > \tag{1.12}$$

$$(\hbar = \text{Planck's constant}/2\pi)$$

$$\mathscr{L}_z \mid nlm > = m\hbar \mid nlm > \tag{1.13}$$

Thus, for each ψ_{nlm} there are specific eigenvalues of $\mathscr{L}^2$ and $\mathscr{L}_z$ associated with it, and described by the numbers l and m. We can thus use the numbers l and m to classify the wave functions. In passing, one may note that the operators $\mathscr{L}_x$ and $\mathscr{L}_y$ do not commute with $\mathscr{L}_z$ (or each other), and consequently that the eigenfunctions of $\mathscr{L}_z$ are not those of $\mathscr{L}_x$ and $\mathscr{L}_y$; we can only know the value of one component of the angular momentum associated with the wave function.

The representation of the wave function as $R_{nl}(r) \cdot Y_{lm}(\theta,\phi)$ is extremely convenient for many calculations – for example, if the hydrogen atom is placed in a magnetic field **H**, then the magnetic dipole moment due to the electronic angular momentum will be proportional to the component of the angular momentum parallel to **H** – we may choose the z axis to be parallel to **H** and use $\mathscr{L}_z$:

$$\mu_{\text{mag.}} \propto < nlm \mid \mathscr{L}_z \mid nlm > = m\hbar < nlm \mid nlm > = m\hbar \tag{1.14}$$

the energy of the interaction is then $\mathbf{H}\cdot\mu \propto m\mathbf{H}$; this splitting of levels according to the m value in a magnetic field is known as the Zeeman effect. There is, however the disadvantage that the functions $Y_{lm}(\theta, \phi)$ are complex if $m \neq 0$. In representing these angular wave functions it is thus common practice to take linear combinations which are real and orthogonal. For $l = 1$ (p-type functions) these functions are:

$$\begin{aligned}
p_z &= R_{nl}(r) \cdot Y_{10}(\theta,\phi) &&= (\sqrt{3}/2\pi)\cos\theta \cdot R_{nl}(r)\\
p_x &= R_{nl}(r) \cdot (1/\sqrt{2})\,[Y_{11}(\theta,\phi) + Y_{1\text{-}1}(\theta,\phi)] &&= (\sqrt{3}/2\pi)\sin\theta\cos\phi \cdot R_{nl}(r)\\
p_y &= R_{nl}(r) \cdot i(1/\sqrt{2})\,[Y_{11}(\theta,\phi) - Y_{1\text{-}1}(\theta,\phi)] &&= (\sqrt{3}/2\pi)\sin\theta\sin\phi \cdot R_{nl}(r)
\end{aligned} \tag{1.15}$$

These new functions have well-defined directional properties, and this is of great use in discussing chemical bonding. They are no longer eigenfunctions of the operator $\mathscr{L}_z$, but they are still eigenfunctions of $\mathscr{L}^2$ and the Hamiltonian. The angular part of the p_z function is shown in Fig. 1.2 (i); the p_x and p_y functions are similar, but are directed along the x and y axes respectively. It should be noticed that there is a nodal plane between the two lobes, and that the wave function changes sign on crossing this plane.

A similar operation may be undertaken for the functions $Y_{2m}(\theta, \phi)$, giving the angular variation of d type functions ($l = 2$). The combinations are:

$$\begin{aligned}
&d_{z^2} = Y_{20}\ ;\ d_{xz} = \sqrt{\tfrac{1}{2}}(Y_{21} + Y_{2\text{-}1})\ ;\ d_{yz} = -i\sqrt{\tfrac{1}{2}}(Y_{21} - Y_{2\text{-}1})\ ;\\
&d_{xy} = -i\sqrt{\tfrac{1}{2}}(Y_{22} - Y_{2\text{-}2})\ ;\ d_{x^2\text{-}y^2} = \sqrt{\tfrac{1}{2}}(Y_{22} + Y_{2\text{-}2})
\end{aligned} \tag{1.16}$$

The angular variations are sketched in Fig. 1.2 (ii) for the $d_{x^2\text{-}y^2}$ and d_{z^2} functions. The functions d_{xy}, d_{xz}, and $d_{x^2\text{-}y^2}$ have four lobes and two nodal planes, the lobes pointing between the axes x and y, x and z, y and z for xy, xz, and yz, and along the x and y axes for $d_{x^2\text{-}y^2}$. The d_{z^2} function is somewhat different, having cylindrical symmetry about the z axis, with two large lobes pointing along the z axis, and a torus

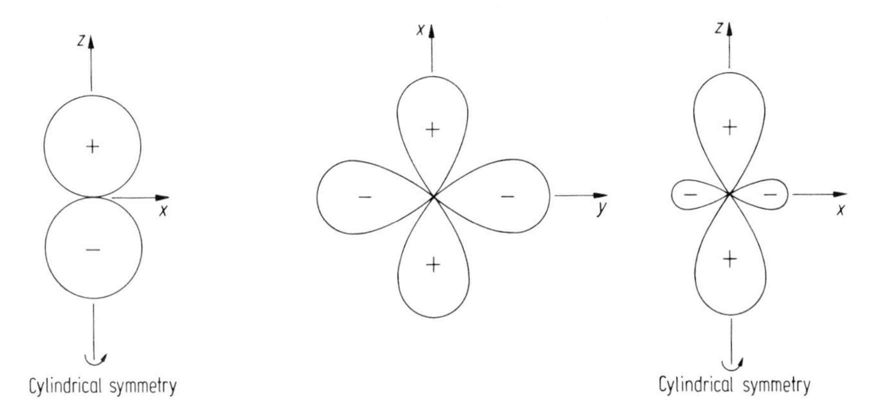

Fig. 1.2. The angular functions for (i) $l = 1$ the p_z function in the xz plane, (ii) $l = 2$ – the $d_{x^2-y^2}$ and d_{z^2} functions in the xz and xy planes respectively.
The distance from the origin of a point on the curve represents the value of the angular function in that direction

(or 'doughnut') in the x-y plane around the z axis. A similar treatment may be extended to the f-type functions ($l = 3$) which have more lobes and nodal planes, but this does not concern us here. It is useful to note that s and d functions are symmetric with respect to inversion in a centre of symmetry, and that p and f functions are anti-symmetric.

The final form of some of the hydrogen wave functions is shown in Fig. 1.3.

Fig. 1.3 The value of the hydrogen wave functions in the xz plane. (i) The 1s function, (ii) the 2s function, (iii) the 3s function, (iv) the $2p_x$ function, (v) the $3d_{z^2}$ function. Reproduced by kind permission of Dr. N. Magnenat-Thalmann.

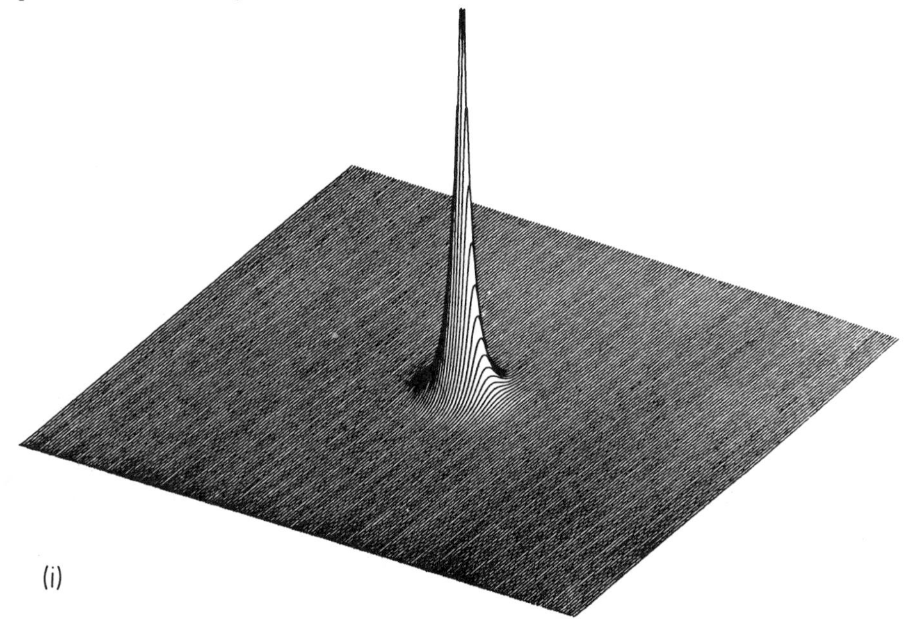

Fig. 1.3. (i) The 1s function

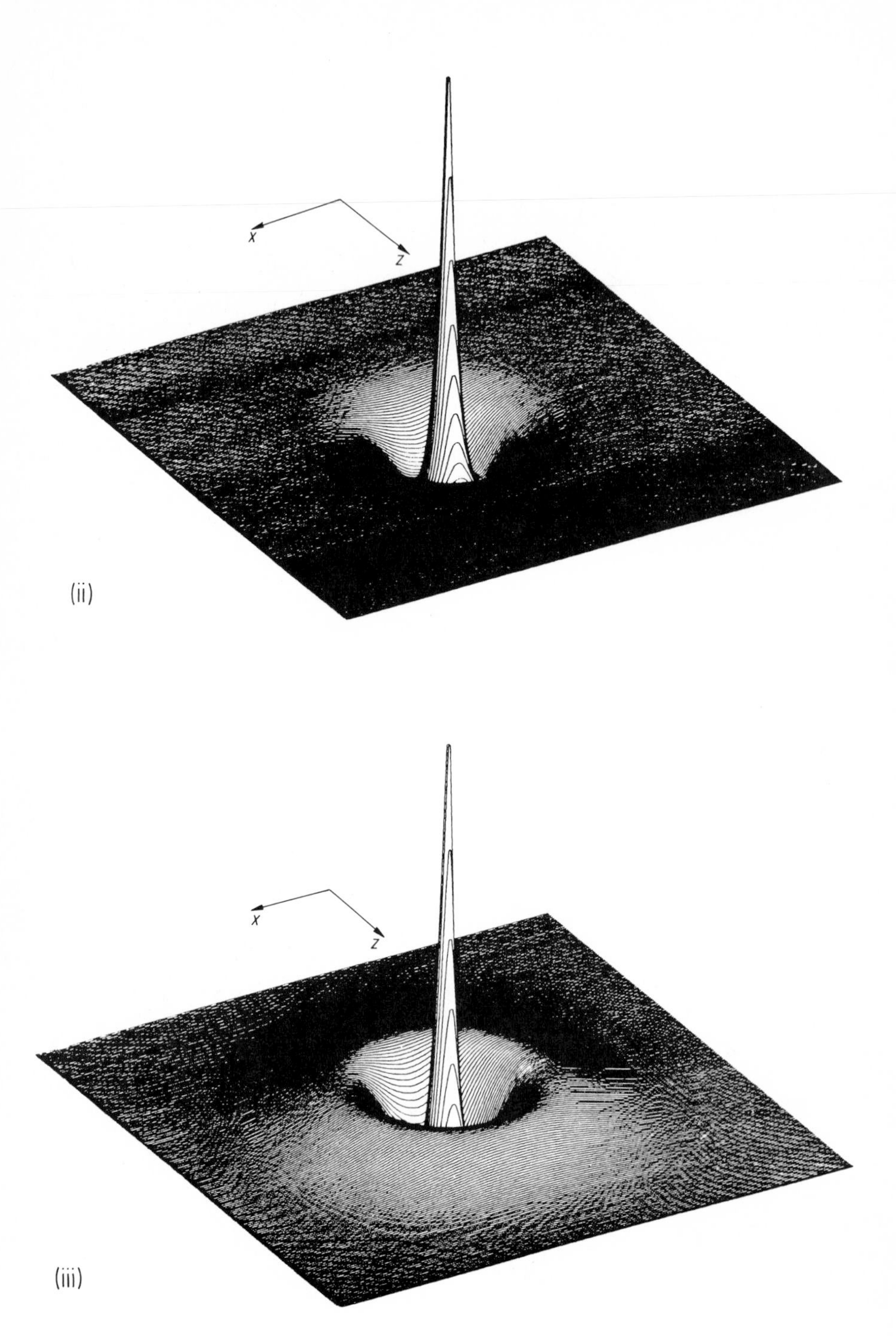

Fig. 1.3. (ii) The 2s function
(iii) The 3s function

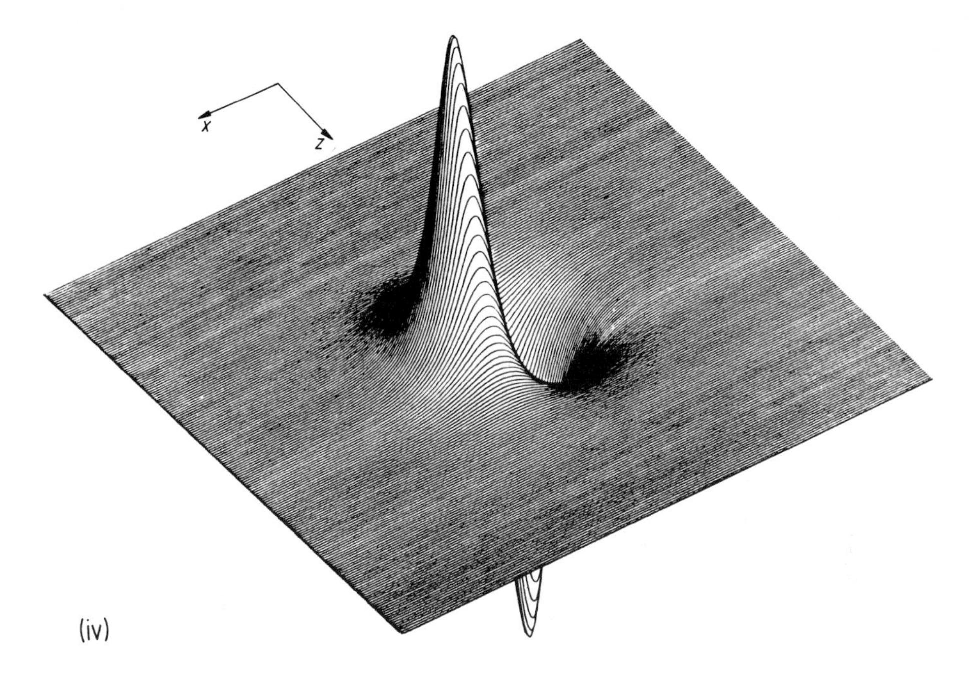

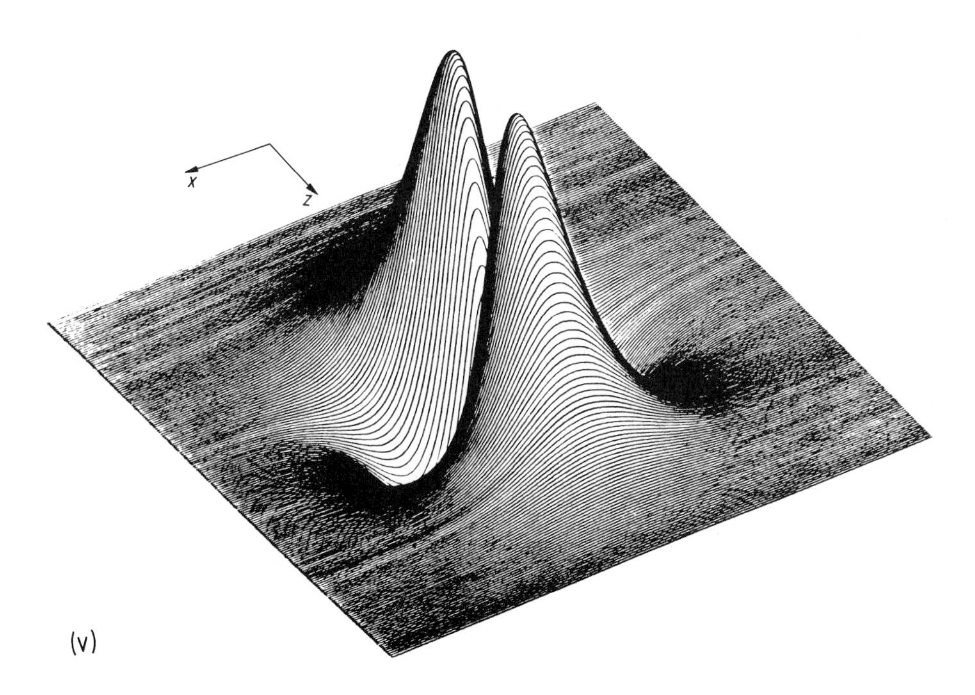

Fig. 1.3. (iv) The $2p_x$ function
(v) The $3d_{z^2}$ function

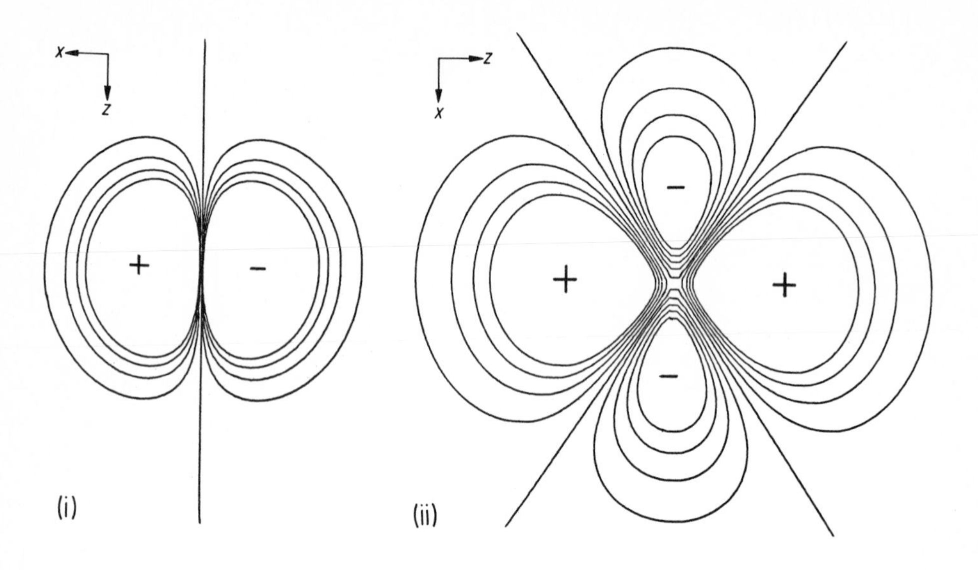

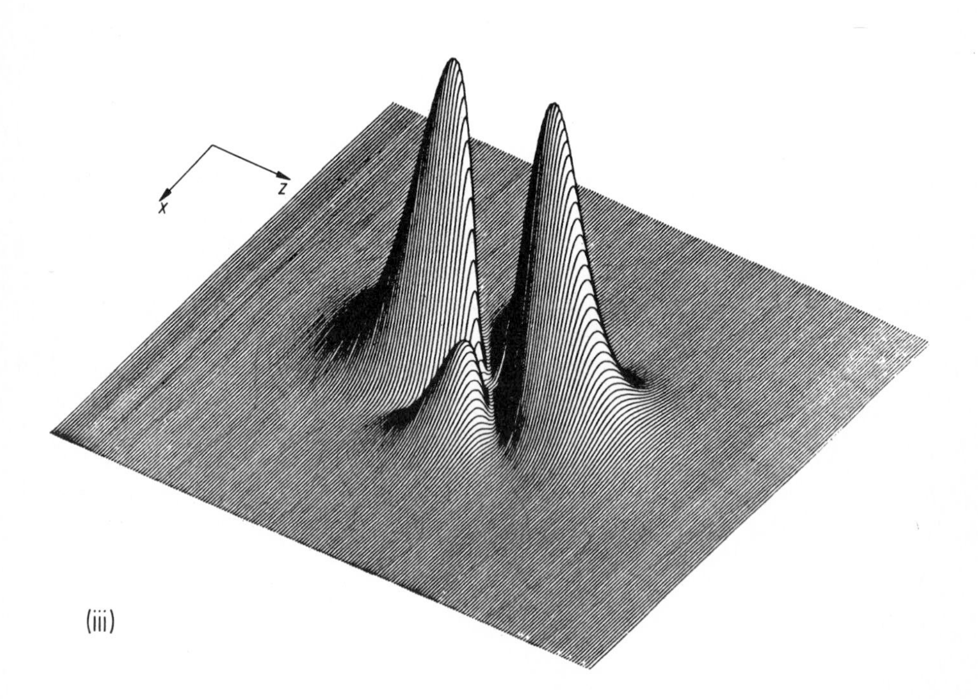

Fig. 1.4. Contour maps of the wave function.
(i) The $2p_x$ function, (ii) the $3d_{z^2}$ function, (iii) shows the value of $|\psi^2|$ in the xz plane for the $3d_{z^2}$ function. Reproduced kind permission of Dr. N. Magnenat-Thalmann

Figure 1.4 shows a 'contour map' for the 2p and $3d_{z^2}$ wave functions and shows the electron density (i.e. $|\psi|^2$) for a $3d_{z^2}$ wave function. We have now discussed almost all the important features of the wave functions of the hydrogen atom, and combine our discussion of the most important remaining property with the last topic of pure quantum mechanics introduced in this section.

b) Spin and Perturbation Theory

It is an experimental observation and also a prediction from Dirac's relativistic theory that an electron possesses spin angular momentum which may be described by a spin quantum number S equal to one half. Spin angular momentum may be treated similarly to electronic angular momentum discussed above, and the operator $\mathcal{S}^2$ leads us to the square of the magnitude of the spin angular momentum, $S(S+1)\hbar^2 = 3/4\,\hbar^2$ for $S = \frac{1}{2}$. If a particular direction z is defined (for example, by application of a magnetic field) the spin angular momentum may have one of two possible values for its component along the z axis, corresponding to $\pm\frac{1}{2}\hbar$. If we denote the eigenfunctions for the two different values by $|\alpha>$ (for $+\frac{1}{2}\hbar$) and $|\beta>$ (for $-\frac{1}{2}\hbar$), then

$$\begin{aligned}
\mathcal{S}^2 |\alpha> &= S(S+1)\hbar^2 |\alpha> = 3/4\,\hbar^2 |\alpha> \\
\mathcal{S}^2 |\beta> &= S(S+1)\hbar^2 |\beta> = 3/4\,\hbar^2 |\beta> \\
\mathcal{S}_z |\alpha> &= \frac{1}{2}\hbar |\alpha> \\
\mathcal{S}_z |\beta> &= -\frac{1}{2}\hbar |\beta>
\end{aligned} \qquad (1.17)$$

The eigenvalue of $\mathcal{S}_z$ is generally referred to as the m_s quantum number.

Associated with this spin is a magnetic dipole; since an electron may also possess a magnetic moment as a result of the angular momentum of its motion, we may expect an interaction or coupling between these two dipoles – this interaction is called *spin-orbit coupling.* Clearly, if we were being exact, we should allow for this in constructing our Hamiltonian, but this would make the calculation more difficult, and we can use an approximate method called perturbation theory.

The principle of perturbation theory is very simple: if we have a set of eigenfunctions of a system close to the one we are considering, we can use them to derive the exact wave function we seek, or at least a better approximation to it, if we know the form of the interaction that is 'perturbing' our original system. In the case of spin-orbit coupling, the magnitude of the interaction is very small, and our original wave function is very nearly exact. We can construct the spin-orbit coupling operator $\mathcal{H}'$, the perturbing Hamiltonian which operates on our unperturbed system: it can be shown to be $\lambda\,\mathcal{L}\cdot\mathcal{S}$ where λ is a constant (the spin orbit coupling constant) and $\mathcal{L}$ and $\mathcal{S}$ the electronic and spin angular momentum operators respectively.

If the initial energy of the system was E_0 and the initial wave function $|a>$, then the first order correction to the energy of the system is

$$E = E_0 + <a|\mathcal{H}'|a> = E_0 + E_1 \qquad (1.18)$$

This essentially corrects the energy for the interaction of the perturbation with the old wave function.

To correct the wave function, we consider the possibility that the old wave function may be mixed with some of the other eigenfunctions of the unperturbed Hamiltonian. Let these other eigenfunctions be $| b >$. Then the first order correction to the wave function is

$$\psi = | a > + \sum_b \frac{< a | \mathcal{H}' | b >}{E_b - E_0} \cdot | b > \tag{1.19}$$

where the summation is over all the eigenfunctions $| b >$ apart from $| a >$. Note that the wave function will be increasingly perturbed as $< a | \mathcal{H} | b >$ increases, and as $| b >$ and $| a >$ move closer in energy. If $E_b = E_0$ it is necessary to use a different approach (see, for example, Atkins), and the mixing may be quite considerable.

Finally we may consider the second order correction to the energy:

$$E = E_0 + E_1 + \sum_b \frac{< a | \mathcal{H}' | b > < b | \mathcal{H}' | a >}{E_b - E_0} = E_0 + E_1 + E_2 \tag{1.20}$$

This considers the change in energy of the system as a result of the change in wave function, that is, the mixing of the other eigenfunctions $| b >$ with $| a >$. It is possible to continue to higher orders of approximation, but those described here are generally sufficient; if they are not, then the system has probably been perturbed so much that the approximations made are invalid.

To recapitulate the principles of the approach: we take a set of eigenfunctions for a system very similar to that we are interested in, and establish the perturbation that will give us a better model of the system. We then calculate the change in energy that this would give if the wave function was unchanged. Next we consider the changes to the wave function due to mixing in of other eigenfunctions, and the changes in energy that this may produce. The mixing effect becomes increasingly important as the functions mixed approach each other in energy. The other term affecting the mixing is $< a | \mathcal{H}' | b >$ – this is often called the matrix element of $\mathcal{H}'$ between $| a >$ and $| b >$ and written $\mathcal{H}'_{ab}$. If $| a >$ and $| b >$ are orthogonal eigenfunctions of $\mathcal{H}'$ then $\mathcal{H}'_{ab}$ is zero and there is no mixing. Expressions of the type $< a | \mathcal{H}' | b >$ are extremely important in quantum chemistry: for example, it can be shown that the probability of the absorption of radiation producing a transition of an electron from one wave function to another (i.e. spectroscopic excitation) is proportional to $| < a | \hat{\mu} | b > |^2$ where μ is the electric dipole moment operator and $| a >$ and $| b >$ are the initial and final wave functions. This quantity, the *transition probability,* must be non-zero for normal optical absorption to take place, and this is the basis of selection rules in spectroscopy. We may also note that if $\mathcal{H}'$ is the unit operator, then $< a | \mathcal{H}' | b > = < a | b > = S_{ab}$, and this integral S_{ab} is known as the *overlap integral* between ψ_a and ψ_b. If ψ_a, ψ_b are orthogonal, then $S_{ab} = 0$; if $a = b$, then $S_{aa} = 1$, that is, the overlap is complete.

Perturbation theory is of great importance in quantum chemistry, and not only as a means of correcting wave functions for small effects such as spin-orbit coupling. If we seek to associate a set of chemical properties with a particular element, then clearly these properties are related to the atomic wave function of that element. We can regard the changes the atom undergoes in forming a compound as a perturbation of this wave function, albeit a fairly drastic one. By starting with the unperturbed atomic wave function, we may derive our new wave function in terms of the atomic wave function of that particular element. We shall discuss this in more detail in Chap. 2.

To return to our discussion of spin-orbit coupling, the strength of the interaction depends on the spin orbit coupling constant λ; it can be shown that λ increases rapidly with the nuclear charge Z, and also (rather less rapidly) with the closeness of the electron to the nucleus. Thus for hydrogen, $\lambda_{2p} > \lambda_{3p}$. λ_{1s} will of course be zero, as there is no angular momentum for an s-type function ($l = 0$). There will be a slight difference in energy between the function $R_{nl}(r)\, Y_{lm}(\theta, \phi) \mid \alpha >$ and $R_{nl}(r)\, Y_{lm}(\theta, \phi) \cdot \mid \beta >$ as a result of the different relative orientations of spin and orbital angular momentum. Thus the perturbation (spin orbit coupling) lifts the degeneracy of two otherwise degenerate orbitals.

The functions $R_{nl}(r)\, Y_{lm}(\theta, \phi) \mid \alpha >$ or $\mid \beta >$ are solutions to the Schrödinger equation for the hydrogen atom characterised by the quantum numbers *n, l, m* and the spin quantum number m_s ($= \frac{1}{2}$ or $-\frac{1}{2}$ corresponding to α or β), and are described as spin orbitals. The concept of an orbital is related to Bohr's idea of an electron orbiting around the nucleus. From every orbital described by *n, l and m* we may obtain two *spin-orbitals* $\psi_{nlm\alpha}$ or $\psi_{nlm\beta}$. The use of spin-orbitals to describe the wave function is highly convient, and we shall make considerable use of this term; we shall also use the simple term orbital, with the tacit understanding that this involves two orthogonal spin-orbitals.

If the spin-orbit coupling constant was larger, then we would need to consider the first order correction to the wave function, and to allow for the corresponding mixing of wave functions; if λ is very large, then the mixing of functions is such that our wave functions are no longer anything like eigenfunctions of the angular momentum operators $\mathscr{L}^2$ and $\mathscr{L}_z$; in such cases, it is no longer possible to classify our wave functions by l and m values, but one can use j values, where j is the quantum number describing the total angular momentum, the vector sum of orbital and spin angular momentum. A purist might argue that we ought to include a second perturbation for the interaction of the total electronic angular momentum with any magnetic dipole due to nuclear spin, but the magnitude of this 'hyperfine interaction' between nucleus and electron is so small that it is generally ignored.

The Schrödinger equation for the hydrogen atom is the only case that we shall discuss that can be solved exactly. In future, all solutions discussed will be approximate, although some may be very accurate approximations. In the rest of this book, we shall be concerned more with the general form of the solutions, and we shall endeavour to predict as much as possible from necessary properties of the wave function (such as its symmetry). We should always try to distinguish those properties which are exact and necessary, and those which are deduced from approximations, and we should also seek as much support as possible from experiment.

c) Summary

Quantum mechanics discusses systems in terms of a wave function ψ. For a system to have a well defined energy this function must obey the Schrödinger equation $\mathscr{H}\psi = E\psi$ where $\mathscr{H}$ is the Hamiltonian operator and E is the energy of the system. ψ is then said to be an eigenfunction of $\mathscr{H}$, and E an eigenvalue. Different ψ which have the same E value are said to be degenerate. The electron density at any point (or the probability density of the electron) is given by $|\psi|^2$ evaluated at that point. The Dirac notation for quantum mechanical systems was introduced, and for two functions, $|a>$ and $|b>$, if $S_{ab} = <a|b> = 0$, they are said to be orthogonal, if $S_{ab} \neq 0$, they are said to overlap and S_{ab} is the overlap integral.

For the hydrogen atom the wave functions may be separated into the product of a radial part R_{nl} (r) and an angular part $Y_{lm}(\theta, \phi)$ where n, l and m are quantum numbers which may be used to classify these functions. These wave functions are also eigenfunctions of two angular momentum operators, and the angular momentum properties may be classified by l and m. As the electron also has spin angular momentum, and there are two values for the spin angular momentum quantum number m_s, then for each solution (or orbital) R_{nl} (r) $Y_{lm}(\theta, \phi)$ there are two possible solutions (or spin-orbitals) corresponding to $m_s = +\frac{1}{2}, -\frac{1}{2}$. The hydrogen wave function is thus completely specified by the values of n, l, m and m_s.

The shape of the wave functions was discussed, and it was shown that for increasing n, R_{nl} (r) becomes more spread out, and the energy of the function increases. If $l \geqslant 1$, then the wave function has well defined directional properties for the Y_{lm} (θ, ϕ) term. Finally we discussed the weak interaction between spin and orbital angular momentum (spin-orbit coupling) in terms of perturbation theory; this theory allows us to improve almost exact solutions by giving a correction to the energy, and shows that distinct solution can be mixed together by this small perturbation $\mathscr{H}'$ to give new solutions. The extent of this mixing is dependent on the integral $<a|\mathscr{H}'|b>$, and the reciprocal of the energy difference between the two mixed wave functions $|a>$ and $|b>$.

B. The Use of Symmetry

a) Representations

It is an experimental observation that many chemical molecules exhibit high symmetry, and of course, isolated atoms exhibit spherical symmetry. If we cannot solve the Schrödinger equation exactly for these species, we may perhaps be able to gain some ideas about the wave functions by considering their symmetry; it is clear that the wave function must exhibit the same symmetry properties as the molecule or atom since the energy is invariant under symmetry operations. We can in fact use symmetry properties to classify wave functions. The relation of wave functions to symmetry is accomplished by the use of group theory. This subject represents a sizeable fraction of pure mathematics, and even to discuss its chemical applications would require at least one whole book. Fortunately several such books exist, and our intention here is only to provide a vocabulary with which we can follow the group theoretical discussions which play such a large part in contemporary inorganic chemistry. For a proper understanding of group theory, the reader is strongly encouraged to read one of the many introductory texts for chemists (see Bibliography, page 37).

A set of quantities or elements form a group if they obey the following conditions:

(i) They multiply associatively: for three elements **A**, **B**, and **C**:

A x (**B** x **C**) = (**A** x **B**) x **C** where x symbolises an operation of multiplication

(ii) If **A**, **B** are elements of a group, then **A** x **B** is also an element of the group

(iii) The inverse, $\mathbf{A}^{-1}$, of any element **A** exists and is also a member of the group

(iv) The identity element **E** exists as an element of the group. The properties of **E** are such that:

$$\mathbf{A} \times \mathbf{E} = \mathbf{E} \times \mathbf{A} = \mathbf{A}$$

$$\mathbf{A}^{-1} \times \mathbf{A} = \mathbf{A} \times \mathbf{A}^{-1} = \mathbf{E}$$

Only a little thought will show that the set of symmetry operations applicable to a given molecule or figure form a group – for example, two successive symmetry operations will leave the figure in a position which could be reached by one single symmetry operation (condition (ii)).

Now let us consider an example, a series of points ±1 unit from the origin of two-dimensional cartesian axes.

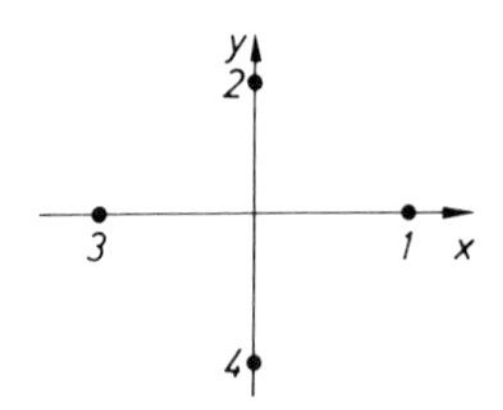

These four points show quite a high symmetry. The effect of the symmetry operations will be to interchange some or all of these points. The identity element leaves them unchanged, but a rotation of 90° clockwise about the origin causes 1 → 4, 4 → 3, 3 → 2, and 2 → 1. Its inverse, a rotation of 90° anticlockwise, causes 4 →1, 1 → 2, 2 → 3, 3 → 4. A reflection in the y axis leaves 2 and 4 unchanged, but interchanges 1 and 3.

We may express the interchanges by a matrix with elements a_{ij} such that $a_{ij} = 1$ if point j is changed to point i by the operation, and $a_{ij} = 0$ if it is not. Thus the identity

element is $\begin{pmatrix} 1&0&0&0 \\ 0&1&0&0 \\ 0&0&1&0 \\ 0&0&0&1 \end{pmatrix}$ and the reflection $\begin{pmatrix} 0&0&1&0 \\ 0&1&0&0 \\ 1&0&0&0 \\ 0&0&0&1 \end{pmatrix}$

These matrices obey the same multiplication laws as the symmetry operators and thus also form a group. We say that these matrices form a *representation* of the group. The points on which they operate are called the *basis functions* of the representation.

Suppose now that we put a point halfway between the origin and each of the points P1 and P4, and take all eight points as basis functions:

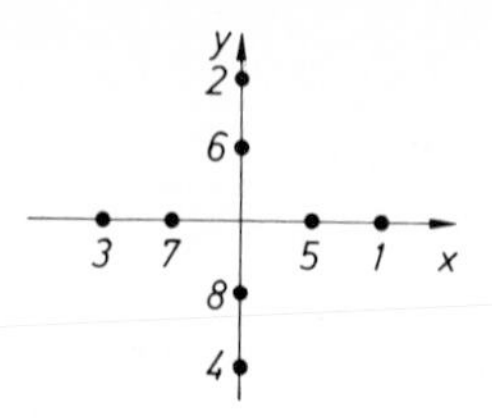

The matrices for the operations will now be 8 x 8, but will still obey the same multiplication laws; we thus have a new representation which is eight dimensional, and clearly we can go on ad infinitum putting in more sets of symmetry related points and using them as basis functions – there is thus an infinite number of representations. However, no symmetry operator will be able to mix points 1 to 4 and points 5 to 8, and we might expect to be able to break up each matrix into two halves, one operating on P1 – P4, the other on P5 – P8. A little thought will convince one that this matrix will have the form

$$\begin{pmatrix} \mathbf{M_1} & 0 \\ 0 & \mathbf{M_2} \end{pmatrix}$$

where $\mathbf{M_1}$ operates on points 1 to 4, and $\mathbf{M_2}$ on points 5 to 8. Furthermore the two matrices $\mathbf{M_1}$ and $\mathbf{M_2}$ will be the same since there will be an exact correspondance between the effect of the symmetry operators on P_n and P_{n+4} (n = 1 to 4).

This may all be expressed in mathematical language as follows:

We have *reduced* the 8 dimensional representation to two four dimensional representations. The matrix where all elements which would mix $P_{1\text{-}4}$ and $P_{5\text{-}8}$ are zero is said to be in diagonal form, and to be the direct sum of $\mathbf{M_1}$ and $\mathbf{M_2}$. We denote representations by the Greek letter Γ, and put $\Gamma^{(8)} = 2\,\Gamma^{(4)}$. It is important to note that although the two $\Gamma^{(4)}$ are identical, they have different basis functions ($P_{1\text{-}4}$; $P_{5\text{-}8}$).

We have taken simple points as our basis functions, but it should be evident that we could equally have taken something rather more complicated – for example, hydrogen 1s orbitals centred at points 1 to 4. As these functions would show the same symmetry properties as points 1 to 4 they would equally be basis functions for the same representation. The important point is that points or functions showing a certain symmetry can form basis functions for a representation of the symmetry group. We may express this in a different way by saying that a series of functions (e.g. P_1 to P_4) transform as a given representation of the symmetry group (in this case $\Gamma^{(4)}$).

b) Irreducible Representations

Can we further reduce our 4 x 4 representations? If we consider the combinations $(P_1 + P_2 + P_3 + P_4)$, $(P_1 + P_3 - P_2 - P_4)$, $(P_3 - P_1)$, and $(P_4 - P_2)$, we find that the first two are unchanged or simply multiplied by -1 by the symmetry operations of the group, and the third and fourth are mixed, but only with each other. They thus form the basis for three representations, the first two $(P_1 + P_2 + P_3 + P_4; P_1 + P_3 - P_2 - P_4)$

one-dimensional and the second two combinations ($P_3 - P_1$; $P_4 - P_2$) a two-dimensional representation. The four dimensional representation is thus the direct sum of three other representations – let us call them Γ_x, Γ_y, Γ_z; then

$$\Gamma^{(4)} = \Gamma_x + \Gamma_y + \Gamma_z \tag{1.21}$$

It is impossible to reduce Γ_x, Γ_y or Γ_z further; these representations are *irreducible representations* of the symmetry group, while the original representation $\Gamma^{(4)}$ is *reducible.* The irreducible representations of a group (often abbreviated to I.R.s) have a set of well defined properties, and are, for finite groups (those with a finite number of operators or elements) limited in number; for the symmetry group we have discussed there are only 10 I.R.s. There is a convention (due to Mulliken) for naming these I.R.s:

1) The dimensionality of the group defines the first part of the name. One dimensional I.R.s are given the label A or B, two dimensional E, three dimensional T, four dimensional U, and henceforth alphabetically (V, W, . . .). Thus the third of our I.R.s above is E type.

2) The letter A is used for representations which are symmetric with respect to the principal rotation axis of the group, and B for I.R.s which are antisymmetric. In our example, the principal axis is the fourfold rotation axis: ($P_1 + P_2 + P_3 + P_4$) is symmetric to a rotation of 90° and is thus A type, but ($P_1 + P_3 - P_2 - P_4$) gives ($P_2 + P_4 - P_1 - P_3$) – hence it is antisymmetric and B type.

3) If there is a centre of symmetry, then I.R.s symmetric with respect to inversion are labelled with a subscript g (German – gerade or even), and those antisymmetric with a subscript u (ungerade). Thus Γ_x and Γ_y are A_g and B_g, but Γ_z is E_u.

4) If there are any I.R.s not distinguished by this label, they are further numbered by subscripts; in our case, following the conventional assignment:

$$\Gamma_x = A_{1g},\ \Gamma_y = B_{2g},\ \Gamma_z = E_u$$

Every group has an I.R. for which all the matrices are the scalar 1. This is the *totally symmetric* I.R., and is labelled A, A_1, or A_{1g} (as needed to distinguish it from other irreducible representations).[4]

c) Direct Products

If we consider the series of functions $P_1 \cdot P_5, P_1 \cdot P_6, P_1 \cdot P_7, P_1 \cdot P_8, P_2 \cdot P_5 \cdots\cdot P_4 \cdot P_8$, these functions must also form the basis for a representation of the symmetry group, which will (in general) be reducible. This representation is the direct product, $\Gamma_p \oplus \Gamma_q$ of the two representations Γ_p and Γ_q for which $P_1 - P_4$ and $P_5 - P_8$ are the basis functions. In our example the direct product is $\Gamma^{(4)} \oplus \Gamma^{(4)}$. Since the direct product is reducible, it can be broken up into a direct sum of I.R.s, and we can state two useful rules about the components of this direct sum:

1) If, and only if, $\Gamma_p = \Gamma_q$, then the totally symmetric I.R. (A_{1g}) will be one of the components of $\Gamma_p \oplus \Gamma_q$.

   (If Γ_p and Γ_q are reducible, then they must both have one component I.R. which is identical.)

4 The newcomer to group theory and symmetry may find it useful to read Appendix I in conjunction with the rest of this section.

2) If Γ_p (or Γ_q) is the totally symmetric I.R., then $\Gamma_p \oplus \Gamma_q$ is Γ_q (or Γ_p). The totally symmetric I.R. thus acts as a kind of identity.

d) Application to Quantum Chemistry

We stated above that the wave functions of a system must reflect its symmetry. We may actually impose a more stringent condition on the wave functions: they must be basis functions for I.R.s of the symmetry group. We may classify wave functions by the I.R.s for which they are basis functions. More useful, however, are the symmetry restrictions of the values of the integrals over all space which we encountered earlier. The integral

$$\int_{\text{all space}} f(\tau)\, d\tau$$

will be zero unless f (τ) transforms as (i.e. can act as a basis function for) the totally symmetric I.R. If we now consider an overlap integral $< a \mid b >$, the function we wish to integrate $(\psi^*_a \cdot \psi_b)$ must transform as the totally symmetric I.R. Rule 1 for direct products tells us that this requires ψ_a and ψ_b to transform as the same I.R. of the symmetry group. Consequently, if we know the symmetry properties of two functions, and the irreducible representations with the same transformation properties, we may immediately predict whether or not they will overlap. This can simplify calculations very considerably.

We may extend this to a consideration of integrals of the form $< a \mid \textit{operator} \mid b >$; this will be zero unless the direct product $\Gamma_{operator} \oplus \Gamma_b$ contains a term which transposes as Γ_a. This type of integral is extremely common in quantum chemistry (for example in perturbation theory), and the symmetry requirements can greatly reduce the number of integrations in a calculation.

As an example of all these results, we may re-examine the solutions of the Schrödinger equation for the hydrogen atom. The symmetry of the hydrogen atom is spherical, and there are an infinite number of symmetry operations. However, the spherical symmetry group ($R3$) has a well defined (although infinite) set of I.R.s, and we find that the spherical harmonics $Y_{lm}(\theta, \phi)$ are in fact basis functions for these I.R.s. The functions Y_{lm} ($m = l, l-1, l-2, \ldots, -l$) are the basis for a $(2l + 1)$–dimensional I.R. of $R3$. The Mulliken system of nomenclature is not used for $R3$, the I.R.s being labelled S, P, D, F, G, . . . as $l = 0, 1, 2, 3, 4, \ldots$. The totally symmetric representation is S. Note in passing that the I.R.s are g-type for $l = 0, 2, 4, \ldots$ and u-type for $l = 1, 3, 5, \ldots$ This immediately gives the result that $Y_{lm} Y_{l'm'} = 0$ if $l' \neq l$; furthermore, the spherical harmonics are chosen so that the integral is zero unless $m = m'$.

e) Transition Probabilities

It was mentioned earlier that the probability of a spectroscopic transition is proportional to $\mid < a \mid \hat{\mu} \mid b > \mid^2$; it is therefore interesting to see how symmetry determines the allowed transitions. If $< a \mid \hat{\mu} \mid b >$ is non-zero the transition will be allowed by the electric dipole mechanism. Now the operator $\hat{\mu}$ has similar transformation properties to the three p orbitals, and in spherical symmetry transforms as Y_{1m} ($m = \pm 1, 0$); we may see immediately that μ will transform as a u-type I.R. Since the final product $< a \mid \hat{\mu} \mid b >$ must have g-type symmetry, $\mid a >$ and $\mid b >$ cannot both be u or both be g. The electric dipole mechanism thus requires $|a>$ and $|b>$ to have different parities (i.e. one u-type, the other g-type). This is the *Laporte* selection rule, which forbids transitions such as s → d or d → d. The complete selection rule requires that, for a transition from ψ_{nlm} to $\psi_{n'l'm'}$, $l - l' = \pm 1$, $m - m' = 0, \pm 1$. The Laporte rule is very frequently used in molecular spectroscopy for molecules with centres of symmetry where the wave functions must either be u or g-type. There are other mechanisms which may excite an electronic transition (notably the magnetic dipole and electric quadrupole mechanisms) but these have a much lower probability, and although the different symmetry properties of the operators 'allow' the transitions, they are too weak to be seen. This illustrates a weakness of symmetry arguments: they may only predict when a given integral is identically zero; an integral may still be zero, even though symmetry does not predict this. Thus the 1s and 2s wave functions of the hydrogen atom are orthogonal even though they both transform as the S I.R. Similarly, the symmetry allowed electric quadrupole transitions are nonetheless too weak to be seen.

It should be noted that all functions or properties of a system which are affected by symmetry may be classified according to their symmetry properties. Let us consider the example of molecular vibrations: the vibrational ground state always transforms as the totally symmetric I.R., and will have g symmetry in a molecule with a centre of symmetry. We may vibrationally excite the molecule by two different methods: (i) direct absorption of energy resulting from the interaction of the oscillating electric dipole moment of the radiation with the vibrating molecule (infra-red spectroscopy), and (ii) absorption of energy by inelastic scattering of incident visible light (Raman spectroscopy). For infra-red spectroscopy the transition probability depends on the electric dipole moment operator, but for Raman spectroscopy the interaction depends on the change in polarisability associated with the vibration, and the appropriate operator has symmetry g. The selection rules are therefore:

$$< \psi_{vib\ excited} \mid \hat{\mu}_u \mid \psi_g > \neq 0 \quad \text{(infra-red)} \qquad (1.22)$$

$$< \psi_{vib\ excited} \mid R_g \mid \psi_g > \neq 0 \quad \text{(Raman)} \qquad (1.23)$$

(R = interaction operator for Raman transitions)

Clearly infra-red spectra will invole excitation only of u vibrations, whilst Raman spectra will only show excitation of g type vibrations. This mutual exclusion rule is of use in identifying centro-symmetric molecules.

The effects of electron spin may also be treated by group theoretical methods, using 'spinor' or double groups which take into account the extra symmetry operations that consideration of spin implies. The two spin eigenfunctions $|\alpha>$ and $|\beta>$ are basis functions for different I.R.s and are orthogonal. Only perturbations which involve an interaction with the electron spin (most commonly magnetic perturbations or spin-orbit coupling operators) will give a non-zero value for integrals of the type $< \alpha \mid \mathscr{H} \mid \beta >$. Thus the electric dipole moment operator, which produces normal electronic transitions, is not able to produce a change of spin state during a transition, unless there is very considerable spin-orbit coupling mixing spin and orbital angular momentum. This is the basis of the $\Delta S = 0$ selection rule of spectroscopy.

f) Degeneracy

In our first example we saw that the two dimensional I.R. had the basis functions $(P_3 - P_1)$ and $(P_2 - P_4)$, and that symmetry operations (such as rotation by 90°) will mix these two functions. Now, a symmetry operation cannot change the energy of a system, so, if two wave functions are basis functions for the same I.R., and are mixed by the symmetry operations, they must have the same energy, or, in other words, they must be degenerate. Thus, in *spherical* symmetry the five 3d orbitals of hydrogen must be degenerate since they all transform as the D I.R. and would be mixed by rotation. If we lower the symmetry by putting six other charges or atoms (which we shall call *ligands*) at the vertices of an octahedron centred on the origin of the d orbitals, then, in the new symmetry, the d orbitals transform as two different I.R.s of the symmetry group O_h, the e_g and the t_{2g} I.R.s, and there is no longer any need for them to be completely degenerate, but only for the two e_g and three t_{2g} orbitals to have the same energy as the other members of that set. If the symmetry is lowered further, for example, by stretching the octahedron along one axis to give the new symmetry group D_{4h}, the d orbitals now transform as $a_{1g} + b_{1g} + e_g + b_{2g}$ and only the two orbitals transforming as e_g need be degenerate. If we carried on lowering the symmetry we would eventually completely remove the degeneracy of these orbitals; if there is no symmetry at all present, every function will transform as the totally symmetric I.R. and there will be no degeneracies required.

We may further apply symmetry arguments to the discussion of perturbation of the d orbitals by the ligands – the d orbitals will only overlap with orbitals which transform as the same I.R.; thus for an octahedral complex, the e_g d orbitals will only interact with e_g symmetry orbitals of the ligand. Since the Hamiltonian is always totally symmetric, the first order correction to the wave function as given by perturbation theory will only involve mixing of orbitals of the same symmetry. The strength of this interaction with the ligands is a function of their distance from the d orbitals: when the ligands are at a great distance the d orbitals will be virtually unaffected; when the ligands are very close, it may be inappropriate to talk of d orbitals at all, as they will be very greatly distorted from their form in the unperturbed atom. Symmetry and group theory cannot help us with this distinction; for this we must use our chemical knowledge and any experimental knowledge. The use of symmetry arguments in such cases is to tell us that some interactions must be zero, and may be ignored.

This does, however, emphasise the fact that symmetry arguments are unable to predict by how much the degeneracy has been lifted, only whether it is possible. If the symmetry is lower than spherical, but the d orbitals are in fact still degenerate or very nearly so, we say that the *microsymmetry* is spherical. In general, the microsymmetry is determinded only by the immediate environment (within ~3 Å) of a given atom. Microsymmetry is a useful concept, since it is clear that, if one goes far enough away from the central atom, the universe is sufficiently random for the total symmetry to be very low. Furthermore, since the symmetry near the atomic nucleus is spherical (since it is dominated by the nuclear charge), we may discuss the chemical effects of surrounding atoms in terms of low symmetry (e.g. octahedral) perturbations of the spherical atomic orbitals.

g) Summary

We have seen that the wave functions of a species with a given symmetry must form the basis for a representation of the symmetry group. This representation can generally be reduced to a sum of I.R.s (irreducible representations) of the group. We may classify wave functions by I.R.s for which they are basis functions. For the integral $\int f(\tau)$ over all space to be non-zero. $f(\tau)$ must transform as the basis function for the totally symmetric I.R., and we can use this rule, and the rules for the formation of direct products of I.R.s to predict whether such integrals will be zero or non-zero. This technique allows the prediction of overlap integrals, selection rules, etc. Group theory does not allow us to predict the size of non-zero integrals. We introduce the concept of microsymmetry to describe the symmetry of chemically important features, rather than the generally low symmetry of a species surrounded by the rest of the universe.

In this short description, we have not discussed the characterisation of I.R.s, but this is discussed in most introductory texts on the subject. For this book, the I.R.s for which the most important functions are the basis, together with some useful direct products, are listed in Appendix I for some of the more important symmetry groups.

C. Polyelectronic Atoms

In the rest of this book we will be concerned with systems containing more than two charged particles, polyelectronic atoms in this section, molecules in the other chapters. The potential energy term in the Schrödinger equation will now contain a large number of interactions of the form $A/\mathbf{r}_{ij}$ where $\mathbf{r}_{ij}$ is the distance between particles i and j. It is not possible to solve the Schrödinger equation exactly for these many body systems, even when there are as few as three bodies, and consequently we must use approximate methods. The first two sections of this chapter were different in that we were able to solve the full equation for the hydrogen atom exactly (at least if we ignore electron spin), and the laws of group theory are exact. We can still use the laws of group theory to give ourselves precise details about certain features of a wave function (e.g. orthogonality), but we *may* subconsciously have made an approximation in assuming that a given system actually has an exact symmetry (see the discussion of microsymmetry).

Since approximation is necessary, it is perhaps worthwhile mentioning that the level of approximation can vary: in chemistry we find approximations that are conceptually clear and useful, and are fairly accurate; those which are inaccurate but qualitatively correct; and finally those which are poor both qualitatively and quantitatively, but are easy to apply. At all times our criterion for acceptance must be agreement with experiment, but the placing of an explanation in the categories of approximation described above is often a matter of personal opinion. The application of quantum mechanics to chemistry demands the application of approximations in a critical fashion. In the remainder of this book we will try to make clear the level of approximation involved, the agreement with experiment, and the sections where the author is exercising his own prejudices; the reader should try to develop his own critical faculties, and not lose sight of the physical nature of the approximations involved.

Consider a polyelectronic atom with n electrons; clearly the wave function will have the form $\psi(\mathbf{r}_1, \theta_1, \phi_1, \ldots, r_i, \theta_i, \phi_i, \ldots, r_n, \theta_n, \phi_n)$ where the coordinates are those of the n electrons. There will be interactions between all the n electrons, and the position of each electron will be affected by the position of the nucleus and all the other electrons. In order to simplify the wave function we may make the *central field approximation:* we can consider an electron α_i as moving independently in the average potential field due to the nuclear and electronic potentials, and that this potential can be used to find the one electron wave function for α_i. We split the wave function $\psi(\mathbf{r}_1, \mathbf{r}_2, \mathbf{r}_3 \ldots)$ into the product of one electron wave functions: $\phi(\mathbf{r}_1) \cdot \phi(\mathbf{r}_2) \cdot \phi(\mathbf{r}_3)$ etc. Each wave function ϕ_j is thus associated with a given electron a_j; this is not only a mathematical simplification but also a conceptual one – we can now identify each electron with its own wave function, and consequently the one electron jumps excited in optical spectra (for example) may be regarded as the change of one electron wave functions in the total wave function. This we must recognise as an assumption as the energies of all the electrons are dependent on the other electrons. However, it turns out to be a good assumption for the weaker bound electrons.

We assume that the polyelectronic atom has spherical symmetry, as does the hydrogen atom, and indeed we find many similarities between the spectra of polyelectronic atoms and that of hydrogen. We find that we can use hydrogenic type functions of the form $f_{nl}(r) \cdot Y_{lm}(\theta, \phi)$ for our one electron wave functions, where the $Y_{lm}(\theta, \phi)$ are the spherical harmonics, but the $f_{nl}(r)$ are now different from the hydrogenic wave functions as a result of electron repulsion interactions. However we can still classify our one electron wave functions by n, l, m to give orbitals, and n, l, m, m_s to give spin orbitals – this means of classification is extremely convenient. We regard our one electron wave function as an eigenfunction of the one electron Hamiltonian, the potential energy terms in this Hamiltonian operator containing terms corresponding to repulsion from all the other electrons. The eigenvalue of the one electron wave function and its corresponding Hamiltonian is the *one electron energy*.

We have given labels 1, 2, 3 to our electrons in orbitals a, b, c, etc. but what would happen if we permuted electrons 1, 2, 3 . . . . amongst orbitals a, b, c . . .? Since the labels 1, 2, 3 . . . are artificial ones we have created ourselves, we would also expect the permuted total wave function to be a solution to the Schrödinger equation. However, it is found experimentally that all electronic wave functions must be antisymmetric with respect to such permutations (i.e. the wave function changes sign on exchange of the labels of two electrons). This is the *Pauli exclusion principle;* the simple product wave function we introduced above does not have this property of antisymmetry and so the total wave function is represented by the *Slater determinant:*

$$\psi = \frac{1}{\sqrt{n!}} \begin{vmatrix} \phi_a(1) & \phi_a(2) & \phi_a(3) & \phi_a(4) \ldots & \phi_a(n) \\ \phi_b(1) & \phi_b(2) & \phi_b(3) & \phi_b(4) \ldots & \phi_b(n) \\ \phi_c(1) & \phi_c(2) & \phi_c(3) & \ldots\ldots & \\ \ldots\ldots\ldots & & & & \\ \ldots\ldots\ldots & & & & \\ \phi_n(1) & \ldots\ldots\ldots & & & \phi_n(n) \end{vmatrix}$$

$$\psi = | \phi_a \cdot \phi_b \cdot \phi_c \cdot \phi_d \cdots\cdots\cdots \phi_n \tag{1.24}$$

The factor $1/\sqrt{n!}$ is a normalisation constant. This determinantal wave function has all the properties required by the Pauli exclusion principle. In discussing the total wave function of an atom or molecule, it is important that we use the *determinantal* wave function. One of the most important results of the Pauli principle is that no two electrons may occupy the same spin orbital (or be characterised by the same 4 quantum numbers), and only two electrons may occupy each orbital, corresponding to $m_s = +\frac{1}{2}, -\frac{1}{2}$.

Let us now recapitulate our remarks on polyelectronic atoms. We have split up our total wave function into one electron wave functions by the central field approximation. We then use functions similar to the wave functions of the hydrogen atom for our one electron functions, and can thus talk about one electron spin orbitals. The Pauli principle requires that these one electron functions are combined to give an antisymmetric total wave function using a device such as the Slater determinant, and further requires that only one electron may occupy each spin orbital (and two electrons each orbital). This gives the framework of our quantum model for polyelectronic atoms. We now consider the effects of interelectron repulsion.

The most important general feature of the wave functions is that the degeneracy of orbitals with different l disappears. For hydrogen the orbital energy was a function of n only; for many-electron atoms it is a function of l as well. Thus we can draw up orders of energies: $E_{3s} < E_{3p} < E_{3d}$; $E_{4s} < E_{4p} < E_{4d} < E_{4f}$. Orbitals with the same n, l but different m are still degenerate (in the absence of a magnetic field).

We are now of course in a position to picture the electronic structure of atoms – if we assume that the ground state of an atom is that with the lowest energy, then we can fill up the orbitals with the lowest energy as we increase the number. Thus helium with two electrons has the two lowest spin orbitals $1s\alpha$, $1s\beta$ occupied; lithium with 3 electrons has $1s\alpha$ $1s\beta$ $2s\alpha$ boron (5 electrons) $1s\alpha$ $1s\beta$ $2s\alpha$ $2s\beta$ $2p\alpha$ and so on. We describe these neutral atoms as having their configurations given by these one electron functions. Helium has the configuration $1s^2$, oxygen $1s^2 2s^2 2p^4$, argon $1s^2 2s^2 2p^6 3s^2 3p^6$. It is common to write a configuration by reference to that of the previous noble gas i.e. oxygen is [He] $2s^2 2p^4$, chlorine [Ne] $3s^2 3p^5$.

One notices that elements with the same number and disposition of electrons outside their noble gas "core" occur below each other in the periodic table. Thus all the alkali metals have the electron configuration (noble gas) ns^1. The assumption that the chemical behaviour of an element is governed by the outer (or valence) electrons is thus confirmed experimentally by the observation that chemical periodicity is in correspondance with the periodicity of orbital filling predicted by quantum mechanics.

If we continued to follow the building up or *Aufbau* principle we would expect to fill the 3d orbitals after filling the 3p orbitals (in argon). Calculations, chemical, and spectroscopic evidence all indicate however that the 4s orbitals are filled next, then the 3d, and so on as shown in Table 1.1. The effects of electron repulsion are sufficiently large to raise the energy of the 3d above that of the 4s; similar effects are found for the 4d (filled after the 5s), the 5d (after the 6s), and the 4f (after the first electron in the 5d shell).

This explanation of periodicity is well known, and we do not wish to labour it here, but will note only a few points of interest:

1) The chemical behaviour of elements depends strongly on their valence shells, in particular the last subshell (i.e. set of n, l orbitals) to be filled.

2) Calculations and experimental evidence generally agree on the ordering of orbitals.

3) The difference between levels is not always clear and definite – the 4s/3d energy difference is very small as the first transition series is crossed, and for ions, or for elements heavier than the first transition series E_{4s} may be higher than E_{3d}. We discuss this later, but the order given here and in Table 1.1. is correct for neutral atoms only.

This model of the atom agrees well with experimental evidence. Electrons ejected from atoms by X-rays show well separated, clearly defined, binding energies corresponding to orbital subshells in the atom. X-ray spectra involving transitions between a vacancy in a tightly bound orbital (caused by electron bombardment), and higher filled orbitals, and optical and U.V. spectra showing the nature of the valence electron configuration give considerable support for the model. Finally we may mention that the picture of electron distribution within an atom shown by electron diffraction agrees well with the calculation.

In the rest of this chapter, we will consider the effects of electron repulsion in greater detail, and, after a brief discussion of atomic spectra, consider the validity of our model of the atom. We then give a more quàlitative discussion of ionisation energies and electron affinities, and conclude with a note on atomic structure calculations.

a) A Perturbation Treatment of Electron Repulsion

Consider a simple two electron system corresponding to one electron in each of the spin orbitals ϕ_a ϕ_b calculated as for hydrogen, ignoring electron repulsion, and let us treat the repulsion between the electrons as a perturbation with the Hamiltonian $\mathscr{H}_{rep}$. The Slater determinant

$$\psi = \begin{vmatrix} \phi_a(1) & \phi_a(2) \\ \phi_b(1) & \phi_b(2) \end{vmatrix} \cdot \frac{1}{\sqrt{2}} = [\phi_a(1)\,\phi_b(2) - \phi_a(2)\,\phi_b(1)] \cdot \frac{1}{\sqrt{2}} \tag{1.25}$$

The first order perturbation energy is given by Eq. (1.18) as:

$$\begin{aligned} E_1 = \langle \psi \mid \mathscr{H}_{rep} \mid \psi \rangle = \frac{1}{2} \; & \langle \phi_a(1)\,\phi_b(2) \mid \mathscr{H}_{rep} \mid \phi_a(1)\,\phi_b(2) \rangle \\ + \; & \langle \phi_a(2)\,\phi_b(1) \mid \mathscr{H}_{rep} \mid \phi_a(2)\,\phi_b(1) \rangle \\ - \; & \langle \phi_a(1)\,\phi_b(2) \mid \mathscr{H}_{rep} \mid \phi_a(2)\,\phi_b(1) \rangle \\ - \; & \langle \phi_a(2)\,\phi_b(1) \mid \mathscr{H}_{rep} \mid \phi_a(1)\,\phi_b(2) \rangle \end{aligned} \tag{1.26}$$

If we examine the 1st and 2nd integrals we see that they differ only in the position of the labels 1 and 2 for the electrons; the repulsion operator $\mathscr{H}_{rep} = e^2/4\pi\,\epsilon_o\,r_{12}$ is unchanged by exchange of these labels. Thus the numerical values of the first two integrals are equal; a similar argument holds for the third and fourth terms. We write:

$$J_{ab} = J(a,b) = \langle \phi_a(1)\,\phi_b(2) \mid \mathscr{H}_{rep} \mid \phi_a(1)\,\phi_b(2) \rangle \tag{1.27}$$

$$K_{ab} = K(a,b) = \langle \phi_a(1)\,\phi_b(2) \mid \mathscr{H}_{rep} \mid \phi_a(2)\,\phi_b(1) \rangle$$

Then:

$$E_1 = \frac{1}{2}(2J(a,b) - 2K(a,b)) = J(a,b) - K(a,b) \tag{1.28}$$

We have thus expressed our correction for electron repulsion in terms of two integrals J_{ab} and K_{ab}. What are the physical interpretations of these two integrals? The J integral is a measure of the interaction between the charge densities associated with the two occupied spin-orbitals a and b, and is referred to as the *Coulomb integral.* The K integral is not readily assignable to a classical interaction as its existence is a result of the Pauli principle and its requirement of antisymmetry. Because it involves an "exchange" of electrons between the two orbitals a and b it is referred to as the *exchange integral.* It has one very important property: the electron repulsion operator $\mathcal{H}_{rep}$ is a function of interelectronic distance only, and not of spin; if ϕ_a and ϕ_b have different m_s values (e.g. ϕ_a is α and ϕ_b is β) then the K integral is identically zero. The K integral thus represents an interaction exclusively between electrons of parallel spin; an immediate corollary of this is that if a and b are the two spin-orbitals of the same orbital then $K_{ab} = 0$.

For the helium atom with an electron in each of the ls spin orbitals the total energy is thus: $E = 2E_{1s} + J_{1s,1s}$, where E_{1s} is the calculated energy for the 1s orbitals considering the nuclear potential only. It is possible to express the total energy E of a poly-electronic atom in terms of one electron energies ϵ_i such that

$$E = \Sigma_a \epsilon_a - \Sigma_{a<b} (J(a,b) - K(a,b)) \tag{1.29}$$

(the summation of repulsion terms is carried out only for b > a, avoiding counting them twice) where $\epsilon_a = E_a + \Sigma_b (J(a,b) - K(a,b))$ (1.30)
The theorem of Koopmans (so important in photoelectron spectroscopy (Chap. 8)) postulates that the ionisation energy is the negative of the one electron energy ϵ_a. This 'theorem' is only valid in so far as the central field, one electron wave function model is valid; although it fails quantitatively in many cases, the concept of a one electron energy is very useful.

The integrals J and K are difficult to evaluate, although by expanding the perturbation operator $\mathcal{H}_{rep}$ ($= e^2/4\pi\epsilon_0 r_{12}$) in terms of spherical harmonics and a radial function (Griffith 1971, Brink and Satchler 1968), and consequent use of symmetry rules, they may be simplified. For our purpose, it suffices to note that they are both positive, and that consequently, in our two electron example above, the energy of the system is raised by the Coulomb repulsion energy, but lowered by the exchange repulsion energy. It follows from the positive value of K, and its stabilising effect, that, all other things being equal, a system with spins parallel (and consequently $K \neq 0$) will be stabilised relative to a system with spins paired ($\Sigma m_s = 0$). This phenomenon is sometimes described by saying that electrons with parallel spin avoid each other and consequently electron repulsion integrals are smaller and the system is stabilised; this is called spin correlation. The electron is regarded as surrounded by a 'Pauli hole' from which other electrons of the same spin are automatically excluded. This is certainly a convenient way of looking at the phenomenon, but we should be careful not to look for mystic forces involved in this mutual evasion of electrons; rather should one recognise it as a necessary result of the experimentally observed antisymmetry requirement of the Pauli principle.

This treatment tells us in principle how to treat electron repulsion effects, but it is clear that if it is necessary to evaluate the repulsion integrals between each electron and all the other electrons the calculations will become impossible for a human being and tedious for a computer if we are going to consider atoms with up to 100 electrons. We thus note that any full subshell (i.e. the $(4l + 2)$ degenerate spin-orbitals associated with a given n, l) will transform as the totally symmetric representation of the spherical symmetry group. We thus approximate all full subshells by a spherical charge density. This leaves the problem of partially filled subshells. Although the $(4l + 2)$ spin-orbitals associated with a given n, l level are degenerate, there is no reason to assume that if q electrons ($q < 4l + 2$) are distributed in these spin-orbitals, all the arrangements will have the same energy. We could expect, for example, that as the spins of the electrons are parallel or antiparallel (paired), the K integral would be zero or non-zero, thereby changing the energy. We will thus briefly consider the varying energies of the states arising from different arrangements of electrons in partially filled shells.

b) Atomic Spectra

Most of the data pertaining to the possible energy states of partially filled shells comes from atomic spectroscopy. It is found experimentally that we may associate with a given energy state of a polyelectronic atom, well defined values of the magnitude of the total angular momentum and its component in the z direction; that is, the total wave function is an eigenfunction of the operators $\mathcal{J}^2$ and $\mathcal{J}_z$ (which now operate on *all* the electrons). We also find in almost all cases that the spin-orbit coupling is sufficiently weak to enable us to describe the total wave function as an eigenfunction of the total electronic orbital and spin angular momentum operators $\mathcal{L}^2$, $\mathcal{L}_z$, $\mathcal{S}^2$, $\mathcal{S}_z$. We use this observation to classify the energy states arising from the various arrangements of electrons in partially filled shells. If it is possible to separate the spin and orbital components, the wave function is said to show Russell-Saunders coupling between spin and orbital angular momenta. The classification of states by J, L, and S values is of course a classification of states according to the spherical symmetry group, allowing for spin. The selection rules for electric dipole allowed transitions are $\Delta L = \pm 1$, $\Delta S = 0$. If we have to couple the spin and orbital angular momenta of each electron first, we speak of *j-j* coupling – this is only important for heavy elements where the high Z value gives rise to very high spin-orbit coupling parameters.

The method of determining the various states arising is detailed in many books (Herzberg for example) and we mention it very briefly. There are two basic methods – the vector coupling and the group theoretical approaches. The vector coupling method takes as its basis that the total angular momentum quantum numbers M_L (sometimes called M) and M_S will be the sums Σm_L (or $\Sigma\, m$) and $\Sigma\, m_s$. Thus each component of a state with a particular M_L, M_S will account for one combination (or microstate) of the individual m_L and m_S values. It then suffices to draw out all the possible combinations of the various m_L and m_S values (remembering the Pauli principle) and then strike out one combination for each corresponding component of the final state. The final state is denoted as follows: the total L value is denoted by a capital letter, S, P, D, F, G, H, ... as $L = 0, 1, 2, 3, 4, \ldots$. etc., the spin state is denoted by the multiplicity $(2S + 1)$ – thus, for $S = 0$ (two paired electrons) the multiplicity is 1 – a *singlet.* For two parallel, unpaired electrons $S = \frac{1}{2} + \frac{1}{2}$, the multiplicity = 3. The state is represented by ${}^{2S+1}L_J$ where J is the total, coupled angular momentum. This expression is frequently referred to as a *Russell-Saunders term.* In the absence of spin-orbit coupling all the values of J will be degenerate; all the values of M_S and M_L will be degenerate in the absence of a magnetic field.

A set of rules due to Hund gives the ground state (i.e. the term with the lowest energy) for a configuration $(n\, l)^q$:

1) The term or terms with the maximum multiplicity will be lower than the others
2) given a choice of terms with maximum multiplicity, that with the maximum value of L is chosen

3) In this term the J value $L - S$ will be lowest for $q < 2l + 1$, $L + S$ lowest for $q > 2l + 1$ (for $q = 2l + 1, J = S$).

Thus for nd^2 there is a choice of terms ^1G, ^3F, ^1D, ^3P, ^1S. The first rule requires that the ground state is either ^3F or ^3P and the second rule that the ground state is ^3F. If the spin-orbit coupling is great enough to lift the degeneracy of the J values then as $q < 2l + 1$, the lowest J value will be $L - S = 3 - 3/2 = 3/2$. We may note that the first rule might be expected from the fact that the K integral acts as a stabilising influence if there are a large number of unpaired, parallel, spins. Hund's rules are extremely reliable for predicting the ground states in atoms and ions arising from configurations where only one subshell is partially filled – they do however fail for predicting the lowest term in excited configurations.

The group theoretical approach to the energy terms is in principle quite simple – if we have two electrons in orbitals transforming as Γ_a, Γ_b (two irreducible representations of the symmetry group) then the direct product $\Gamma_a \oplus \Gamma_b$ gives the terms ($\Gamma_i + \Gamma_j + \Gamma_k + \ldots\ldots$) and we reduce this representation to give the individual terms Γ_i. This method has the advantage that it can be applied to systems with lower than spherical symmetry (e.g. molecules). The mathematical treatment of electron repulsion in partially filled shells is elegant but complicated – for full details the reader should refer to the works of Condon and Shortley (1953), Griffith (1961) or Brink and Satchler (1968). Suffice it to say that the distances between various terms can be expressed as sums of rational multiples of a relatively small number of integrals. Racah introduced a series of combinations of these integrals which enable the distances to be expressed in terms of only 3 parameters (A, B and C) for d shells and 4 parameters for f shells, assuming Russell-Saunders coupling – happily Russell-Saunders coupling is sufficiently accurate for these systems. If not, it is necessary to use the double groups mentioned in Sect. B.e. This approach enables a theoretical justification of Hund's first rule to be given (Slater 1968, Jørgensen 1971) and we can abstract the useful qualitative concept of spin pairing energy – the energy associated with the change from q parallel spins to (q – 1) parallel spins and one antiparallel spin.

The treatment of electron repulsion outlined above is reliable in predicting the observed terms and their order, but not in predicting actual energy values – the reason for this is the neglect of electron correlation.

c) *Correlation Effects

Correlation may be regarded as the expression of the failure of the one electron orbital approximation, and arises from the fact that each electron is really dependent on the position of all the others; for example in the 1s orbital one electron will endeavour to be on the opposite side of the nucleus to the other electron, thereby decreasing the electron repulsion, even if the time-averaged one-electron density is spherical. Rather than seek an unreal qualitative picture of correlation effects, it is better simply to regard correlation as the cause of the difference between experimental electron repulsion parameters, and the (larger) calculated values. The consequent error in the calculated energy of a system is called the correlation energy, and is generally defined as the difference between the experimental energy and the best calculated value. Correlation can be allowed for by writing the wave function as a mixture of the groundstate with small amounts of excited states of the same symmetry – the energy of the system is slightly lowered by the matrix elements of the electron repulsion terms between the ground and excited states. This method is known as *configuration interaction* (C.I.); it gives much better energies, but involves many excited states, and in many electron systems can involve immense calculations. A necessary result of C.I. is that the final wave function cannot really be described by one simple determinantal wave function or configuration, but as a sum of several, one of which (the preponderant configuration) may account for 99 percent of the wave function.

Correlation effects are generally fairly small in proportion to the total energy of an atom (~1%) but as the total electronic energy of an atom rises rapidly with the number of electrons, and the valence shell energies remain very roughly constant, the correlation energy reaches values greater than the energies of chemical bonding.[5] Ab initio calculations on poly-electronic systems give very poor results for total energies, and consequently heats of formation. Fortunately for the qualitative explanations in this book we may regard correlation as an additional stabilising effect, and assume it to be roughly constant for a given element.

The reader may be beginning to wonder where all this discussion of electron repulsion is leading, especially since correlation effects appear to be likely to render the arguments useless. The situation is not, however, as bad as it might seem – the model predicts the type and order of the various terms in atomic spectra very well. If we take the electron repulsion parameters (e.g. the Racah parameters, A, B, C) from experimentally observed spectra, the correlation effects are allowed for, and we can predict energy terms quite accurately. This suggests that our one electron spin-orbital model is at least usable, and our picture of atomic structure satisfactory. This fact alone is of enormous importance for chemists.

In most chemical compounds it must be admitted that the effects of electron repulsion within partially filled shells are swamped by the effects of neighbouring atoms. There are however two exceptions of considerable importance to inorganic chemists: in many first row transition metal complexes, and almost all lanthanide complexes, there is a partially filled shell which is not too greatly affected by the surrounding atoms (or *ligands*). An enormous amount of chemical information about such complexes has been derived from the optical spectra, in which transitions from one term to another of a given electron configuration are observed. Since the chemical effects here are smaller than the electron repulsion effects, we can treat the chemical effects as a perturbation of the levels given by atomic structure theory as above – it is clear then that we need to know at least the fundamentals of electron repulsion theory to discuss such spectra usefully.

Most elements have partially filled shells in their ground states, and if we want to discuss the energy of these shells we should consider the various terms associated with the partially filled shell – this will normally be quite difficult, and the ordering of terms will be completely changed by the chemical interactions (with the exceptions of the transition metal and rare earth complexes where the partially filled shell is little affected). It is thus convenient to consider an average value corresponding to a particular configuration l^q – this value, the *baricentre,* is the average of all the term ener-

[5] An example of this is the fact that for sodium ($Z = 11$) the correlation energy is greater than its first ionisation energy. If we compare total energies E_{Na^+} (obs) $< E_{Na}$ (calc.).

gies, weighted by their degeneracy – thus a triply degenerate level counts three times as much as a non-degenerate level.

When a chemist talks of the energy of a subshell of electrons l^q, he normally implies that this energy is the baricentre, and that each spin-orbital is associated with $1/q$ of this energy. This is an approximation of course, but as ionisation potentials and chemical bonding effects are generally much greater than the electron repulsion effects the approximation ignores, the error involved is not too serious, and for qualitative discussions quite justified.

A useful expression relates the energies of baricentres of different multiplicity

$$E_S - E_{S'} = D\,\{S'(S'+1) - S(S+1)\} \tag{1.31}$$

where E_S is the energy of the baricentre of states with multiplicity $2S + 1$ and D is the spin-pairing parameter which is related to an average value of the K integral. On going from the baricentre of the states with spin S' (i.e. $2S'$ unpaired electrons) to a state with spin S ($2S$ unpaired electrons) we must supply $E_S - E_{S'}$ to overcome the greater electron repulsion resulting from the pairing of spins (clearly if $S' < S$, energy will be liberated). The concept of spin pairing energy is further discussed by Jørgensen.

d) Ionisation Energies and Electron Affinities

The ionisation energy (*I.E.*) of any substance is the energy of the reaction S(gas) → S^+(g) + e^- (g). Historically it has often been measured in units of electron volts and referred to as an ionisation potential. The electron affinity (*E.A.*) is the energy released in the reaction S (gas) + e^- → S^-(g). It is fairly clear that *I.E.* (S) = −*E.A.* (S^+). These two quantities are of considerable interest to the chemist since the *I.E.* of an element (or rather of its various orbitals) gives some indication of how strongly the electrons are bound, and their readiness to take part in chemical bonding, while the *E.A.* gives an indication of the readiness of an element to accept electrons into its empty orbitals in a chemical bond. We can rationalise a good ideal of chemistry by consideration of ionisation potentials and electron affinities, and we can qualitatively relate their variation to our picture of atomic structure. It is important in such discussion to realise exactly which energy is being discussed: an electron with a high *ionisation energy* will have a low (i.e. very negative) *orbital energy* but a high *binding energy.*

Although it is possible to ionise strongly bound electrons with X-rays, the most important ionisation process will be related to the most weakly bound orbital, the nature of which will be determined by the electron configuration. We introduce two qualitative ideas for the discussion of electron repulsion – *penetration* and *screening.* Consider the atomic shell with $n = 3$. Referring to Fig. 1.1 we see that the 3s orbital has three rings of electron density while the 3d has only one. If there are other electrons around the nucleus (in the $n = 1, n = 2$ shells) we might expect that the 3s orbital will gain greatly in stability from the region of electron density close to the nucleus while the region furthest from the nucleus will experience very little attraction.

The 3d orbitals will be largely outside the $n = 1$ and $n = 2$ shells and thus will be screened from nuclear attraction, giving them a higher orbital energy. This is a simple qualitative explanation of the lower orbital energy of the 3s orbitals compared with the 3d – the 3s orbital is said to be more *penetrating* than the 3d; conversely we may say that the 3d is more heavily *screened.* In general the order of penetration for a given n shell is $s > p > d > f$. We may also consider the screening effects – s orbitals screen outer electrons more effectively than p, which are more effective than d and so on. Thus the screening of the 3d, and the penetration of the 4s causes the 4s shell to be filled before the 3d. The strong screening of the 4f causes the 5s, 5p, 6s, and one orbital of the 5d shell to be occupied before the 4f shell is filled. Another feature of screening is that an electron in a given nl orbital will screen other electrons in the same subshell very poorly. Thus as we move across the periodic table filling for example the p subshell, the nuclear charge Z increases by one unit, but the extra screening by the new electron is fairly small – the *I.E.* thus rises. The change in *I.E.* for increasing Z is shown in Fig. 1.5 for the filling of the p shell; the *I.E.* of Al ([Ne] $3s^2 3p^1$) is a third of the *I.E.* of Ar ([Ne] $3s^2 3p^6$). We would expect a similar increase in electron affinity as Z increases from Mg ([Ne] $3s^2$) to Cl ([Ne] $3s^2 3p^5$); here the experimental evidence is much weaker as electron affinities are difficult to measure, but such evidence as there is justifies this conclusion, and indirect evidence supports it very strongly.

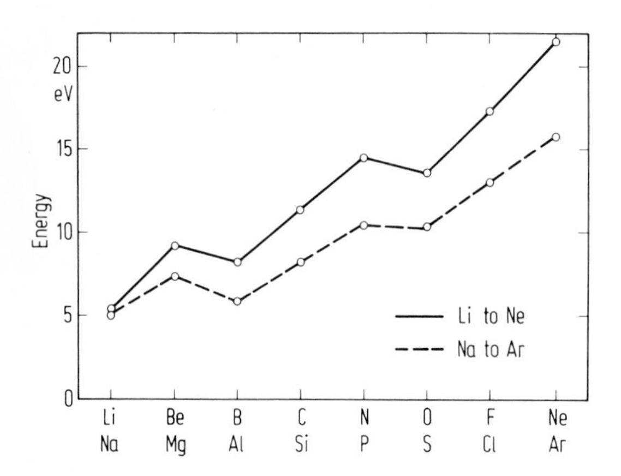

Fig. 1.5. The ionisation energies of the first and second row elements

Let us consider the change in *I.E.* across the first row of the periodic table, from Li to Ne (Fig. 1.5). Lithium has a relatively low *I.E.* as the weakest bound electron is in a new shell (n = 2) and is strongly screened by the 1s electrons. The second *I.E.* of lithium is much higher, as we are now breaking into the 1s shell – successive ionisation potentials in fact give strong support for the shell model of the atom, there being sharp increases as a new shell is ionised. The *I.E.* of Be is higher than Li as the nuclear charge has increased (by 33%) but the 2s electrons screen each other weakly. There is a fall to boron, since the 2p subshell is fairly strongly screened by the 2s and 1s subshells. As Z increases the *I.E.* rises through carbon to nitrogen; there is a fall to oxygen and then a rise again to neon. After neon it is necessary to fill the n = 3 shell – there is thus a consequent fall in *I.E.* for sodium ([Ne] $3s^1$). Why is there a discontinuity at oxygen? This is a result of our rather cavalier treatment of electron repulsion – the reader can show without difficulty that for a p^4 system it is essential that at least one electron has an antiparallel spin and thus there is a destabilisation, with consequent fall in *I.E.*, as a result of the spin pairing energy. It is quite common for there to be a discontinuity in *I.E.* when it is necessary to start pairing electrons – the *I.E.* of Mn^{2+} ($3d^5$) is higher than Fe^{2+} ($3d^6$) in spite of the higher charge of the Fe nucleus. This has led to invocations of the "stability of the half filled shell" perhaps somewhat confusingly since the real cause is the spin pairing energy for the case (half filled shell + 1 electron) – an example is the gadolinium atom where an electron prefers to enter the 5d subshell at a higher energy than pair with an electron in the 4f shell to give [Xe] $4f^8 6s^2$ as would be expected from the configurations of the other lanthanides. If electron pairing effects are allowed for, the *I.E.*s show no discontinuities within a subshell.

On traversing a period of the periodic table we expect to see a rise during the filling of each subshell (with a discontinuity when the shell is half full), a drop in *I.E.* when a new subshell is reached. We may also note that as the *n*d shell is filled, the energy of the $(n + 1)$s shell is lowered, since the *n*d shell screens this subshell only weakly. Thus the *I.E.* of calcium ([Ar] $4s^2$) is much less than that of zinc ([Ar] $3d^{10}4s^2$). A similar effect is found before and after the filling of the poorly screening 4f shell. This effect is also shown up in the size of the atoms[6]; the s^2 elements after the d shell is filled (Zn, Cd, Hg) are much smaller than those before the d shell is filled (Ca, Sr, Ba) as a result of the "shrinkage" due to poor screening by the d electrons. With the lanthanides this is even more apparent – the very poor screening properties of the 4f subshell cause the atomic radius to shrink as the f shell is filled, and the shrinkage is such that the mean size of elements with a partially filled 5d shell is very similar to that of those with a partially filled 4d shell. These phenomena are frequently referred to as the d-block and lanthanide *contractions*.

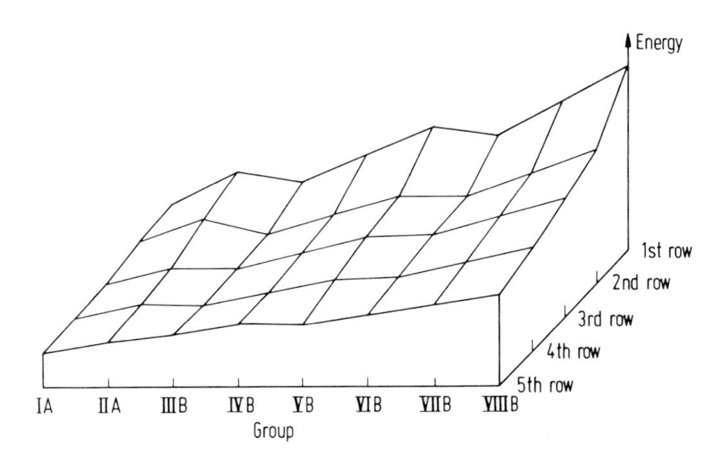

Fig. 1.6. The ionisation energies of s and p block elements plotted against group and row of the periodic table

Figure 1.6 shows the first ionisation energies of all the s and p block elements: the dip in *I.E.* for first electron in the p subshell (group IIIB) is much smaller for the 3rd and later rows where a d subshell is filled between groups IIA and IIIB. It will be noticed that the *I.E.* rises on crossing the periodic table, and falls on descending a

6 Atomic size is a rather ill-defined concept as a result of the diffuseness of electrons, but we may take as a more concrete example the interatomic distances in the solid elements.

column of the table; the irregularities due to electron repulsion effects are much less marked for the heavier elements.

What of electron affinity? The experimental data is much less complete here, but in general the *E.A.* rises as one progressively fills a subshell, but if the extra electron must enter a new subshell, the *E.A.* is negative. Thus the halogens (np^5) have relatively high *E.A.*s, but the noble gases (np^6) have negative *E.A.*s. Some values for the s and p blocks are shown in Fig. 1.7.

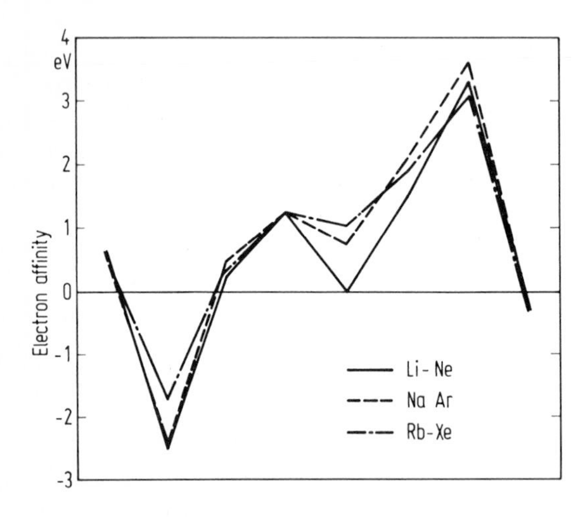

Fig. 1.7. Electron affinities for the s and p block elements

It seems from the general values of *I.E.* and *E.A.* that the neutral atom is fairly well balanced – it always requires a fair attraction of energy to remove an electron from an atom, and electron affinities are very small and sometimes negative. The highest known *E.A.* (for chlorine) is less than the lowest known *I.E.* (caesium) – hence isolated atoms of Cs and Cl will not undergo spontaneous ionisation (although if the atoms are brought close together other effects will stabilise the ionisation). Addition of a second or third electron is unfavourable in all known cases.

The chemical use of *I.E.*s and *E.A.*s is considerable, and will become increasingly clear throughout the book – however even without proposing any precise theories of chemical bonding we may anticipate that the noble gases, with high *I.E.* and low *E.A.* will not be very susceptible to any rearrangement of electrons resulting from chemical bonding. The halogens with high *I.E.* and high *E.A.* will probably form their most stable compounds if there is a net gain of electron density or chemical bonding, while the alkali metals with low *I.E.* may be prevailed upon to lose electron density without too great a loss of stability.

The considerations of electron repulsion effects which have brought us this far cannot be applied so easily to ions. The effect of ionising an atom is generally to reduce electron screening effects. Thus the groundstates of the positive ions of the first transition series have the configuration [Ar] $3d^n$; the ionisation has reduced electron screening so that the 3d subshell now lies lower than the 4s. It has also been argued that in compounds where an atom might be expected to have a high positive charge (e.g. sulphur in SF_6) the energy of the nearest d orbitals may be lowered sufficiently for them to have appreciable electron affinity. As orbitals move away from the valence shell, screening effects are less important, thus the 3d in potassium has a higher energy than the 4s, but in the core of e.g. Cs (Z = 55) the 4s subshell lies above the 3d shell (which can be said to have shrunk into the core). In using the concept of orbital energies it is important to realise that they vary with the charge on the atom. We return to this subject in Chap. 7 (page 250).

e) *Theoretical Calculations of Atomic Structure

It is of interest to discuss briefly the methods used to calculate atomic orbital wave functions and energies, especially as atomic orbital wave functions are the starting point for many calculations of molecular properties. The simplest form is the Slater orbital where the wave function is given as ψ_{nlm} (r, θ, ϕ) = $\mathrm{Nr^{n-1}}$ exp ($-$ ζ_{nlm}r) $Y_{lm}(\theta, \phi)$ where N is a normalisation constant, $Y_{lm}(\theta, \phi)$ is a spherical harmonic and ζ_{nlm} is the orbital exponent (the same for all m given n and l). Slater related this to an effective atomic number which he defined by assuming that orbitals become more shielding as n decreases, and that d and f orbitals screen more weakly than s or p orbitals. The rules are given and discussed in many books (e.g. Huheey; Murrell, Kettle and Tedder).

This wave function can be improved by undertaking a self-consistent field (S.C.F.) calculation. If we take a trial wave function, we can calculate the electron repulsion integrals to give a measure of the potential acting on a given electron – we now resolve the Schrödinger equation for that one electron to give a new solution – this is then used to calculate the potential acting on the other electrons. We resolve the equation for each electron in turn, until we find that the potential fields are unchanged in the new solutions. The field is then said to be self-consistent. S.C.F. wave functions have been obtained for all atoms up to radon (Z = 86) using Slater type orbitals (S.T.O.s) similar to Slater orbitals but with different values of ζ_{nlm}.

The main disadvantage of S.T.O.s is that they have no nodes, having only one region of electron density, and they are unsuitable for calculations which involve electron-nucleus interactions (e.g. in n.m.r. and n.q.r., Chap. 8). However in the valence region of an atom they are not too bad an approximation, and have the very useful property that overlap integrals $\int \psi_a^* \psi_b$, where a and b are on different atoms, can easily be computed.

More accurate wave functions are given by approximate numerical solutions of the wave equation. The initial work by Hartree used a simple product wave function $\psi = \phi_1 \times \phi_2 \times \phi_3 \times \ldots$ – which takes account of the J integrals of electron repulsion, but not of the K integrals; the Hartree-Fock system uses a determinantal wave function and thus includes K terms. The Hartree-Fock method is currently regarded as the most accurate method, although it must always be remembered that the correlation energy is not allowed for. This has led some chemists to maintain that the Slater determinant wave function considering only occupied orbitals of the ground state is not a justifiable representation of the atom, as the excited states mixed in by configuration interaction should also be considered. However, the Hartree-Fock values are still used, primarily because of the complexity of C.I. calculations but also because the correlation energy is generally regarded as a constant source of error, not changing greatly for a given atom upon incorporation into a molecule.

A good introduction to Hartree-Fock methods, together with a useful compilation of wave functions is given by Herman and Skillman. Because the functions are numeric, quantities such as overlap integrals must be calculated numerically; this is tedious, and is sometimes avoided by expanding the Hartree-Fock wave function as a sum of Slater type orbitals – the case where two S.T.O.s are used is referred to as the double zeta method, and enables overlap integrals to be calculated easily for a better atomic wave function than a single S.T.O. Wave functions are also sometimes expanded as sums of Gaussian functions as this simplifies the calculation of electron repulsion integrals.

f) Summary

The Schrödinger equation cannot be solved exactly for a many electron system, and consequently we introduce the concept of one electron wave functions and energies, which are combined to give the total wave function of the atom. The *Pauli principle* requires that the one electron functions be combined so that the final product is antisymmetric, and also requires that no two electrons can occupy the same spin orbital. The one electron functions are similar to those of hydrogen ($f_{nl}(r)\, Y_{lm}(\theta, \phi)$) but are no longer degenerate for a given n, the functions with higher l having a lower binding energy. By progressive filling of the lowest energy spin orbitals (the Aufbau principle) we may explain the structure of the periodic table in terms of the number of electrons in the valence shell.

The perturbation treatment of electron repulsion leads us to two forms of integral, one (the Coulomb, J) corresponding to a simple repulsion between charge densities, the other (the exchange, K) resulting from the Pauli principle, and leading to an increase in stability for systems with parallel (rather than paired) spins. These integrals may be used to give expressions for one electron energies, total energies, and ionisation energies. In general, closed subshells (all orbitals with a given n, l, filled) behave as spherical charge densities, but with a partially filled shell, the various distributions of electrons possible within this shell have different energies, giving different energy states of the atom which may be classified, according to their angular momentum properties, into *Russell-Saunders* terms.

Calculations of energy are generally in error as a result of *correlation effects,* which are due to the inaccuracy of the one electron model. The error involved is however roughly constant for each element, and if the parameters of the atomic structure are derived experimentally, the one electron model is quite accurate for describing atomic properties.

A simple qualitative model in which the energy of a given *nl* orbital is determined by its penetration effects (decreasing with increasing l) can explain the disposition of the periodic table, and can also explain the variation in ionisation energies and electron affinities. Finally the various methods of calculating atomic wave functions were briefly mentioned.

Bibliography

Texts marked with an asterisk present a more advanced treatment.

Atkins, P.W.: Molecular Quantum Mechanics. London: Oxford University Press 1970

Atkins, P.W., Child, M.S., Phillips, C.S.G.: Tables for Group Theory. London: Oxford University Press 1970

Ball, M.C., Norbury, A.H.,: Physical Data for Inorganic Chemists. London: Longman 1974

*Brink, D.M., Satchler, G.R.: Angular Momentum, 2nd Ed. London: Oxford University Press 1968

Chen, E.C.M., Wentworth, W.E.: Electron Affinities and Their Determination.J. Chem. Ed. *52*, 486 (1975)

*Chisholm, C.D.H.: Group Theoretical Techniques in Quantum Chemistry. London: Academic Press 1976

*Clementi, E., Raimondi, D.L.: J. Chem. Phys. *38*, 2686 (1963)

*Clementi, E., Raimondi, D.L., Reinhart, W.P.: J. Chem. Phys. *47*, 1300 (1967)

*Condon, E.U., Shortley, G.H.: The Theory of Atomic Spectra. Cambridge: Cambridge University Press 1970

Cotton, F.A.: Chemical Applications of Group Theory, 2nd Ed., London, New York, Sydney: J. Wiley 1971

Coulson, C.A.: Valence, 2nd Edn. London: Oxford University Press 1961

*Griffith, J.S.: The Theory of Transition Metal Ions. Cambridge: Cambridge University Press 1971

Hanna, M.W.: Quantum Mechanics in Chemistry, 2nd Ed. New York: Benjamin 1969

*Herman, F., Skillman, S.: Atomic Strucure Calculations. New Jersey: Prentice-Hall, Englewood Cliffs 1963

Herzberg, G.: Atomic Spectra and Atomic Structure. New York: Dover 1943

Huheey, J.E.: Inorganic Chemistry. London, New York: Harper and Row 1972

Jaffe, H.H., Orchin, M.: Symmetry, Orbitals, and Spectra. London, New York, Sydney: J. Wiley 1971

*Jørgensen, C.K.: Modern Aspects of Ligand Field Theory. Amsterdam: North Holland, and New York: Elsevier 1971

McGlynn, S.P., Vanquickenbourne, L.G., Kinoshita, M., Carroll, D.G.: Introduction to applied quantum chemistry. New York: Holt, Rinehart, and Winston 1972

Murrell, J.N., Kettle, S.F.A., Tedder, J.M.: Valence Theory, 2nd Ed. London, New York, Sydney: J. Wiley 1970

Newman, D.H., Leech, J.W.: How to Use Groups. London: Methuen-Science Paperbacks 1969

*Slater, J.C.: Spin-pairing energy. Phys. Rev. *165*, 655 (1968)

Urch, D.S.: Orbitals and Symmetry. London: Penguin Books 1970

Problems

1. For the following operators, identify which of the functions listed below are eigenfunctions, and give the eigenvalues.

$$\text{Operators: } \frac{d}{dx}\ ;\ \frac{d^2}{dx^2}\ ;\ x \rightarrow -x\ ;\ x \rightarrow x + \pi$$

(The operator $x \rightarrow -x$ changes x to $-x$ in any function of x)

$$\text{Functions: } e^{nx}\ ;\ \sin(nx);\ \cos(nx)\ ;\ x^3$$

2. Consider the basis functions of the I.R.s discussed in connection with the symmetry of the points P_1 to P_4 (page 17). Can you think of any atomic orbitals (or a combination of atomic orbitals) which might also be basis functions for these I.R.s?

3. If the operator $\mathscr{P}$ transforms as the I.R. E, the wave function $|a\rangle$ transforms as the I.R. A, and the wave function $|b\rangle$ transforms as the I.R. E, will $\langle a|\mathscr{P}|b\rangle$ necessarily be zero?

4. By considering the integral $< a \mid \mathscr{H} \mid a >$, where $\mid a >$ is any wave function, convince yourself that the Hamiltonian $\mathscr{H}$ must be totally symmetric.

5. Classify the following functions as symmetric or antisymmetric with respect to the operator $x \rightarrow -x$: x^2; $1/x$; x^5; e^{-x^2}. For which functions is the integral $\int_{-\infty}^{+\infty} f(x)\,dx = 0$? Comment.

6. The ionisation energy of copper is less than that of zinc, but copper has a higher electron affinity than zinc. Can you explain this?

7. It requires less energy to remove one electron from aluminium than from magnesium, but more to remove two electrons. Comment.

8. Calculate the possible values of S for a configuration [Ar] $3d^5$. What will the differences in energy between the different S values as given by 1.31?

9. The configuration $4f^2$ (as in Pr^{3+}, [Xe] $4f^2$) has the following possible Russell-Saunders states: ^1I, ^1G, ^1D, ^1S, ^3H, ^3F, ^3P. What will be the groundstate? Compare your result with Fig. 4.3. Tm^{3+} ([Xe] $4f^{12}$) has the same possible Russell-Saunders states. Will the groundstate be the same?

10. The Complex $[PtCl_4]^{2-}$ is planar and has four equal Pt-Cl bond lengths (a)

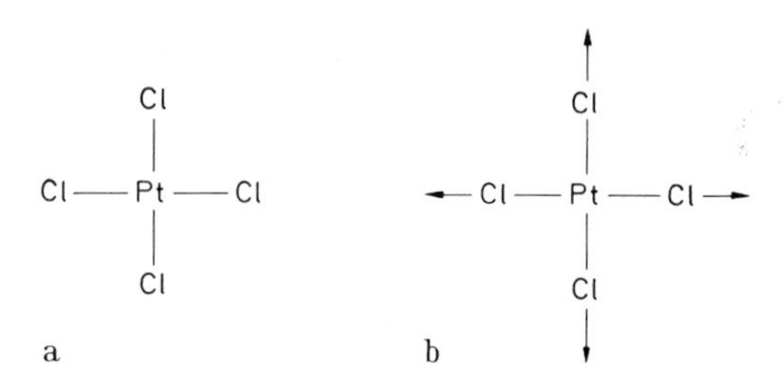

(i) The four Pt-Cl bonds form the basis for a representation of the symmetry group. Will this representation be reducible? If so, into which I.R.s? Would it still be reasonable to talk about four independent Pt-Cl bonds?

(ii) Consider the Pt-Cl stretching vibrations shown in (b). Is the representation for which they are basis functions reducible? If so, sketch the combinations of vibrations which are basis functions for the I.R.s, and predict which may be observed using Raman or infra-red spectroscopy.

2. Simple Molecular Orbital Theory

In the previous chapter we saw that the quantum theory of the atom gives a fairly good explanation of atomic behaviour, and can form the basis for quite accurate ($\sim$ 99%) calculations. Molecular systems are much more complicated, with many many-body interactions, and, in general, fairly low symmetry; it is not possible, for example, to give a general expression for an orbital wave function such as $f(\mathrm{r}) \cdot Y_{lm}(\theta, \phi)$. We seek a qualitative theory that we can easily apply to almost all chemical compounds, rather than a theory which will enable us to perform quite accurate calculations on a limited range of compounds.

Our aim is thus a theory which is simple, and will classify types of compounds and predict trends in their behaviour, rather than give numerical predictions. The theory should be compatible with the quantum theory of the atom, but its qualitative nature requires us to justify our assumptions by comparison with experimental results. The approach we shall use is that most commonly applied in inorganic chemistry today, the *Linear Combination of Atomic Orbitals* (LCAO) version of Molecular Orbitals (M.O.) theory.

A. Molecular Orbitals

For atomic structure we found that the one electron wave function (spin-orbital) with its one-electron Hamiltonian

$$\mathscr{H}_{\substack{\text{nuclear}\\\text{attraction}}} + \mathscr{H}_{\substack{\text{kinetic}\\\text{energy}}} + \mathscr{H}_{\substack{\text{electron repulsion}\\\text{energy}}}$$

was a very useful concept. If we study the electronic spectra of molecules, we find that they have many similarities to those of atoms, and we can again describe their spectra in terms of one electron jumps from an occupied to an unoccupied spin-orbital. We picture the total wave function of the molecule as a set of one electron wave functions combined in a Slater determinant, and again use the 'building-up principle' of filling the orbitals of lowest energy to predict the molecular structure. We find from X-ray spectra that the more tightly bound an electron in a molecule, the closer is its energy to that of the equivalent electron in an isolated atom. This leads us to assume that the tightly bound subshells of an atom are not appreciably perturbed by the chemical bonding, and consequently take little or no part in the bonding process.

The pattern of energy levels of the weakly bound electrons is completely changed, and these electrons are assumed to play a very important role in chemical bonding. We make the distinction between core shells, atomic orbitals left unchanged by any chemical bonding (apart from repulsive effects if two atomic cores are pushed too close together in a molecule), and valence shells, which dominate chemical behaviour. Although the distinction between core and valence shells is not absolutely clear cut, it is an empirical fact that chemical behaviour is shown by the Periodic table to be closely related to the composition of the valence shell of the most weakly bound electrons. Thus, for the alkali metals, the valence shell is the outermost s subshell, while for the first row transition metals it is composed of the 3d and 4s subshells.

The electronic spectra of the weakest bound electrons in a molecule show pronounced vibrational structure in their spectra, implying that they are strongly affected by the relative motions of nuclei in a molecule, and their energies change very considerably from one compound to another. We thus consider these electrons to be described by one-electron wave functions which are peculiar to a given molecule, the *valence molecular orbitals.*[1] To describe these valence shell M.O.s we shall use the *linear combination of atomic orbitals* model, in which $\psi_{M.O.}$ is written as a combination of (or an expansion in terms of) atomic orbitals ϕ_i:

$$\psi = \sum_i c_i \phi_i$$

There are other approaches, some of which are discussed in Chap. 5, but this one has two obvious advantages: 1) by discussing molecular behaviour in terms of functions of the isolated atom, we are relating the behaviour of molecules to that of their constituent atoms; and 2) the theory of atomic structure is sufficiently simple for us to have a fairly good understanding of the nature of atomic orbitals, and hence their behaviour in molecules; furthermore, the orthogonality properties of atomic orbitals will be very useful in calculation.

Let us consider where this leads us: for the hydrogen molecule, it is clear that the ls orbitals will be involved in bonding, and that the molecular wave function will be of the form

$$\psi = c_1 \phi_1 + c_2 \phi_2$$

where ϕ_i is the 1s orbital of hydrogen atom *i.* If we consider an alkali metal molecule such as K_2, we would expect that the only atomic orbital important for the bonding would be the outermost s orbital, in this case the 4s orbital, and the M.O. will have the form:

$$\psi K_2 = c_1 \phi(4s)_1 + c_2 \phi(4s)_2$$

It should be noticed that we have been using orbitals rather than spin-orbital wave functions. It is general in M.O. theory to assume that for every molecular orbital there are two molecular spin-orbitals of the same energy. For hydrogen, the two electrons are placed in the two molecular spin-orbitals with the lowest energy. In most cases we find that the stabilisation of the M.O. with respect to the atomic orbitals is much

---

1 The core orbitals, even if they are unchanged from the isolated atom, are still orbitals in a molecule, and thus formally molecular orbitals.

greater than the spin pairing energy required to doubly fill a given M.O. We will return to this topic later.

We can gain considerable insight into chemical bonding if we study the form of the Schrödinger equation for L.C.A.O. molecular orbitals.

Solution of the Schrödinger Equation

Our molecular orbital is given as a combination of a series of n atomic orbitals ϕ_i such that $\psi = \Sigma_i^n c_i \phi_i$.

Thus:

$$\mathcal{H}\psi = \Sigma_i^n c_i \mathcal{H}\phi_i = E \Sigma_i^n c_i \phi_i \tag{2.1}$$

we may use the variation theorem which requires the approximate wave function closest to the true value to be that with the lowest energy, to yield the requirement that

$$\Sigma_i^n \Sigma_i^n c_i c_j \mid \mathcal{H}_{ij} - ES_{ij} \mid = 0 \tag{2.2}$$

where $\mathcal{H}_{ij} = \langle \phi_i \mid \mathcal{H} \mid \phi_j \rangle$, $S_{ij} = \langle \phi_i \mid \phi_j \rangle$. The determinant $\mid \mathcal{H}_{ij} - ES_{ij} \mid$ is known as the *secular determinant,* and clearly it will be considerably simplified if symmetry arguments allow us to put some of the $\mathcal{H}_{ij}$ and S_{ij} to zero. The secular determinant has n values of E such that Eq. (2.2) holds; thus, if we start with n atomic orbitals, we obtain n molecular orbitals, one corresponding to each value of E, the energy of the molecular orbital.

For the hydrogen molecule, if we take the 1s orbital on each of two hydrogen atoms 1 and 2, we can form two molecular orbitals. The integrals we shall need are:

$$\langle \phi_1 \mid \phi_2 \rangle = S \tag{2.3}$$

$$\langle \phi_1 \mid \mathcal{H} \mid \phi_2 \rangle = \langle \phi_2 \mid \mathcal{H} \mid \phi_1 \rangle = \beta \tag{2.4}$$

$$\langle \phi_1 \mid \mathcal{H} \mid \phi_1 \rangle = \langle \phi_2 \mid \mathcal{H} \mid \phi_2 \rangle = \alpha \tag{2.5}$$

(the first equality in Eqs. (2.4) and (2.5) is evident from the symmetry of the system). The secular determinant is now

$$\begin{vmatrix} \alpha - E & \beta - ES \\ \beta - ES & \alpha - E \end{vmatrix} = 0 ,$$

yielding:

$$E_+ = \alpha + \frac{(\beta - \alpha S)}{1 + S} \tag{2.6}$$

$$E_- = \alpha - \frac{(\beta - \alpha S)}{1 - S} \tag{2.7}$$

The two corresponding unnormalised wave functions are:[2]

$$\psi_+ = (\phi_1 + \phi_2) \tag{2.8}$$

$$\psi_- = (\phi_1 - \phi_2) \tag{2.9}$$

The two functions are the molecular orbitals. We could equally have obtained them by symmetry arguments, since the two ls orbitals, in the linear symmetry of H_2,

---

[2] For the question of normalisation, see Problem 7, page 65.

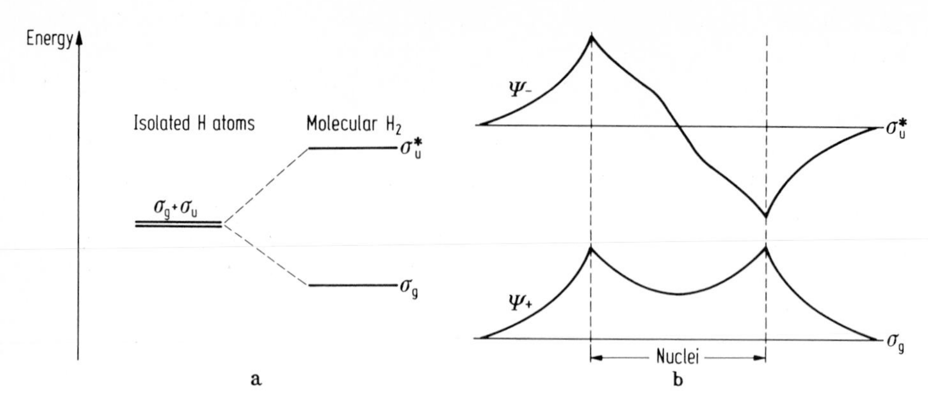

Fig. 2.1. (a) A molecular orbital diagram for H_2, showing the relationship between the atomic and molecular orbitals, **(b)** the wave functions ψ_+ (σ_g) and ψ_- (σ_u) along the internuclear axis. It will be seen that ψ_+ has a greater electron density between the nuclei

transform as the I.R.s $\sigma_u + \sigma_g$ of the point group $D_{\infty h}$ (see Appendix I). The function ψ_+ transforms as σ_g, while ψ_- transforms as σ_u. We may represent this by a M.O. diagram (Fig. 2.1 a) showing the splitting in energy of the two degenerate ls levels (on the left) as the hydrogen atoms are brought closer. The purpose of these diagrams is to represent the molecular orbitals of a species and their energy, and to show their relationship to the atomic orbitals of the isolated atoms. We label the orbitals by the I.R.s of the symmetry group for which they are basis functions.

When they are far apart, the two atoms hardly interact, and the effective symmetry (i.e. the microsymmetry) is spherical around each atom, but as the internuclear distance falls, it becomes more important to consider the true cylindrical symmetry.

The values of ψ_+ and ψ_- along the internuclear axis are shown in Fig. 2.1 b. It will be seen that ψ_+ will have increased charge density in the internuclear region, but that ψ_- has a nodal plane between the nuclei, and will have decreased charge density in this region. It turns out that ψ_- has a higher energy than ψ_+, and that ψ_+ is the more stable, or bonding, orbital; at least some of the stability of ψ_+ is associated with the increase in charge density in between the two nuclei, lowering the potential energy of the system as a result of the attraction due to both nuclei felt by the electrons. The symmetry of ψ_+ and ψ_- is cylindrical about the internuclear axis; since linear molecules are somewhat different from others in having an infinite axis of rotational symmetry, the Mulliken notation is not used for the I.R.s, but one adapted from spherical symmetry is used instead. Orbitals with no angular momentum about the axis are classed as σ, those with one unit as π, those with two as δ. If we consider an atom in a cylindrically symmetrical environment, with the z axis along the internuclear axis, then an s orbital has σ symmetry, a p_z orbital is invariant under the rotation operation and has σ symmetry, but the p_x and p_y may be mixed by rotation, and both have symmetry π. Similarly the d orbitals give one σ (d_{z^2}), two π (d_{xz}, d_{yz}), and two δ ($d_{x^2-y^2}$, d_{xy}) orbitals. If the atomic orbitals have this symmetry, then the molecular orbitals they form must also have this symmetry – we can thus classify our M.O.s by this treatment, giving σ_g and σ_u for the M.O.s of the H_2 molecule, the g and u subscripts referring to the change of sign on inversion in the centre of symmetry of the molecule; thus ψ_+ is g-type, ψ_- is u-type.

For linear molecules, this classification is exact, but for conceptual reasons, it is often convenient to regard a molecular orbital as localised between two atoms, even for molecules where the symmetry is not cylindrical. In such a case, the orbitals are classified according to the σ, π, δ convention, although the distinction is made because of the nature of the wave function, and not from the angular momentum properties. A molecular orbital is σ, π, or δ type according to whether it has 0, 1 or 2 nodal planes passing through the internuclear axis. A system such as the hydrogen molecule, where there are two electrons in the lowest (σ_g) M.O. is said to be σ-bonded. The σ, π, δ .. classification of M.O.s has considerable use, but it is worth noting that it is based on two assumptions:

1) That the M.O. can be regarded as concentrated in the region between two atoms only – the localisation condition.
2) That the symmetry in this region approximates cylindrical symmetry, i.e. cylindrical microsymmetry.

We shall discuss condition 1 later in the chapter.

Let us now reconsider the Schrödinger equation and the terms which contribute to the energiy of the system. The molecular Hamiltonian operator may be regarded as the sum of four terms:

$$= \mathscr{H}_{\text{atom}} + \mathscr{H}_{\text{kin.}} + \mathscr{H}_{\text{rep.}} + \mathscr{H}_{\text{ligand}}$$

$\mathscr{H}_{\text{atom}}$ represents the electrostatic potential energy due to the nucleus and core electrons of the atom on which a given atomic orbital ϕ_i is located; $\mathscr{H}_{\text{kin.}}$ is the kinetic energy term, and $\mathscr{H}_{\text{rep.}}$ allows for the electron repulsion effects of other electrons in the valence shell. $\mathscr{H}_{\text{ligand}}$ is the electrostatic potential energy term due to the surrounding atoms which we shall henceforth refer to as ligands.

We may now discuss the numerical importance of the three parameters α, β, and S in the secular determinant. $\alpha = \langle \phi_i \mid \mathscr{H} \mid \phi_i \rangle = \mathscr{H}_{ii}$ is sometimes referred to as the Coulomb integral.[3] It corresponds to the energy of the pure atomic orbital ϕ_i in the molecule; since the dominant potential acting on ϕ_i is that due to the electrons and nucleus of the atom on which ϕ_i is situated, $\mathscr{H}_{ii}$ is close to the one electron energy of ϕ_i in the isolated atom, and in calculations is frequently approximated by the I.E. of ϕ_i in the free atom. α is always negative (i.e. the orbital ϕ_i is stable with respect to spontaneous ionisation). However, $\mathscr{H}_{ii}$ must be corrected for the potential of surrounding ligands, and is also sensitive to the charge on the atom. A build-up of electron density on an atom in a molecule will cause a rise in orbital energies (and thus $\mathscr{H}_{ii}$) as a result of increased electron repulsion.

The overlap integral S, or, more generally, $S_{ij} = \langle \phi_i \mid \phi_j \rangle$, gives a measure of the overlap of the two orbitals, and thus of the extent to which an electron in the internuclear region will feel the attraction of both nuclei. The absolute value of S is dependant on the internuclear distance, but some overlaps will always be zero from symmetry considerations. Thus orbitals on the same atoms are orthogonal, and orbitals of different symmetry will also be orthogonal. The 1s orbital on hydrogen atom will overlap with the 1s, 2s, and $2p_z$ orbitals of the other (all σ symmetry), but not with the $2p_x$ and $2p_y$ orbitals (π symmetry). S varies from 0 to 1, but is rarely greater than 0.5.

---

[3] This name unfortunately leads to confusion with the J integral introduced in Chap. 1.

The final integral β (frequently called the resonance integral) is rather harder to visualise. The form of $\beta = \mathcal{H}_{ij} = \mathcal{H}_{ij} = <\phi_i | \mathcal{H} | \phi_j>$ recalls the expression for mixing of wave functions in perturbation theory (Eq. (1.19)), and we may crudely visualise it as the energetic consequence of the overlap of the atomic orbitals. The stabilisation due to build up of charge in the internuclear region thus appears in the β term. Since $\mathcal{H}$ is totally symmetric, $\mathcal{H}_{ij}$ will necessarily be zero if ϕ_i and ϕ_j have different symmetries. This condition recalls that for non-zero overlap of orbitals, and β is in fact very roughly proportional to S. Coulson and Blyholder discuss how the terms due to $\mathcal{H}_{\text{atom}}$, $\mathcal{H}_{\text{rep.}}$, and $\mathcal{H}_{\text{ligand}}$ are nearly proportional to S, while the term $<\phi_i | \mathcal{H}_{\text{kin.}} | \phi_j>$ is much more sensitive to S, being large for large S, but falling away rapidly as S decreases. It has been claimed by some authors (notably Ruedenberg and Jorgensen) that this kinetic energy term has considerable importance in the stabilisation of chemical bonds. For the rest of the book we will however assume a proportionality between β and S.

We may now reexamine Eqs. (2.6) and (2.7), using the knowledge that α and β are negative, and S is positive. As deduced from the form of the wave function, E_+ will have a lower energy than E_-, and, if the difference $(E_- - E_+)$ is greater than the spin pairing energy, the two electrons will occupy the σ_g orbital, to give a molecule with configuration $(\sigma_g)^2$ which is more stable than the isolated atoms. The stability of σ_g (and thus of the bond) will increase with increasing S and β. σ_g is said to be a *bonding* molecular orbital. σ_u has a higher energy than the isolated atomic orbitals, and is *antibonding*. It is traditional to mark antibonding orbitals with an asterisk $\sigma_u{}^*$.

It will be seen that the different denominators in Eqs. (2.6) and (2.7) result in E_- being more unstable with respect to the isolated atoms than E_+ is stable. Thus $H_2{}^{2-}$ with both σ orbitals occupied $((\sigma_g)^2 (\sigma_u{}^*)^2)$ is unstable with respect to dissociation into H^- ions. He_2, which would also have both σ orbitals occupied is unknown. This is an important result: *antibonding orbitals are more antibonding than bonding orbitals are bonding;* a molecule where all the bonding and antibonding M.O.s are occupied will be unstable to dissociation. The noble gases all have configurations $ns^2\ np^6$ in the valence shell; if we try to form a diatomic molecule with two noble gas atoms, we will start with 4 A.O.s on each atom, giving 8 M.O.s, but there will be 16 electrons to fill these – thus all bonding and antibonding orbitals will be filled and the molecules will be unstable. Atoms with filled shells will not form chemical bonds with each other unless one ore more electrons are excited into a low-lying empty orbital; it is then sometimes possible to form M.O.s which compensate for the excitation by the stability of the chemical bond so formed. The noble gases have no low lying orbitals to which an electron might be excited, and consequently cannot use this method to stabilise diatomic molecules.

It is a general observation that systems with full shells repel each other when brought into contact. We mentioned previously that core orbitals are ignored in the formation of molecular orbitals – this is tantamount to requiring that $\mathcal{H}_{ij}$ and S_{ij} are zero for ϕ_i or ϕ_j a core orbital. In the solution of the secular determinant, the energy of the orbital is left invariant, and it is mixed with no other orbital. If the internuclear distance is decreased so that $\mathcal{H}_{ij}$ and S_{ij} are no longer negligible, then any M.O. formed by the core orbitals will have both bonding and antibonding levels filled – the net effect will consequently be destabilising. This effect is one of the reasons for the non-implosion of compounds. The effects of repulsion between the cores are balanced with the chemical bonding effects of the valence shells, since Eq. (2.6) would lead us to expect that the stability of the bond would increase on decreasing the internuclear distance, and consequently increasing S. For the hydrogen molecule there is no core repulsion, but the two nuclei repel each other at short bond lengths.

The effects of core repulsion are, however, difficult to calculate, especially for heavy atoms with many electrons, and the inorganic chemist usually ignores them, apart from their limitation of the maximum value of the overlap integral, and discusses the bonding in term of the valence electrons.

The salient points of our treatment of H_2 are thus:

(i) From the two atomic orbitals we obtain two molecular orbitals, one stabilising, the other destabilising the molecule. The antibonding orbital is more antibonding than the bonding orbital is bonding.

(ii) The stabilisation is due to the build up of charge in the internuclear region, and is represented by the β ($\mathscr{H}_{ij}$) term. It is roughly proportional to overlap.

(iii) The stability of the molecule is dependent on the occupation of bonding and antibonding orbitals.

We may also notice that the electron density ($\psi^*. \psi$) is symmetric about the centre of the bond and that the bond is therefore non-polar. It is in fact the classic example of the covalent bond whose stability arises from the 'sharing' of electron density in the intermolecular region.

We can extend our treatment of the hydrogen molecule to the alkali metal vapours M_2. The alkali metals all have valence shells ns^1, and consequently we can draw the same sort of M.O. diagram as for H_2 with two σ orbitals giving a σ_g (bonding) and a σ_u (antibonding) orbital. The bonding M.O.s of the alkali metals are simply a doubly filled σ_g M.O., assuming that the core orbitals take no part in the bonding. The bonds in the alkali metals are very weak as the overlap falls sharply as the atomic size increases, and the atoms cannot move close together because of the repulsion between the nonbonding atom cores at low internuclear distance. In general, the stability of the molecules falls as the atomic number rises, as a result of falling overlap and increasing core-core repulsion.

B. Heteronuclear Diatomic Molecules

So far, we have only considered symmetric diatomic molecules. If we consider gaseous LiH as an example of the heteronuclear molecule AB, the plane of symmetry between the nuclei has disappeared, and the symmetry is now cylindrical $C_{\infty v}$. The valence orbitals are the hydrogen 1s (ϕ_1) and the lithium 2s (ϕ_2), each containing one electron. Both orbitals have σ symmetry, and may interact to give two σ molecular orbitals σ_1 and σ_2 (Fig. 2.2a). With two valence electrons only the bonding orbital will be filled, and the molecule will be stable.

The higher binding energy of the hydrogen 1s orbitals leads us to expect that the bonding pair of electrons will be more strongly attracted to the hydrogen atom, and that the bond will be *polar*. If we solve the secular determinant with $\mathscr{H}_{11} < \mathscr{H}_{22}$,

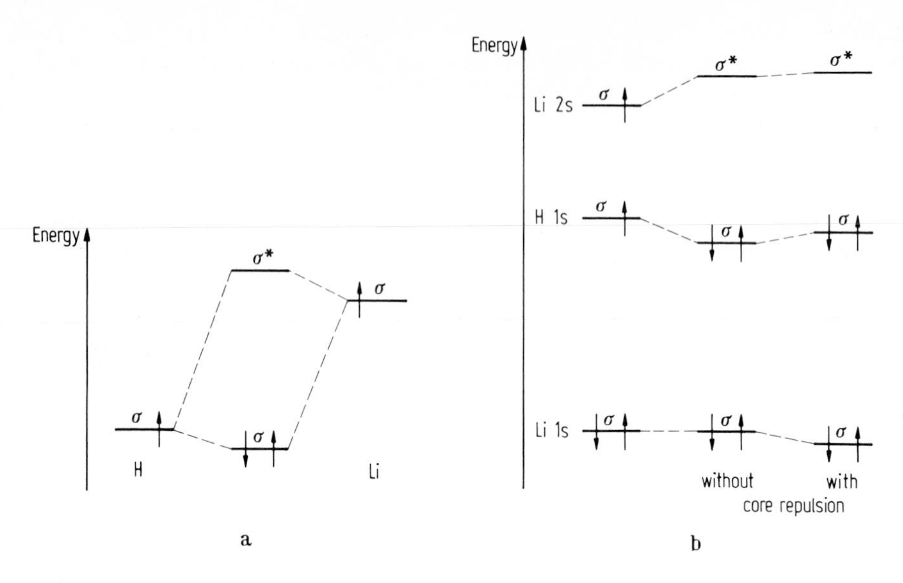

Fig. 2.2. (a) The M.O. diagram for gaseous LiH, **(b)** the same diagram allowing for core repulsion from the lithium ls orbital

we find that the σ_1 bonding M.O. lies at lower energy than $\mathscr{H}_{11}$ (hydrogen) and that the antibonding $\sigma_2{}^*$ M.O. lies above $\mathscr{H}_{22}$ (lithium). As our qualitative argument suggested, $|c_1| > |c_2|$ for the bonding M.O., and it is somewhat localised on the hydrogen, implying a negative charge on the H.

It is difficult to define exactly the charge on an atom in a molecule, since it is impossible to say where one atom stops and another begins in the region of the bond. The idea of charge on a atom is however very useful, and Mulliken suggested the following relationship: if the molecular orbital is given by $\psi = c_1\,\phi_1 + c_2\,\phi_2$, then the normalised wave function $\psi = \frac{1}{N}(c_1\,\phi_1 + c_2\,\phi_2)$ where $N = (c_1{}^2 + c_2{}^2 + 2c_1c_2S_{12})^{\frac{1}{2}}$.
The electron density on atom 1 is then $2/N^2\,(c_1{}^2 + c_1c_2S_{12})$ i.e. the population actually localised on atom 1 $(c_1\,\phi_1)^2$ and half the overlap population $(c_1c_2S_{12})$; this treatment has been criticised as a method for obtaining charges, but in view of its simplicity and the fictitious character of the charge, it is perhaps the best method available. For H_2 this gives a zero charge on each atom $(c_1 = c_2)$, but for LiH $(|c_H| > |c_{Li}|)$ this gives a negative charge on the hydrogen atom. This illustrates one of the attractive features of LCAO M.O. theory, that it is able to describe simply the variation between pure covalent bonds $(c_1 = c_2)$, polar covalent $(c_1 \neq c_2)$ and pure ionic $(c_1 = 1, c_2 = 0$ or $c_2 = 1, c_1 = 0)$; even if it is difficult to calculate energies accurately, it is capable of a simple description of electron distribution in molecules.

We may represent the effects of core repulsion on our M.O. diagram (Fig. 2.2b). The lithium ls orbital has σ symmetry also, and consequently $\mathscr{H}_{ij}$ and S_{ij} terms are not necessarily zero. The orbital is clearly too strongly bound to allow appreciable interaction with the H atom, but clearly some interaction will exist. We proceed as before – the lower σ orbital (Li ls) is lowered slightly, the higher σ orbital (bonding

M.O.) is raised slightly – the remarks made earlier about the interaction between two filled shells (σ(ls) and σ_{bond}) lead us to expect that the total effect will be slightly anti-bonding i.e. there is a slight repulsion between the ls core and the bonding M.O. Henceforth we will ignore this slight mixing of the core orbitals and concentrate our discussion on the qualitative features of the valence orbitals.

Two factors now affect the stability of the bond: electrostatic attraction of the electron density towards the lower energy orbital showing up in the $\mathscr{H}_{ii}$ term, and the covalent 'electron sharing' term $\mathscr{H}_{ij}$. In the case of LiH, the value of $\mathscr{H}_{11}$ will be raised by the increased electron repulsion as the hydrogen atom acquires a negative charge, but slightly lowered by the electrostatic potential of the nearby, positively charged lithium atom. Part of the stability of gaseous LiH comes from the double occupation of an orbital close in energy to the hydrogen atomic orbital, compared with single occupation of the hydrogen and lithium orbitals in gaseous Li and H.

The $\mathscr{H}_{ij}$ term is responsible for the mixing of atomic orbitals; we thus expect a covalent stabilising contribution due to $\mathscr{H}_{ij}$. The size of this contribution is dependent on the energy difference $|E_i - E_j|$ (roughly equal to $|\mathscr{H}_{ii} - \mathscr{H}_{jj}|$) between the two atomic orbitals. This effect is shown in Fig. 2.3. If ϕ_i is the lower energy orbital, then the bonding orbital is lowered much more relative to ϕ_i if ϕ_j has only slightly greater energy than if ϕ_j has much greater energy. The greater the energy of ϕ_j (for fixed $\mathscr{H}_{ij}$), the closer the bonding orbital will be to ϕ_i in energy and in character. This implies that $c_i \rightarrow 1$, $c_j \rightarrow 0$ for the bonding orbital. Our discussion of c_i implies that the bonding becomes more and more ionic as $|E_j - E_i|$ becomes greater. For large $|E_j - E_i|$ it can be shown that the bonding and antibonding orbitals have energies given approximately by:

$$E_i\,(\text{bonding}) = \mathscr{H}_{ii} - \frac{\mathscr{H}_{ij}^2}{|\mathscr{H}_{jj} - \mathscr{H}_{ii}|}$$

$$E_j\,(\text{antibonding}) = \mathscr{H}_{jj} + \frac{\mathscr{H}_{ij}^2}{|\mathscr{H}_{jj} - \mathscr{H}_{ii}|} \tag{2.10}$$

This is an important result since it tells us that the chemical bonding energy will be almost entirely dependent on $\mathscr{H}_{ii}$ when the orbital energies of i and j are far apart, no matter how large the overlap and $\mathscr{H}_{ij}$. This is the basis of the ionic model which sets out to calculate the values of $\mathscr{H}_{ii}$ from data on free gaseous ions and the potential energy effects due to the charged ligands. In fact, our example of LiH is very well described by the ionic model in the solid state. The ionic model is discussed further in

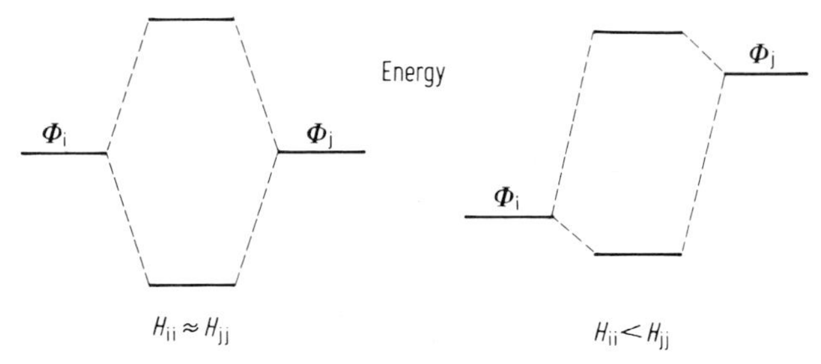

Fig. 2.3. The effect of orbital energy on the formation of molecular orbitals

Chap. 5. Another useful result is that the antibonding character of ϕ_j due to covalence is given by $\mathscr{H}_{ij}^2 / | \mathscr{H}_{ij} - \mathscr{H}_{ii} |$; this is the basis of the angular overlap model for antibonding effects of d orbitals in transition metal complexes, where the antibonding nature of an orbital is taken as proportional to the square of its overlap integral with low energy ligand orbitals (the overlap integral is assumed proportional to $\mathscr{H}_{ij}$).

The foregoing discussion enables us to draw some useful conclusions on the nature and stability of chemical bonds:

1) If $E_i = E_j$ the bonding will be essentially covalent, increasing in strength with $\mathscr{H}_{ij}$ and by extension, with increasing overlap. Examples: H_2, K_2.
2) If $E_j \gg E_i$, $\mathscr{H}_{ij} \approx 0$ the bonding will be essentially ionic with the bonding orbital consisting mainly of ϕ_i. Example: NaF.
3) If E_j is only slightly greater than E_i and $| \mathscr{H}_{ij} | > 0$, then the bonding will be covalent, but polarised, with a greater component of ϕ_i in the bonding orbital. Examples: LiH, HCl.

We would predict a non-bonding state if:

1) $E_j \gg E_i$, and ϕ_i is already doubly occupied, as for a tightly bound core orbital. If $E_j \approx E_i$ and ϕ_i is doubly occupied (e.g. the lone pair of a donor nitrogen atom in NH_3), bonding is still possible as a result of sharing and partial transfer of large with ϕ_j (e.g. an acceptor orbital in BF_3) if $\mathscr{H}_{ij}$ is sufficiently large i.e. if there is a good overlap.

2) $E_i \approx E_j$, $\mathscr{H}_{ij} \approx 0$. The approximately equal energies of i and j do not favour charge transfer, and the value of $\mathscr{H}_{ij}$ is too small to allow much orbital mixing and covalency. This situation arises frequently with compounds of the rare elements and some lighter transition metals.

We may use these deductions to explain the comparative rarity of paramagnetic molecules (i.e. radicals) since the unpaired electrons will be in the valence shell (the aufbau principle), and consequently not very strongly bound. If we take two molecules of the same compound, then the orbital (with the unpaired electron) on each atom will have the same energy, and if the overlap energy is greater than the spin pairing energy of the bonding M.O. the dimerisation will be favourable (Fig. 2.4). For most compounds the overlap is sufficiently great for the bonding energy to exceed the spin pairing energy. Consequently undimerised radicals are rare, and one finds $Mn_2(CO)_{10}$ rather than $Mn(CO)_5$, N_2H_4 rather than NH_2, and NO_2 dimerises readily to give N_2O_4. NO and O_2 are stable as radicals with unpaired spins since the overlap in the dimer is poor. Paramagnetic compounds are very common in transition metal and rare earth chemistry where the overlap between two metal atoms is poor if not negligible, and the unpaired electrons are in orbitals which do not project greatly from the core of the atom.

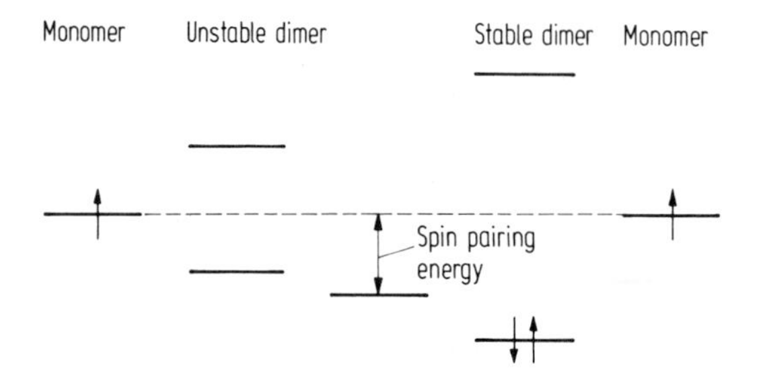

Fig. 2.4. The dimerisation of radicals

C. Further Applications of M.O. Theory

We now move to molecules with more than two atoms. MgF_2 is a solid in normal conditions, but in the gas phase is a stable linear molecule. We will assume to begin with that the $2p_z$ and $2p_y$ orbitals of the fluorine take no part in the bonding. If we solve the secular equation for the two fluorine $M2p_z$ orbitals ϕF_1 and ϕF_2 and the magnesium 3s orbital we obtain a bonding orbital, an antibonding orbital, and an orbital $\frac{1}{\sqrt{2}}$ $(\phi F_1 - \phi F_2)$ whose energy is close to that of the original fluorine orbitals[4] – this orbital is said to be non-bonding and also has symmetry σ. This is the situation shown in Fig. 2.5 i. We could equally well have done this by noting than the ϕF_1 and ϕF_2 transform as σ_u $(\phi F_1 - \phi F_2)$ and σ_g $(\phi F_1 + \phi F_2)$ and that only σ_g will interact with the magnesium 3s (σ_g). It is clear that with four electrons (two from the Mg, one each from the fluorine atoms) the molecule will be stable.

We may note that the magnesium $3p_z$ orbital energy is only slightly above the 3s σ_g, (about 25000 cm^{-1}). This orbital points along the internuclear axis and we would expect the overlap with the fluorines would be quite good from size considerations, and its symmetry σ_u will enable it to overlap with the nonbonding fluorine M.O. There will be a consequent stabilisation of the molecule by mixing in a little Mg $3p_z$ character with σ_u. We now have four M.O.s – two bonding and two antibonding, and the molecule will be more stable (see Fig. 2.5 ii). This shows an important result – that unoccupied atomic orbitals can take part in bonding provided:

1) they are not too far away energetically
2) the overlap with the surrounding atoms is sufficiently large. (and consequently that they are of correct symmetry)

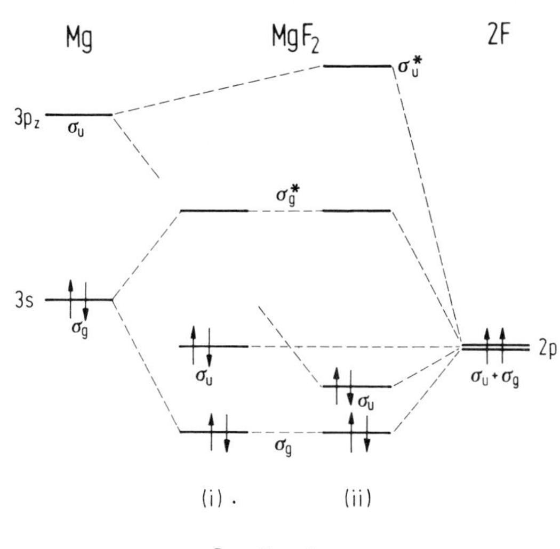

Fig. 2.5. The M.O. diagram for MgF_2. In part (i) the magnesium $3p_z$ orbital plays no part in the bonding; in the second part, it stabilises the σ_u orbital

[4] It is exactly equal if there is no overlap between ϕF_1 and ϕF_2 and electron repulsion is ignored.

Thus the magnesium 4s orbital can be ignored for the first reason, and the magnesium $2p_x$ for the second. The fluorine 2s orbital probably overlaps with the magnesium orbitals almost as much as the 2p orbital, but its strong binding energy makes it unlikely to indulge in any real extent in bonding. Finally we may note that the bonding orbitals are delocalised over all three atoms, each bonding orbital having a component from each atom. One may therefore describe the molecule as showing three centre, four electron bonding. Both bonding orbitals have σ symmetry type, and because there are two bonding orbitals filled and both antibonding orbitals empty, we may say that there are two σ bonds in the molecule.

For more complicated molecules it is much simpler to use symmetry arguments from the beginning. Thus for methane, a tetrahedral molecule with symmetry T_d, Appendix I tells us that the 2s orbital of carbon transforms as a_1 and the 2p orbitals as t_2. The hydrogen valence shell is the 1s orbital, and along the C–H internuclear axis it has σ type symmetry; Appendix I states that the 4 σ type ligands orbitals in T_d symmetry transform as $(a_1 + t_2)$. We are thus led to the M.O. diagram in Fig. 2.6. Again we have all bonding orbitals filled and antibonding orbitals empty, a stable bonding situation; methane is in fact a very stable molecule. Confirmation of our M.O. diagram comes from ultraviolet photoelectron spectroscopy (Chap. 8), which shows two low ionisation potentials of methane, one at 14 eV, and another, less intense, at 23 eV; it is easy to assign these to the t_2 and a_1 orbitals respectively, the more strongly bound M.O. associated with the carbon 2s having the higher ionisation energy.

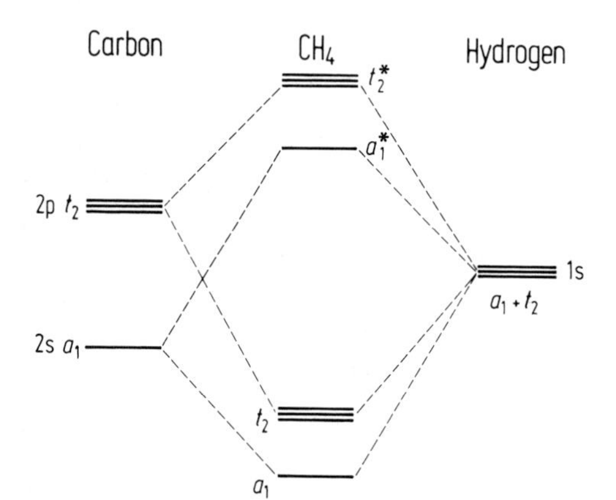

Fig. 2.6. The M.O. diagram for methane, CH_4. Only the t_2 and a_1 M.O.s are occupied

It is clear that the tetrahedral ions BH_4^- and NH_4^+ will have the same number of electrons, and the molecular orbital diagram will be qualitatively similar. On crossing the periodic table however, the 2s orbital becomes progressively lower in energy (as a result of its strong penetrating powers and the increasing nuclear charge), and from the general arguments given earlier we would expect to find the EH_4 system less stable. In fact NH_4^+ is less stable thermally than BH_4^- and CH_4, and in the second row PH_4^+ is very unstable, dissociating to PH_3 which has much less s character in its bonding, while AlH_4^- and SiH_4 are relatively stable, H_4O^{2+} is unknown.

Compounds with the same number of electrons such as NH_4^+, CH_4, and BH_4^- are said to be isoelectronic and it is frequently found that they have the same structure.

This has been used as evidence for a method of structure prediction based on the number of electrons in the *valence shell* (Chap. 5), and we can see that with the same number of electrons one might expect similar M.O. diagrams and hence the same structures to be found. There are also similarities in structure in compounds with the same number of valence electrons (for which compounds Jørgensen has suggested the term isologous) such as SiH_4, CH_4, NH_4^+, PH_4^+, all with 8 electrons in the valence shell and tetrahedral structures.

Once we know the geometrical structure of the compound, we may use symmetry arguments to make predictions about the electronic structure. Unfortunately, the prediction of the geometrical structure is not always straightforward. In theory it is necessary to calculate all the possible configurations and their energy, and then choose the structure with the lowest energy. This is tedious, and in view of the errors in M.O. calculation, quite likely to give the wrong answer. There exist simple methods for predicting structure, and careful consideration of the M.O. diagrams arising from these can tell us when the approximations are likely to fail.

Bond localisation and hybridisation

If we were to calculate the values of $|\psi|^2$ for the various one electron molecular orbitals of CH_4 we would find that they were distributed over the whole molecule (as in MgF_2); the M.O. are highly delocalised. This approach is extremely useful for explaining the molecular spectra, but is not so useful for studying the electron distribution (since we have to sum $|\psi|^2$ for all 4 M.O.s). We may also wish to study the exact nature of a particular CH bond. There is much evidence from the additivity of bond energies, internuclear distances (the sum of covalent radii) and other evidence that a given CH bond does not change very much whatever the other groups bonded to the carbon. This constancy is not really predicted by qualitative molecular orbital theory. The early descriptions and theories of valence regarded a chemical bond as an essentially local link between neighbouring atoms A–B. This independent bond approach is not suitable for all descriptive purposes, but as an approximate approach it has the advantage of simplicity, and for complicated molecules where it would be necessary to look at orbitals delocalised over more than a hundred atoms it is essential.

If we consider MgF_2 again we may consider the combination of Mg σ orbitals $\phi_a = \frac{1}{\sqrt{2}}(\phi_{2s} - \phi_{2pz})$, $\phi_b = \frac{1}{\sqrt{2}}(\phi_{2s} + \phi_{2pz})$. These two orbitals are orthonormal, and their spatial distribution (Fig. 2.7) is interesting as they each have the bulk of their electron density concentrated on only one side of the atom. We may thus expect that ϕ_a will interact strongly with ϕF_1 but not with ϕF_2. Similarly ϕ_b will overlap with ϕF_2 but not ϕF_1. The bonding with ϕF_1 is thus strongly localised between ϕ_a and ϕF_1,

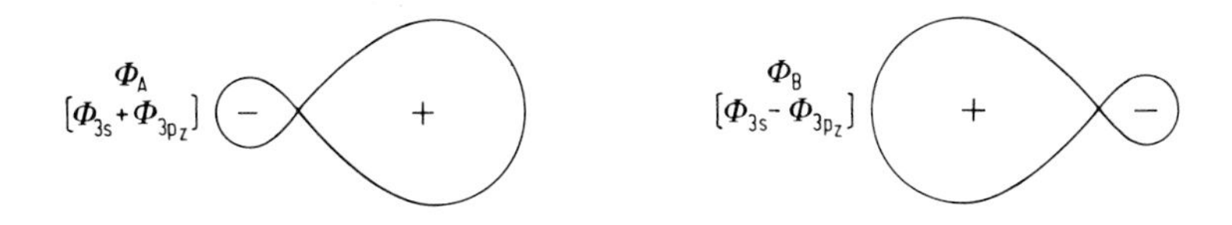

Fig. 2.7. Localised orbitals for MgF_2

and we have achieved a localisation of the bonding. The two orbitals ϕ_a and ϕ_b are said to be equivalent orbitals. They are mixtures of atomic s and p orbitals, and are clearly not themselves atomic orbitals; they are referred to as *hybrid orbitals*. The mixture of atomic orbitals to give equivalent orbitals is called *hybridisation*.

The process of hybridisation can be regarded as a rearrangement of the Slater determinant without changing its value. For MgF_2 before, the bonding orbitals where $c_1\phi_{2s} + c_2(\phi F_1 + \phi F_2)$ and $c_3\phi_{2p} + c_4(\phi F_1 - \phi F_2)$, and the total wave function would be a Slater determinant including these orbitals. We can rearrange the elements *inside* the determinant (according to certain mathematical laws) without changing its value, to give elements of the form $(c'_1\phi F_1 + c_2'\phi_{2s} + c_3'\ \phi_{2p})$ and $(c_4'\phi F_2 + c_5'\phi_{2s} + c_6'\ \phi_{2p})$ where ϕF_1 and ϕF_2 each appear in one column only. The metal parts of these M.O.s $(c'_2\phi_{2s} + c'_3\phi_{2p})$ and $(c'_5\phi_{2s} + c'_6\phi_{2p})$ are the components of the metal hybrid orbitals in the bonding M.O.

We may now see why the hybrid orbital description is unsuitable for interpreting spectral data – the rearrangement, while leaving the *total* wave function unchanged, involves making linear combinations of orbitals which are approximate one electron wave functions. These linear combinations will not themselves be eigenfunctions of the one electron Hamiltonian operator, and consequently unsuitable for the description of one electron changes associated with electronic spectra.[5]

More qualitatively the failure of localised orbitals to explain spectral data may be explained by asking the question why should an electron be excited out of one of the hybrid orbitals rather than the other; the M.O. description of this pictures the excitation as being spread over the whole molecule. This is reassuring, as many compounds have absorption peaks in the ultraviolet at energies greater than their localised bond energies.

Localisation of molecular orbitals, and the use of hybrid orbitals played important roles in the early development of bonding theory, and in some cases these concepts have been misapplied; a fairly detailed discussion is thus useful. Historically, localisation is strongly connected to the idea of the electron pair bond, in which each chemical bond involves an exact number of electron pairs which are responsible for the bond. This idea is quite in accord with the hydrogen molecule with two electrons in one bonding orbital linking the two nuclei. The structures of a very large number of covalent compounds (often quite complicated) can be broken down into localised electron pair bonds, and the simplicity of the idea has rendered it very attractive.

Unfortunately, there exists a large number of compounds for which localised electron pair bonding schemes cannot be drawn (e.g. B_2H_6, see Chap. 3), or can be drawn but obscure some important essential features of the bonding (for example electronic spectra, and the 6 equal bond lengths in benzene). Molecular orbital theory can deal adequately with these problems, but if delocalisation over very large numbers of atoms is considered, M.O.s become very difficult to visualise. In general, a balance is struck, in which some orbitals are regarded as localised, and others are treated by delocalised M.O. methods; we must now consider which orbitals to treat in which fashion.

We have established before that core atomic orbitals overlap feebly with each other,

---

[5] Note also that hybrid orbitals are basis functions for *reducible* representations of the symmetry group – for MgF_2 this representation is $(\sigma_u + \sigma_g)$.

are little changed in energy, and do not mix – quite clearly we can regard all core orbitals as localised on their parent atoms, and concentrate only on the valence shell. The condition for drawing a localised bond structure is that the number of electrons involved in the bonding should be twice the number of bonds; if this is not true then we must expect to have to treat at least some of the bonds by M.O. methods. It is not normally necessary to consider M.O.s delocalised over more than a few atoms.

Even if it is possible to draw a completely localised bonding system, it is important to notice if there is any physical evidence to contradict such a picture: the classical Kekule structure of benzene suggests that the molecule has three short and three long C–C bonds, contrary to the experiment – one should thus be very hesitant about using the Kekule structure to explain molecular properties.

Hybridisation is an attempt to combine the valence shell atomic orbitals of a central atom in such a way as to produce new orbitals which are directed towards surrounding atoms or ligands. The bond between the central atom and a given ligand then only involves overlap between the ligand orbital (or orbitals) and the hybrid (or hybrids) pointing at this ligand. In the case of MgF_2, the hybrid orbital ϕ_a is directed towards ϕF_1 and the Mg-F_1 bond arises from their overlap. For such a localisation to work, it is clear that we require that the interaction between ϕF_1 and ϕ_a is much greater than the interaction of ϕ_a with ϕF_2 and of ϕF_1 with ϕ_b or ϕF_2.

Describing the exact nature of the hybrid is rather more difficult; the sp hybrid orbitals $\lambda\,(\phi_{2s} + \phi_{2p}), \lambda'\,(\phi_{2s} - \phi_{2p})$ have the directional properties shown in Fig. 2.7, but also require that the 2s and 2p orbitals participate equally in the bonding. The energy difference of ϕ_{2s} and ϕ_{2p}, and the probable difference in their overlap integrals makes this highly unlikely, and it is thus unwise to regard the hybrid orbital as an equal mixture of s and p. Clearly, the closer the hybrid is to either pure s or pure p, the less it will be concentrated on one side or the other of the central atom – we expect the localisation into hybrids to be less succesful in these cases since a pure s (or pure p) orbital overlaps equally with both ligands. A requirement for successful hybridisation is thus approximately similar energies of the atomic orbitals, and radial functions which are not too different, giving approximately equal overlap integrals with the ligands. In the case of magnesium this is probably not too far from the case; if, however we had attempted to hybridise the fluorine 2s and 2p orbitals, the great difference in the energies of the orbitals (ionisation potentials differing by ~22 eV, or 2000 kJ/

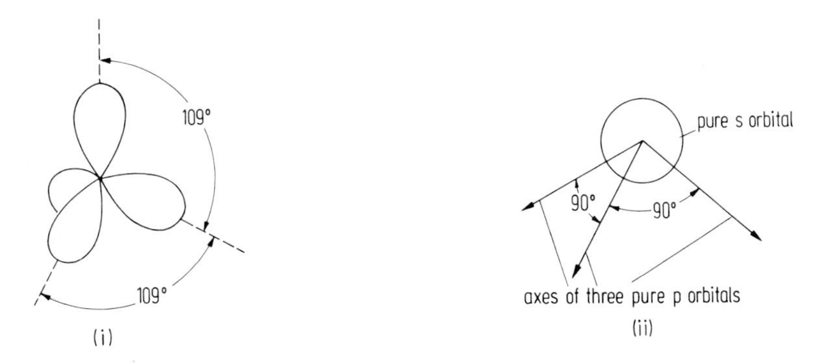

Fig. 2.8. Hybridisation. (i) sp^3 hybrids – four identical tetrahedral orbitals, (ii) no hybridisation – three p orbitals at 90° to each other

mole) would have made this rather unrealistic picture. The element for which hybridisation (and consequent localisation) works par excellence is carbon.

Whatever the relative amounts of s and p orbital in the MgF_2, the two hybrids will have linear symmetry; it has thus become common to associate sp hybridisation with linear symmetry. If we combine two p orbitals with one s orbital the result is three orbitals directed at 120° to each other; while the third p orbital is perpendicular to the plane – this gives planar threefold symmetry (D_{3h}). If the s and p orbitals are imperfectly mixed the two more p like orbitals move closer together, and the more s like becomes more spherical and less "directed". If we combine all three p orbitals with the s orbital we arrive at four directional orbitals pointing towards the vertices of a tetrahedron – if the s orbital participates only slightly, the p-like orbitals move towards a limit where they point down axes at right angles, while, as before, the s becomes more spherical (Fig. 2.8). It is equally possible to construct hybrids from d orbitals (see Problem 10, page 65), notably

dsp^2 – square planar, 4 bonds at 90°
dsp^3 – trigonal bipyramid, 3 bonds at 120°,
two at right angles to this plane
d^2sp^3 – octahedral, six bonds, pointing down Cartesian axes.

In view of our previous remarks about energy and overlaps, we could however expect a hybrid containing d, s and p orbitals to be a rather unrealistic approximation.

Assuming that a useful hybridisation and localisation scheme can be drawn up, it does present certain advantages: the simplicity of the electron pair bond is clear, and, given the structure, we may assign an approximate hybridisation scheme which will allow us to predict the electron distribution; this is particularly true where the valence atomic orbitals are not all localised into bonds. For NH_3, for example, the H-N-H angle is 107°, and the molecule is pyramidal; for pure p orbital bonding we would expect a bond angle HNH of 90°, so there appears to be an appreciable s contribution – the angle is in fact close to the tetrahedral angle of 109°, and we may imagine the two electrons not involved in electron pair bonds to the hydrogen atoms to be in a "lone pair" sp^3 hybrid orbital. The directional properties of this lone pair suggest it may overlap well with an acceptor orbital, and NH_3 is well known for its powers as a Lewis base, or electron pair donor. If the fourth ligand is a proton we have the tetrahedral NH_4^+ in which we would expect to be fairly well described by sp^3 hybrid orbitals.[6] For the heavier group VB hydrides, the bond angles diminish: PH_3 93.3°, AsH_3 91.8°, SbH_3 91.3° implying a much smaller s orbital participation in the bonding and that the lone pair has much more s character; the basicity of these hydrides is much lower, implying, not unexpectedly, that the spherical s orbital is less inclined to act as an electron pair donor. A calculation for NH_3 gives the lone pair 40 percent s character, implying it is not quite a pure sp^3 orbital.

A similar treatment for H_2O with a bond angle of 103.5 predicts two lone pairs, although the bond angle predicts that they have rather more s character than in NH_3. The approximately tetrahedral disposition of lone pairs and bonding pairs is confirmed by the structure of ice, where two hydrogen atoms attached to other oxygen atoms are found close to the oxygen atom giving a roughly tetrahedral coordination. H_2O is however a weaker base than NH_3. The concept of the lone pair is extremely useful in predicting reactivities, and recent very accurate X-ray diffraction data has shown that

[6] See, however, the remarks on s participation on page 50.

there is in fact an increase of electron density in the direction predicted by a simple hybrid orbital. It is worth noting that a complete M.O. calculation should show the same electron distribution as the simple hybrid approach based on bond angles, but the hybrid approach is a great deal easier to apply. The advantages of regarding the lone pair as localised may be clearly seen in triphenylphosphine chemistry: triphenylphosphine $(C_6H_5)_3P$ acts as an electron pair donor in many complexes to give $(C_6H_5)_3PX$; in such cases the C–P–C bond angles change little from the uncoordinated value, and for drawing the molecular orbitals involving the X group, it is much easier to start with one donor lone pair on the phosphorus, than a M.O. delocalised over all 34 atoms. The localisation may be further justified by the very small (though sometimes noticeable) changes in the spectroscopic properties of the C_6H_5 groups on coordination to X. The chlorides $CH_{4-x}Cl_x$ show the partial failure of the localisation condition: on a localised model we would expect the polarity of the C–Cl bond to remain roughly constant, but as shown by n.q.r. spectroscopy (Chap. 8) the polarity falls roughly linearly as x rises from 1 to 4.

The widespread use of hybrid orbitals has led to certain misconceptions which it is as well to point out: hybridisation theory does not predict structures since the hybridisation is only deduced from the known structure; it is possible for some elements (e.g. carbon) to estimate the probability of s–p hybridisation, and given the formula, to propose a series of structures from which other considerations may suggest one particular structure to be stable. It is often claimed that the hybrid is chosen to give the maximum overlap between central atom and ligand – this is not quite true, although clearly the higher the overlap between orbitals the stronger the interaction. For carbon the greatest overlap for a given C–H bond is actually obtained with s–p hybrid rather than an sp^3 which is more commonly found. The energy difference between the constituent atomic orbitals of a hybrid must also be allowed for. The term *promotion energy* is sometimes used to describe the energy necessary to produce the so-called valence state where each orbital constituent of the hybrid is singly occupied e.g. sp^3 is the valence state for a carbon atom in methane while s^2p^2 is the ground state.

The disadvantage of this term is that the configuration sp^3 will have a large number of Russell-Saunders terms, so that a weighted average will need to be taken, and also that the actual degree of participation in bonding is unlikely to be equal for s and p orbitals; the state sp^3 as the valence state is thus rather hypothetical. Promotion energy is probably useful as a qualitative estimate of the probability of participation of otherwise unoccupied orbitals (i.e. in sp^3, the third p orbital), but not as a quantitative concept.

After this rather lengthy discussion, let us summarise the progress we have made so far: localisation of bonds is not always possible, but to localise as many orbitals as possible will simplify our treatment. Hybridisation is also frequently unrealistic as a result of energetic considerations and is not essential for M.O. theory (although it is for V.B. theory, Chap. 5). It is sometimes claimed that an element has a most favourable hybridisation (e.g. sp^3 for carbon) but this idea should be regarded with suspicion: almost all carbon compounds are unstable to combustion to CO_2 which, if anything, is sp hybridised (see later). It is however true that compounds with low coordination number (C.N.) (e.g. 3 in BF_3 – sp^2 hybridised) are often unstable with respect to a species with a higher C.N. (BF_4^-, C.N. = 4, sp^3 hybridised). This is however more a matter of the stability of bonds formed, a subject to which we return later; with oxygen for example boron remains planar and approximately sp^2 hybridised. Henceforth we shall use hybridisation as an approximation, based on information from bond angles etc., to give a rough picture of the electron distribution (e.g. lone pairs), and attempt to localise as many bonds as possible (as for example with PPh_3) in order to simplify the discussion. We will avoid localisation when we wish to discuss electronic spectra or when we have evidence that localisation is not possible (for example unexpected structures or bond lengths).

D. Multiple Bonds

In discussing any bond it is useful to consider the symmetry of the electron density in the region of the bond. We may consider the internuclear axis to be an infinite axis of rotation. In the case of MgF_2 which is linear, we saw all the orbitals involved in the bonding had σ symmetry about the internuclear axis. In one localised MgF bond, the fluorine $2p_z$ orbital and the magnesium sp hybrid both have cylindrical symmetry and are thus σ type. The localised bond between these two orbitals is thus a sigma type bond. The case of methane is a little more complicated, but if we consider a localised CH bond, both the hydrogen ls orbital and the carbon sp^3 hybrid pointing towards it have complete symmetry with regard to rotation about the bond axis, and are thus σ type. The same argument can obviously be applied to the other CH orbitals, and we may say that methane is bonded by 4 localised σ bonds.

If we consider the p_x and p_y orbitals of the magnesium and fluorine atoms, these have a plane of symmetry running down the internuclear axis. Such orbitals transform as the π irreducible representations in cylindrical symmetry and consequently give rise to π type molecular orbitals. Similarly d_{xy} and $d_{x^2-y^2}$ orbitals have two perpendicular planes of symmetry in the z axis, and have δ type symmetry.

It is of course quite straightforward to form bonding and antibonding π orbitals following the same rules as for the σ type orbitals we used for H_2. A M.O. diagram for a diatomic molecule E_2 with s and p orbitals only in its valence shell is shown in Fig. 2.9. The sigma orbitals are much the same as before, although there will be a slight repulsion between $1\sigma_g$ and $2\sigma_g$, $1\sigma_u^*$, and $2\sigma_u^*$, lowering the first and raising the second of each pair (as for the discussion of core repulsion in the LiH molecule).

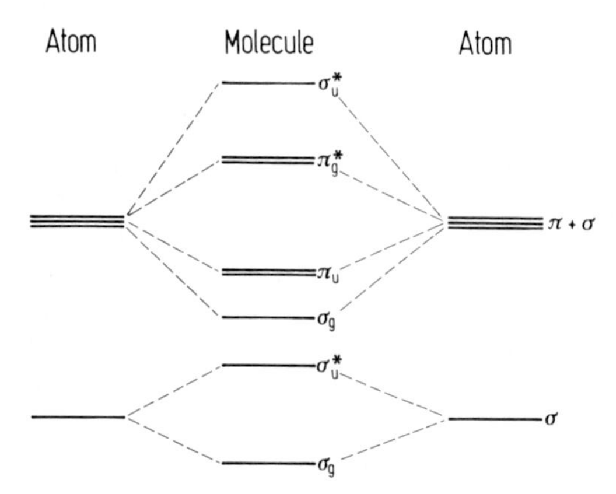

Fig. 2.9. The molecular orbital diagram for a diatomic molecule E_2, for an element E with only s and p orbitals in the valence shell

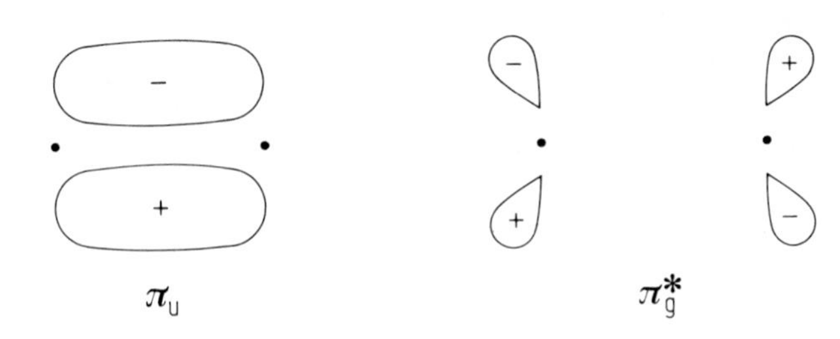

Fig. 2.10. The π molecular orbitals of E_2

The π orbitals are simply visualised (Fig. 2.10.), but now the u combination is bonding and the g combination antibonding.

In most cases π interactions are smaller than sigma interactions since the overlap integrals between π type atomic orbitals are smaller than their sigma counterparts; this may be seen qualitatively from the directional properties of the π orbitals, the overlap being side on rather than end on. A result of this is that compounds with occupied π molecular orbitals are frequently unstable with respect to compounds with occupied σ molecular orbitals only. (For example $C_2H_4 + H_2 \rightarrow C_2H_6$). When atoms have large non-bonding cores the overlap between π orbitals is often quite small; in the first row of the periodic table, however, the core orbitals are only the ls, and physical evidence suggests that the interactions are quite strong.

If we return to the M.O. diagram in Fig. 2.9., we may consider what happens as we gradually increase the number of electrons n in the valence shell of E. For n = 1, we have $1\sigma_g$ occupied and a stable bond; for n = 2 $1\sigma_g$ and $1\sigma_u{}^*$ are filled and the total effect is antibonding, as for helium; however, if the interaction with the $2\sigma_g$ and $2\sigma_u$ levels is sufficiently strong (corresponding to a mixture of p_z orbitals with the bonds, or partial sp hybridisation) the molecule can be stabilised with respect to the free atoms. Consequently the diatomic molecules with n = 2 (i.e. Be_2, Mg_2, Ca_2, the alkaline earths) exist, but are weakly bound. For n = 3 the next bonding orbitals ($2\sigma_g$) are filled and the molecule is again fairly strongly bound. The next two elements involve filling the π bonding orbitals with consequent increasing stability. Nitrogen, with n= 5, has the orbitals $1\sigma_g$, $1\sigma_u{}^*$ $2\sigma_g$ π_u all occupied, i.e. 4 bonding and 1 antibonding M.O. and consequently has a very high stability. We may note here that the *bond order* of a chemical bond is defined as $\frac{1}{2}$ (the number of electrons in bonding M.O. – the number in antibonding M.O.). For nitrogen the bond order is thus 3, and the compound has a triple bond, which we may separate into one σ and two π bonds.

For n = 6, two electrons enter the $\pi_g{}^*$ level; we may note here that, as for atomic orbitals, the energy is lower if the electrons enter different π M.O.s with parallel spin. Thus, in the ground state, molecular oxygen (O_2, n = 6) is paramagnetic as a result of the unpaired spins. The bond order is now only 2. For elements with configuration s^2p^5 (i.e. the halogens) both the π antibonding orbitals are occupied, and the molecule is held together only by 1 σ bond; for fluorine, in the lst row the π antibonding effect is so strong that the dissociation energy of the F_2 molecule is very low. For s^2p^6 elements (the noble gases) the final $2\sigma_u{}^*$ antibonding level is filled, and the molecule is unstable. We can use this model to predict the stabilities of homonuclear diatomic molecules as we cross the s, p group of the periodic table, as shown by their heats of dissociation (Fig. 2.11). The curve is completely in accord with the theory with minima at n = 0, 8, and 2 and a very high value of ΔH_{diss} for n = 5 (e.g. N_2). The heavier elements have similar shaped curves, but lower values of ΔH_{diss}; this presumably results from the lower overlap and increased core-core repulsion as Z increases.

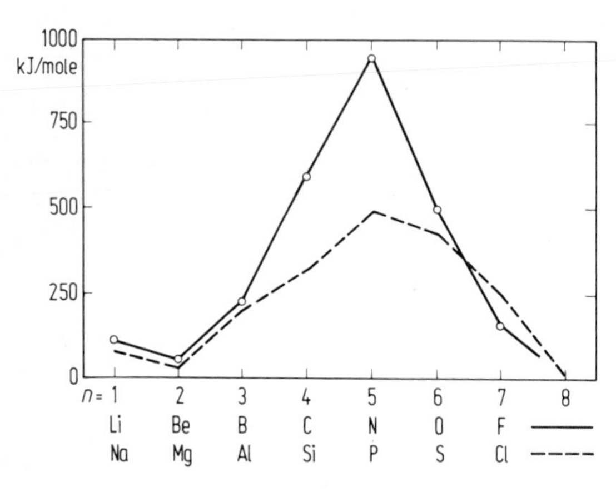

Fig. 2.11. The heats of dissociation of diatomic E_2 molecules for elements with n electrons in the valence shell. Solid Line Li–F, dotted line Na–Cl

For diatomic molecules where the two atoms are different (e.g. CO) a similar diagram to Fig. 2.9 maybe drawn, with the slight difference that there is no longer a centre of symmetry, and the two sets of atomic orbitals will not be degenerate: for CO for example, the carbon atomic orbitals will have slightly higher, and the oxygen slightly lower energy in comparison with nitrogen. The final M.O. diagram will however be very similar, and since CO also has 10 valence electrons, the molecule will have a high stability: in fact the dissociation energy of CO is slightly higher than N_2. As one might expect of two compounds with such similar M.O. diagrams and the same number of electrons, there are certain similarities in chemical behaviour.

A measure of the strength of the binding may be obtained in various ways: for diatomic molecules the heat of dissociation is clearly a very good measure, but for more complicated molecules approximations are necessary: thus for methane, the carbon-hydrogen bond energy is taken as one quarter of the heat of dissociation of CH_4 into one carbon and 4 hydrogen atoms.[7] This is then assumed to be constant, so that the other bonds may be calculated by subtraction e.g. $(CH_3)_2O$ will give the C–O bond energy. This is of course a localisation condition, but for carbon it works quite well. Thermochemical data such as heats of dissociation are not always easy to obtain and for comparative purposes, the bond length (internuclear distance) and vibrational force constant (from the infrared stretching frequency) are often used. Thus for the series O_2^+, O_2 and O_2^- the force constant falls and the bond length rises as the π antibonding shells are filled. The changes are not in general linearly related to ΔH_{diss}, and these methods cannot be used to compare unlike species – thus F_2 has a higher stretching frequency and a shorter bond length than Cl_2, but a bond energy of 158 kJ in comparison with 234 kJ/mole for Cl_2. This can be explained by the smaller core-core repulsions in F_2, and the strongly π antibonding π_g^* orbitals making the potential well very narrow. Although different species cannot be compared directly, the force

---

[7] Bond energies are discussed further in Chap. 5 (page 177).

constant method is useful for studying the changes in a fixed system, and double bonds generally have higher force constants than single bonds but smaller than triple bonds (see Chap. 8, page 286).

The concept of local bond symmetry may equally be applied to π type orbitals, but in general π orbitals cannot be localised as easily as σ orbitals. For a linear molecule, the distinction between σ and π type orbitals is clear; for more complicated molecules the distinctions are made by inspection: if an orbital has a plane of symmetry in a particular internuclear axis, it is said to be of π symmetry type for that bond (we give examples of this in Chap. 3). However it can arise that an orbital on a central atom has σ symmetry for one bond and π symmetry for another – in such cases the distinction between σ and π orbitals is more of a fantasy than a reality. For atoms bond to only one other atom, then the micro-symmetry is effectively linear, and the discrimination is possible – thus for tetrahedral CCl_4 we may talk of σ and π orbitals on the chlorine, but we cannot make this distinction for the central carbon atom. We shall see in the next chapter that the distinction between σ and π M.O.s is useful (if used carefully); for example, it is often highly convenient to take the sigma orbitals as localised, and the π orbitals as fairly extensively delocalised, the classic example being benzene.

It was mentioned above that π bonded systems are often unstable relative to systems with more σ bonding character. The E_2 molecular orbital system given above predicts stability for many diatomic species generally found as solids (e.g. carbon). This is because the solids have a higher coordination number and consequently better overlap with surrounding atoms. We can predict the change in stability of the solids by simple M.O. arguments. If there are n atomic orbitals in the valence shell, then we can form n molecular orbitals, of which n/2 will be bonding, and n/2 antibonding. On filling up the first n/2 M.O. we will be filling only bonding molecular orbitals, but if we fill more than n/2 M.O., some of the antibonding orbitals will be occupied, and the stability of the solid will fall.

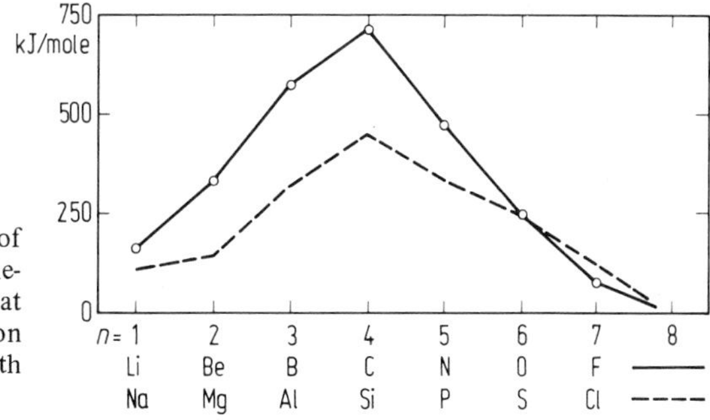

Fig. 2.12. The heats of formation of gaseous monatomic atoms from the elements in their standard states. Note that the maximum is now at n = 4 for carbon (solid line) and silicon (dotted line), both normally found as solids

For elements where the valence shell consists of s and p orbitals, n = 4, and the most stable configuration is obtained with 4 electrons i.e. group IVB. (carbon, silicon, etal). This prediction is well confirmed (Fig. 2.12). Again the lower rows of the periodic table have lower binding energy. If we consider elements of the transition series where the valence shell consists of s and d shells, n = 6 the maximum stability is obtained with 6 electrons. The curve is not quite so smooth in this case, but the maximum value of the heat of dissociation into atoms is close to n = 6, and for the third row is actually found at n = 6 (for tungsten).

We have now established the general features of molecular orbital theory, and in the next chapter we shall apply it to a wide variety of chemical compounds. To conclude this chapter we give a brief discussion of calculations using M.O. theory and the effects of electron repulsion. Since this chapter is concerned with the general principles of LCAO M.O. theory, it seems more appropriate to include this section here, but it may be omitted at first reading without prejudicing the understanding of following chapters.

E. *Electron Repulsion

Electron repulsion in molecules may be treated in the same fashion as for atoms – we have closed shells or sub shells of molecular orbitals which repel each other and raise the energy of the molecular orbitals, and we have the varying terms which may arise from the different arrangements of electrons in partially filled shells. We may define molecular orbital equivalents of the atomic orbital J (Coulomb) and K (exchange) integrals, but now their calculation is made even more complicated by their low (i.e. non-spherical) symmetry. It is consequently even more difficult to make qualitative generalisations, but one or two points are worthy of mention:

(i) Terms. In atoms we stated that if the system has two electrons in orbitals transforming as the irreducible representations M_a, M_b, then the direct product $M_a \oplus M_b$ gives the sum $M_i + M_j + M_k + \ldots$ of the various terms of the configuration. Thus for oxygen, the π_g shell is doubly occupied, and the product $M_\pi \oplus M_\pi$ gives rise to the terms $^3\Sigma + {}^1\Delta + {}^1\Sigma$. The ground state is the triplet $^3\Sigma$ as, if only one molecular orbital shell is being filled, the maximum multiplicity rule still holds, i.e. electrons prefer to have parallel to antiparallel spins. Note that as before, we use capital letters to denote terms or states of the molecule, and small letters to denote the actual orbitals.

It is important to realise that the energy of an absorption in which an electron jumps from orbital *a* to orbital *b* at a higher energy is not necessarily the one electron energy difference between ϕ_a and ϕ_b. Thus for benzene with the highest occupied M.O. $(e_{1g})^4$, an A_{1g} state, a one electron jump gives rise to the configuration $(e_{1g})^3\,(e_{2u})^1$, which has the terms (in order of increasing energy) $^3B_{1u}$, $^3E_{1u}$, $^1B_2$, $^3B_{2u}$, $^1B_{1u}$, $^1E_{1u}$, spread out between +3.5 and 7.0 ev. Needless to say this considerably complicates the interpretation of the spectra.

The measurement of orbital energy differences by absorption spectroscopy is thus difficult unless one can allow for electron repulsion, or make suitable approximations. In some of the most important cases (d–d, f–f transitions in metal compounds) the relevant orbitals are little changed from atomic orbitals, and we may approximate by using pure atomic orbitals with spherical symmetry to express the electron repulsion. The absorption spectroscopy of such compounds has proved a very useful source of information.

The above variations in energy arise from the variations in energy of the total system depending on the arrangement of the valence electrons within a given set (or sets) of orbitals, and is a consequence of the fact that our one electron orbitals are not really independent of the others. It is also important to realise that the orbital energies are very dependent on the orbital occupation. Thus, in the chlorine atom ($Ne3s^2 3p^5$) all the p orbitals will be equal in energy, but if a sixth p electron is added to the chlorine to give Cl^- the increased electron repulsion will give the 6 occupied p spin orbitals a higher mean energy, although they will still be degenerate. If an electron is ionised to give Cl^+, the decreased electron repulsion will lower the p orbital energy appreciably. It has thus to be remembered that in a compound such as LiF, although the energy of the fluorine 2p is much lower than the lithium 2s when separated, the appreciable transfer of electronic density to the fluorine atom in the molecule brings these levels much closer; the effects of the Madelung potential (Chap. 5) are however strong enough to lower the energy of the occupied fluorine orbital sufficiently to stabilise the compound.

This variation of orbital energy with occupation is not linear, and not the same for different orbitals. This can give rise to the apparent paradox in which partially filled orbitals have lower energy than completely filled orbitals. Several examples of this are now known, the best known examples being the lanthanide fluorides which are essentially ionic, $Ln^{3+}3F^-$, with the 4f shell of Ln^{3+} partially occupied. The ionisation of the fluorine 2p electrons requires less energy than that of the 4f although the 2p fluorine shell is completely full. Fig. 2.13 gives a simple explanation of this: in case (i) the partially filled M shell has lower energy than the L shell of the ligand, but if the transfer acutally occurs, the energy of the M orbital rises, and the L orbital falls.

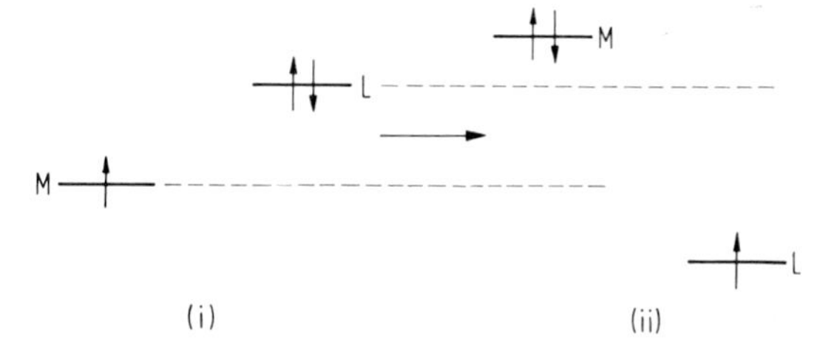

Fig. 2.13. The paradox of an empty or partially filled shell at a lower energy than a filled shell. The change of orbital energies following the transfer of an electron (ii) leads to a higher overall energy

Case (ii) shows the result when L falls sharply and M rises sharply – the system is even more unstable, and the partially filled shell is even more below the occupied shell. This behaviour is particularly common with the lanthanide 4f shell which has a high ionisation potential, but a low electron affinity, giving a large rise in energy on increasing the occupation of the 4f shell. The above paradox has been described by Jørgensen as the "third revolution in ligand field theory".

Even if we cannot calculate electron repulsion effects with any great precision, we can use the theory in a qualitative fashion, and classify the observed bands according to group theoretical predictions. Thus, of the six excited states of benzene mentioned before, we can calculate, as in Chap. 1, which transition will be allowed by the electric dipole mechanisms, and the polarisation that any such bands may have. The qualitative success of such calculations is extremely good, and they play a vital part in the interpretation of electronic spectra.

Correlation energy

In atomic spectra it was found that the electron repulsion terms could not be calculated accurately as a result of correlation in the electronic motion reducing the repulsion.

Correlation exists equally in molecules, and would again affect our calculations of electron repulsion terms. A more important effect is the correlation energy, defined as the difference between the experimental energy, and the calculated energy using a Hartree-Fock method. Although this produced a slight error in the calculation of orbital energies for atoms, the effect is more serious in molecules, where the correlation contribution to the chemical binding energy can be appreciable. For example, for NaCl, the correlation contribution to the heat of dissociation of the molecule is 104 kJ/mole – about 25 percent of the total heat of dissociation. Even worse, for F_2, a Hartree-Fock calculation predicts that the heat of dissociation is actually negative (i.e. the molecule is unstable) implying that the correlation energy actually stabilises the molecule. Calculations of heats of formation are thus likely to be wildly inaccurate, although for heats of reaction the partial cancellation of errors on both sides of the equation may improve matters. This might be thought to throw doubt on the worth of molecular calculations, but, as for atomic calculations, the phenomenon of correlation affects mainly the energy, and calculations of molecular electron distribution and other properties are much more accurate. Thus, a Hartree-Fock calculation of NH_3 gave the following results (A. Rank, L.C. Allen, E. Clementi, J. Chem. Phys. *52,* 4133 (1970))

	Bond length (Å)	Bond angle	Dipole moment (D)	Heat of formation (kJ)
calc.	1.000	107.2°	1.66	–33.22
expt.	1.0116	106.2°	1.48	–46.11

and it can be seen that the bond angles and lengths are very accurate, even if the heat of formation is rather poor, the error coming from the correlation energy. For the planar form of NH_3 we would expect the correlation energy to be about the same (the number of electrons in the molecule remaining the same), and the calculated barrier to "flapping" through a planar form is 21.2 kJ/mole compared with the experimental value of 24.3 kJ.

Correlation energy can be treated, as in atoms, by the method of configuration interaction (C.I.), in which excited states are allowed to mix in to the ground state wave function.[8] Although this gives a much improved result, it considerably lengthens calculations, and is not generally carried out. The proportion of excited state mixed into the ground state is generally of the order of 1 percent, so that we are justified in describing a system by its ground state function alone.

F. *Methods of Calculation

The basis of a Hartree-Fock type linear combination of atomic orbital molecular orbital theory was given by Roothaan in 1951, and the relevant equations and integrals defined. There are however various degrees of approximation which may be applied to these equations, and we give the most important below:

In ab initio or non-empirical calculation one starts with a set of atomic wave functions and calculates the best possible wave function (using the variation principle). This

---

[8] A simple example of this is given by Atkins for hydrogen.

may or may not include configuration interaction in an attempt to allow for correlation. All the integrals are calculated and the amount of calculation involved is formidable. Ab initio calculations have in general been restricted to compounds with fairly light elements, and a fair degree of symmetry.

In the next level of approximation, certain integrals are ignored (i.e. put to zero), used as parameters of the calculation, or approximated by experimental data for the free atoms. This level includes the neglect of differential overlap methods such as NDDO, MINDO, CNDO (in order of increasing approximation). These methods generally involve much less calculation, frequently considering only the valence shells of the atoms considered, and, if the parameters are carefully chosen can give results not much worse than the more complicated ab initio methods. It is even claimed by some practitioners of these methods that the selection of parameters can avoid some of the errors in the ab initio method, although this seems to be more of a matter of personal belief than established fact. Finally, there are the Hückel methods where almost everything is ignored with the exception of the resonance integrals between neighbouring atoms. In general, it is wise not to give too much weight to the results of molecular orbital calculations unless one is fairly certain of the approximations made, and unless there is a fairly good correspondance between some calculated and experimental values. This is particularly true for spectral data; in many cases, calculations are unable to predict the energy order of molecular orbitals. The situation is happier in organic chemistry where the limited number of elements, and the large number of closely related compounds enables quite reliable calculations to be made, but even here the reliability of a method is a personal belief rather than an established fact.

Although one should not blindly accept the use of LCAO M.O.s as the best description possible of a wave function, no more should one treat it with contempt. LCAO molecular orbitals generally show the properties of the true molecular orbitals, even if they are not themselves exact. The calculations, even if not exact, at least achieve order of magnitude accuracy, and provide support for our qualitative description of electron repulsion, overlap and other effects. One should also remember that the total energy of the electrons in a compound is far greater than the energy absorbed or liberated during the formation of a compound, and that a calculated heat of reaction is only a tiny difference between very large quantities; the total electronic energy for Mn is about 3×10^6 kJ/mole, for Th about 6×10^7 kJ/mole, so an error of some 10 kJ in a heat of formation is not at all bad for what we know to be the approximation of one electron orbital functions.

It is possible to expand almost all functions in a power series using the spherical harmonics and consequently we might expect that we can obtain a very good wave function by mixing in enough atomic orbitals of the type $f_{nl}(r)\ Y_{lm}(\theta, \phi)$. This would however be moving away from our initial intention of describing the molecular wavefunction in terms of the valence shell orbitals of the constituent atom. Thus the wavefunction of ammonia is improved considerably by the inclusion of orbitals of d-type symmetry on the nitrogen atom. This represents an improvement of the *wave function* but also the partial abandonment of the LCAO approach.

Within the limits of our approximation we will not worry about the *exact* nature of our wavefunction, and whether it can be improved by addition of a few percent of an empty higher energy A.O., since there are a large number of effects we ignore in our calculation (such as correlation) which ought to be considered at this level of accuracy.

Rather are we interested in a simple picture of the wavefunction which gives many of the observed experimental properties, and for this the LCAO M.O. model is very useful.

Bibliography

Atkins, P.W.: Molecular Quantum Mechanics. London: Oxford University Press 1970

Ballhausen, C.J., Gray, H.B.: Molecular Orbital Theory. New York, Amsterdam: W.A. Benjamin 1965

Blyholder, G., Coulson, C.A.: Theoret. Chim. Acta, *10,* 316 (1968)

Coulson, C.A.: Valence, 2nd edition. London: Oxford University Press 1961

Murrell, J.N., Kettle, S.F.A., Tedder, J.M.: Valence Theory, 2nd Ed. London, New York, Sydney: J. Wiley 1970

Pople, J.A.: Localisation of orbitals. Quart. Rev. *1957,* 273

Purcell, K.F., Kotz, J.C.: Inorganic Chemistry. Philadelphia, London, Toronto: W.B. Saunders 1977

Urch, D.S.: Orbitals and Symmetry. London: Penguin Books 1970

Wagniere, G.H.: Introduction to Elementary Molecular Orbital Theory. Berlin, Heidelberg, New York: Springer 1976

The following texts are mainly concerned with calculations of wave functions:

McGlynn, S.P., Vanquickenbourne, L.G., Kinoshita, M., Carroll, D.G.: Introduction to Applied Quantum Chemistry. New York: Holt, Rinehart, and Winston 1972

Murrell, J.N., Harget, A.J.: Semi-empirical Self Consistent Field Molecular Orbital Theory of Molecules. London, New York, Sydney: J. Wiley 1972

Bingham, R.C., Dewar, M.J.S., Lo, D.H.: Empirical M.O. theory of simple systems. J. Amer. Chem. Soc. *97,* 1285, 1294, 1302, 1311, (1975)

Pople, J.A.: Criticisms of the previous reference. Amer. Chem. Soc. *97,* 5306 (1975) *and* Hehre, W.J., J. Amer. Chem. Soc. *97,* 5308 (1975)

Steiner, E.: Ann. Rep. Chem. Soc. *1970A,* 5

Problems

1. Comment on the following bond dissociation energies in the light of the M.O. description of the bonding:
H_2^+ 270 kJ/mole; H_2 436 kJ/mole; He_2^+ 241 kJ/mole

2.* Using the Heisenberg Uncertainty Principle show that reducing the volume within which an electron is localised will result in an increase in its kinetic energy. Could this provide a resistance to the compression of chemical bonds?

3. Taking the orbital energies as being measured roughly by the first ionisation potential (Figs. 1.5 and 1.6), predict whether the bonding will be ionic, covalent, or covalent polarised for:
NaCl, NF_3, GaAs, $AsBr_3$, PbS, $MgCl_2$.

4. Would you expect a ligand σ orbital to have a large overlap with a central atom d orbital? Does this explain why many transition metals are densely packed solids which vaporise to monatomic gases?

5. The lanthanide metals show relatively small variations in heats of atomisation, but maxima are found for Ce, Gd, and Lu. Can you relate this to the electronic configuration of these elements?

6. (i) The stretching frequency of NO is 1860 cm^{-1}, that of NO^+ 2220 cm^{-1}. Can you explain this?
(ii) The carbide (C_2^{2-}) and cyanide ions are remarkably stable. Can you explain this?

7. The wave function $\psi = c_1\phi_1 + c_2\phi_2$ is unnormalised. Find a value for the normalisation constant $\frac{1}{N}$ such that $\psi' = \frac{1}{N}\psi$ is normalised. Compare your answer with that given on page 46.

8. In a diatomic covalent molecule the bonding orbital is localised between the nuclei. Where is the antibonding orbital 'localised' (i.e. where is the electronic density a maximum)?

9. *Using the form of molecular Hamiltonian given on page 43, consider how the different terms in $\mathscr{H}_{aa}$ will vary as electron density varies on the atom A in a diatomic molecule A–B.

10. Consider the figure on page 17, and imagine that an atom centred at the origin of the axes has four hybrid orbitals directed towards ligands at the point 1 to 4.
(i) Will the hybrid orbitals be basis functions for a representation of the symmetry group?
(ii) If so, is this representation reducible?
(iii) If so, what are the constituent I.R.s?
(iv) Consult the I.R.s of the central atom orbitals for symmetry D_{4h} given in Appendix I. Which atomic orbitals could be used to constitute the hybrids? Compare your conclusions with page 54.

3. Structural Applications of Molecular Orbital Theory

In this chapter we shall apply molecular orbital theory to a wide variety of compounds in order to show that it may be used as a quite general means of description. The subdivisions of the chapter are thus more a matter of convenience than a reflection of changes in the principle of the approach.

A. Simple Polyatomic Species

In this section we will examine the structures of some compounds of elements having only s and p type atomic orbitals in their valence shell.

ML_2 molecules

In discussing MgF_2 we considered only the linear structure known from experiment. We may now consider the M.O.s of a bent ML_2 molecule, assuming that the two M–L bonds are of equal length. This molecule has C_{2v} symmetry, with a twofold symmetry axis along the z axis (Fig. 3.1a). Appendix I gives the symmetry properties of the various orbitals: we see that the ligand b_2 σ orbital may overlap with the p_y orbital, and the ligand a_1 σ orbital with the p_z and the s orbitals of M. The orbital energies will determine which of the s or p_z is more involved in bonding. In Fig. 3.1b a molecular orbital diagram for the system is given, assuming the L σ orbitals to have energy close to the M p orbitals. The b_1 orbital is non-bonding, and the a_1 (bonding) orbital will be raised slightly above the b_2 (bonding) orbital by interaction with the a_1 (s) orbital. The order of the a_1^* and b_2^* orbitals is uncertain.

We may now consider what happens if we increase the L–M–L angle until the molecule is linear along the y axis. The b_2 orbital becomes a σ_u bonding orbital, and the other two p orbitals of M become non-bonding π_u orbitals. The a_1 (s) orbital correlates with the σ_g bonding M.O. As far as the orbital energies are concerned, the p_x orbitals is always non-bonding, and its energy will not change greatly on becoming linear, but the p_z orbital changes from bonding to non-bonding in the linear molecule, and its energy therefore rises. The b_2 (p_y) orbital falls in energy in the linear molecule, since the overlap with the ligands will be greater, and the ligand b_2 orbital is more L–L antibonding in the bent molecule. The a_1(s) orbital will not change greatly in energy on distorting the molecule. We may now draw the correlation diagram shown in Fig. 3.2 relating the molecular orbitals of bent and linear ML_2 molecules.

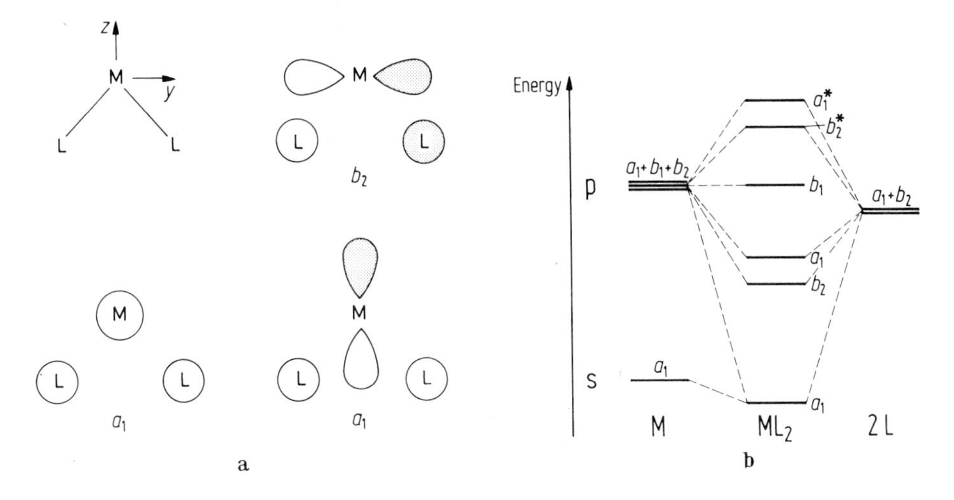

Fig. 3.1. (a) The structure and atomic orbitals for M.O. formation of ML_2 molecules, **(b)** the M.O. diagram for a bent ML_2 molecule

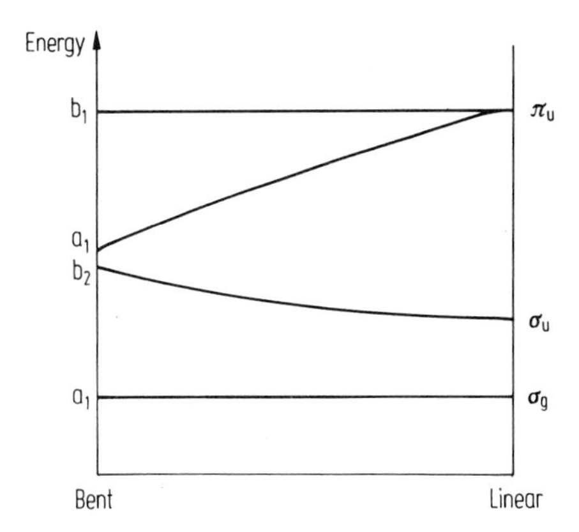

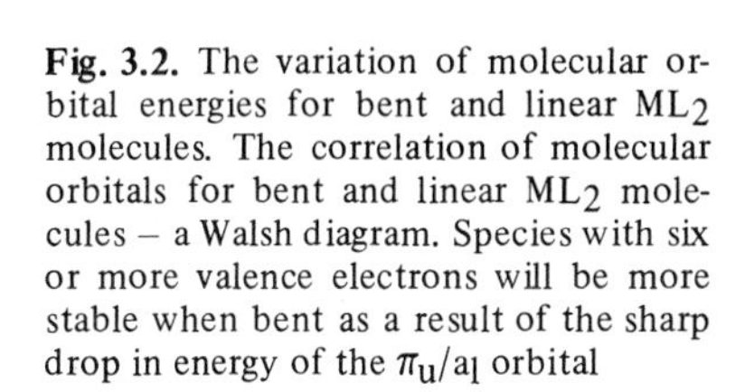

Fig. 3.2. The variation of molecular orbital energies for bent and linear ML_2 molecules. The correlation of molecular orbitals for bent and linear ML_2 molecules – a Walsh diagram. Species with six or more valence electrons will be more stable when bent as a result of the sharp drop in energy of the π_u/a_1 orbital

It will be seen that a molecule with four or less electrons in the valence shell will prefer linear geometry, but that further electrons will enter the π_u/a_1 orbital which is strongly stabilised by bending. Experimental data support this conclusion. BeH_2 with 4 valence electrons is linear, but BH_2 with 5 valence electrons is bent; excitation of one electron to give the excited state of BH_2 $(a_1)^2$ $(b_2)^2$ $(b_1)^1$ reduces the stabilisation of the bending, and the excited state is linear or nearly linear. Singlet CH_2 with two electrons in the a_1/π_u orbital is strongly bent, but the triplet form with one electron in the a_1/π_u orbital and the other in the b_1 orbital is less so. NH_2 and H_2O with respectively 7 and 8 electrons are bent.

Further experimental support for this description comes from the unpaired electron spin in species such as PH_2. The electron spin couples with any nuclear spin present in the molecule, and, if the magnitude of the coupling with a given nucleus can be measured, one may estimate to what extent the unpaired electron is localised on that atom. For PH_2, where the unpaired electron is in the b_1 orbital, the coupling with the phosphorus nuclear spin is much stronger than with the proton spin, confirming it as a non-bonding orbital localised on the central atom.

The treatment given above is similar to that first given by Walsh in a classic series of papers in 1953, and Fig. 3.2 is often referred to as a *Walsh diagram.* Many further examples of this style of discussion are given by Atkins and Symons.

Photoelectron spectra have now shown that the s orbitals of the central atom are rather less important than originally thought and the description given above has taken account of this.

Consideration of M–L multiple bonding does not fundamentally change the approach. In most cases (e.g. MgF_2) the L π orbitals lie at lower energy than the M π orbitals, and, for linear geometry, have the form shown in Fig. 3.3 with the π bonding mainly, and the π non-bonding entirely localised on the ligands. These orbitals change relatively little on bending, and the structure adopted depends mainly on the occupation or non-occupation of the π_u^*/a_1 orbital, as for the σ-bonding only case. The π_u (bonding) and π_g (non-bonding) M.O. lie below the π_u^* and will be filled first. On progressively filling these orbitals, we expect linear geometry until the configuration $(\sigma_g)^2$ $(\sigma_u)^2$ $(\pi_u)^4$ $(\pi_g)^4$ $(\pi_u^*)^1$ is reached, at which point bending will be favoured. Both NO_2 and CO_2^- have this configuration and are bent, whilst CO_2, NO_2^+, and MgF_2 have one less electron and are linear. For simple triatomics where π bonding is possible, up to 12 electrons may be accomodated in the M.O.s described above without bending; it is usual to count as well the two pairs of electrons in the non-bonding 2s shells of ligands such as F, O, and N, and the critical number of valence electrons in these systems is thus 16. 17 and 18 electron species will be bent, as will 19 and 20 electron species since the π_u^*/b_1 orbital above the π_u^*/a_1 orbital is almost unaffected by bending (it is actually slightly stabilised). 21 and 22 electron systems involve filling the σ^* M.O.s which are strongly destabilised by bending, and are consequently linear; 24 electron systems have all the antibonding orbitals filled and are thus unstable.

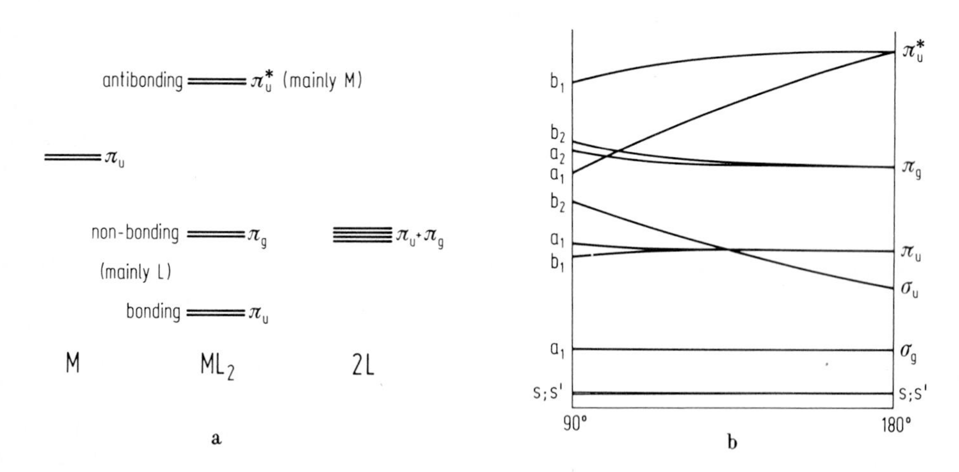

Fig. 3.3. (a) The π M.O. of a linear ML_2 molecule, **(b)** a Walsh diagram for a ML_2 molecule showing the effects of M–L π bonding

Table 3.1. Structures of simple triatomic molecules

Molecule		No. of valence electrons	Bond angle
CO_2		16	180
NO_2^+		16	180
CO_2^-		17	127 ± 8
NO_2		17	134
NO_2^-		18	115
CF_2		18	105
SiF_2		18	101
GeF_2		18	94 ± 4
O_3		18	116.8
SO_2		18	119.5
ClO_2		19	117.5
F_2O		20	103.3
Cl_2O		20	110.9
ClF_2^+	– bent	20	
ClF_2^-	– linear	21	
NNO		16	180
SSO		18	118

Table 3.1 shows the measured bond angles for a large number of ML_2 molecules. The geometric predictions of the Walsh diagram are verified. We may also note that the N–O bond length in the series $NO_2^+ : NO_2 : NO_2^-$ increases from 1.10 → 1.19 → 1.24 Å implying that the new orbital being filled has antibonding character, as predicted by the diagram. Herzberg discusses the electronic spectra of many simple compounds, and relates them to Walsh diagrams. Walsh also pointed out that the loosest bound (and presumably most reactive) electrons in NO_2 and NO_2^- are in the π_u^*/a_1 orbital of the molecule, and that, as this M.O. is localised mainly on the nitrogen atom, reactions of these species will involve the nitrogen atom rather than the oxygen atoms. This is generally observed, and is also found for the sulphur atom in SO_2.

The effect of orbital energy variation on geometry is not easy to predict. One might reasonably assume that, if the s orbital of M is unwilling to take part in bonding, the contribution of the p_z orbital will be more important, and the bending will be more favourable. This will produce a variation in bond angle, as is seen in Table 3.1, but will not cause a change from linear to bent geometry (or vice-versa) for a given number of valence electrons. This is another example of the structure being determined by the number of valence electrons, irrespective of the variation in orbital energy (cf. BH_4^-; CH_4; NH_4^+). Electron counting also works for molecules which have lower symmetry than that assumed above. Thus $IClBr^-$, NNO, or SSO have geometry and spectra similar to those predicted for molecules which are symmetric about the central atom. It appears that the lower symmetry perturbs the molecular orbital levels (and in theory allows more mixing between atomic orbitals) but does not change the pattern of the levels sufficiently to nullify the predictions of the Walsh diagram.

ML_3 molecules

We consider two possible structures: planar (symmetry D_{3h}), and pyramidal (symmetry C_{3v}), and consider only the effects of σ bond formation with the ligands L. For D_{3h} symmetry, Appendix I shows that the orbitals concerned transform as the following I.R.s:

M s	a_1''
M p_x, p_y	e'
M p_z	a_2''
3 L σ	$a_1' + e'$

The M p_z orbital is thus non-bonding, and the interaction will be between the M s, M p_x, and M p_y and the ligand σ orbitals. If we assume the ligand σ orbitals to be intermediate in energy between the M s and M p orbitals, the M.O. diagram is as shown in Fig. 3.4a, and there are three stable bonding M.O.s, and one non-bonding M.O. The structure will be quite stable for up to six valence electrons – it is in fact the stable form for BF_3 with six valence electrons.

The pyramidal structure may be obtained from the planar by pushing the central M atom along the threefold axis, out of the plane of the ligands. In this symmetry (C_{3v}) the orbitals transform as:

M s	a_1
M p_x, p_y	e
M p_z	a_1
3 L σ	$a_1 + e$

The p_z orbital is obviously no longer non-bonding and consideration of Fig. 3.4a shows that it may interact with the $a_1'^*$ orbital, causing one orbital to rise in energy, and the other to fall (as was the case when the σ_u interaction was introduced for MgF_2). The M.O. diagram is given in Fig. 3.4b; the pyramidal structure will thus be more stable than the planar if there are eight electrons involved in the bonding, and the second a_1 orbital is occupied. NF_3 and NH_3 do indeed have this structure. We may also note that this highest occupied orbital (a_1) will have a fair amount of M p_z and s orbital character and will be directed away from the L_3 plane: many pyramidal ML_3 compounds are good electron pair donors (or Lewis bases) as a result of this orbital being occupied.

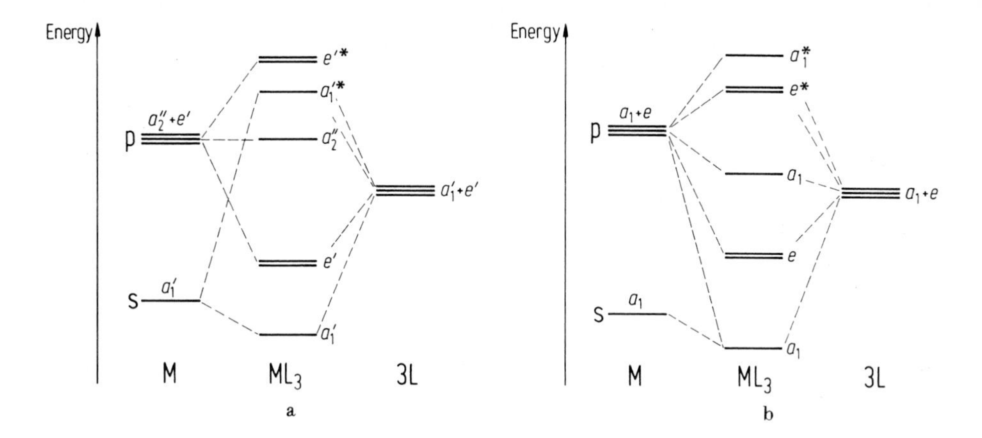

Fig. 3.4. (a) M.O. diagram for a planar ML_3 molecule, **(b)** M.O. diagram for a pyramidal ML_3 molecule

If the M s orbital is virtually non-bonding, either as a result of a low overlap integral with the ligands, or because it is tightly bound, the bonding will involve the M p orbitals only, and the bond angles will be close to 90°. This is observed for the series PH_3, AsH_3 and SbH_3, where the bond angles are much closer to 90° than for NH_3 (see page 54). The planar BF_3 molecule has a low lying, empty, non-bonding orbital (a_2'') localised on the boron atom. This orbital can accept a pair of electrons, functionning as a Lewis acid; however, in accepting a pair of electrons from a Lewis base (such as ammonia), it increases the number of electrons in the valence shell, and distortion to a pyramidal structure is now favoured, as the energy of the acceptor (a_2'' / p_z) orbital will be lowered by the distortion. This is indeed observed, and BF_4^-, the fluoride ion adduct of BF_3, is tetrahedral.

In passing we may note that the planar structure involves only M s, M p_x and M p_y orbitals of the central atom, and is thus close to sp^2 hybridisation. The pyramidal structure uses all s and p orbitals of M, and corresponds to sp^3 hybridisation.

In summary, we see that the comparison of the M.O. diagrams for different structures and the construction of Walsh diagrams enable us to predict the more stable structure for a compound. In the two cases discussed it was seen that the occupation or non-occupation of one M.O. was critical in determining the structure adopted. Some more complicated examples are discussed in a review by Burdett. For compounds where π bonding is relatively unimportant, it is often possible to localise bonds and predict the stereochemistry about a particular atom in quite complicated molecules. Tertiary phosphines and amines (R_3P and R_3N) are pyramidal even if the three organic substituents are different; each carbon atom of the C–N (or C–P) bonds may be assumed to have an identical localised orbital available for bonding.

Molecular orbital energies

The energy of molecular orbitals is not easy to determine: ultraviolet and visible spectra determine only the energy differences between orbitals and not their absolute energies.

Table 3.2

Compound	CH_4	NH_3	H_2O	HF	Ne	H_2S
Structure	tetrahedral	pyramidal	bent	linear		bent
Bond angle	109,5°	106,7°	104,5°			92°
Binding energies (eV):						
p-type	$1t_2$14.2	$3a_1$11.04	$1b_1$12.61	1π16.1	2p21.59	10.48
orbitals		le 16.5	$3a_1$14.73	3σ19.9		13.25
			$1b_2$18.6			15.35
Mean energy	14.2	14.7	15.3	17.4	21.59	
2s-orbital	$2a_1$23.05	$2a_1$27.75	$2a_1$32.6	2σ39.6	48.47	
1s-orbital	290.8	405.6	539.7	693.8	870.3	
Ratio 2s:2p	1.62	1.89	2.13	2.28	2.24	

The recent development of photoelectron spectroscopy (Chap. 8) has enable the vertical ionisation energies of specific molecular orbitals to be measured, and, consequently, an estimate of their binding energy to be made. Table 3.2 shows the values obtained for the eight valence electron hydrides in the series methane to neon.

In all cases the number of bands seen corresponds with the pattern of orbitals predicted by simple molecular orbital theory. The central atom ls orbital is always too tightly bound to have any involvement with chemical bonding. The orbital formed from the hydrogen atomic orbitals and the 2s orbital can be clearly identified in CH_4 ($2a_1$) and the corresponding band in the other compounds rises monotonically in energy to the value for the pure 2s orbital in neon. This suggests that this orbital remains fairly close to a pure 2s orbital, although some mixing with a_1 type p orbitals is possible in molecules of lower symmetry (NH_3, H_2O, HF). The weaker bound electrons must therefore be either non-bonding p orbitals, or p orbitals bonding with the ligand σ orbitals.

It is instructive to take the mean p orbital energies and compare them with the s orbital energies. As one crosses the series, the ratio E_s:E_p become much greater, implying that the 2s orbital rises in binding energy much faster than the 2p. For HF the energiy difference is 22 eV (~2000 kJ/mole), and, as the bond energy is only 556 kJ/mole, extensive s orbital participation in the bonding seems improbable. For NH_3 the s–p gap is much smaller (almost the same as the total NH bonding energy), and interaction with the s orbital seems more probable. The H–N–H angle is much greater than the 90° expected for p orbital bonding only; for H_2O the s type orbital $2a_1$ is more tightly bound, and, in accord with this, the bond angle diminishes, implying less s character.

B. Transition Metal Compounds

All transition metals have a partially filled d subshell in the free atomic state, and chemists were (for a long time) inclined to think that this required a separate theoretical treatment of the bonding in their compounds. One of the purposes of this section is to show that the principles of this treatment are identical to those already established. Reference to the sketches of the d orbitals in chapter one will show that d-orbitals do not have any particularly well established directional properties, with four lobes of electron density pointing in different directions. We are not surprised then to find that elements where the d-orbitals play a strong part in the bonding generally have several atoms bonded to them, or, to put it more elegantly, such elements tend to show high coordination number (C.N.), rarely less than 4, most frequently 6, and sometimes as high as 9. We may note in passing that elements with partially filled f shells often show even higher coordination numbers, although this does not appear to be directly related to the presence of occupied f orbitals.

This tendency to high C.N. is shown not merely in the solid, but also in solution, where central atoms with d-orbitals involved in bonding will keep their ligands (i.e. surrounding atoms or groups) quite strongly bond. This was established by the work of the Swiss chemist, Werner, at the end of the nineteenth century, and the high C.N. species were described as complexes; for many years the high C.N. of the central atom in complexes posed considerable theoretical problems for chemists who had based their bonding theories on elements such as carbon and nitrogen usually showing low C.N. The fact that d-orbitals do not 'point' strongly in one particular direction results in overlap integrals frequently being much smaller than for p orbitals.

Many complexes show quite high symmetry, and even those which do not may be discussed in terms of a high microsymmetry. For example, the complex $[Co(NH_3)_6]^{3+}$ has the ammonia groups octahedrally disposed about the cobalt atom (symmetry O_h). Substitution of one ammonia by a chloride ion gives the complex $[Co(NH_3)_5Cl]^{2+}$, where the disposition of the ligands (5 NH_3, 1 Cl) is still octahedral, but the symmetry is lower (C_{4V}). Although the energy levels are slightly perturbed, the electronic structure can still be discussed usefully in terms of microsymmetry O_h.

A complex is conventionally described as octahedrally coordinated if there are six ligands at the apices of an octahedron centred on the central atom. This does *not* imply that the symmetry is octahedral. Thus $[Co(NH_3)_5Cl]^{2+}$ has octahedral *coordination,* but lower *symmetry.* We may frequently be able to discuss an octrahedrally coordinated species in terms of an octahedral microsymmetry, but this is another assumption. We have already seen an example of this with N_2O and S_2O, which may be discussed in terms of microsymmetry C_{2V}.

The electronic structure of octahedral complexes

Tungsten hexamethyl is a well characterised, if rather reactive octahedral complex of tungsten with symmetry O_h. The methyl ligands have no orbitals suitable for π bonding and we will consider that they have available one sigma orbital each for bonding with tungsten. In view of the closeness of the CH bond angles in methyl groups to the tetrahedral angle of 109.5^o this orbital will probably be close to a sp^3 hybrid, but it is really only its σ bond symmetry which interests us here. We place the six groups on the x, y and z axes, and Appendix I gives the I.R.s spanned by the six ligand orbitals as $a_{1g} + e_g + t_{1u}$.

We now need to consider the metal A.O.s. The only A.O.s available for bonding are the 5d, the 6s and the 6p, although these last may have too high an energy to take much part in bonding. All three sets of orbitals are sufficiently diffuse to overlap well with the ligand groups. Although the 6s and 6p each transform as one I.R. of the symmetry group (a_{1g} and t_{1u}), the 5d orbitals transform as t_{2g} and e_g. The e_g set consists of the d_{z^2} and $d_{x^2-y^2}$ orbitals – reference to Chap. 1 shows that these two orbitals are directed down the axes, directly at the ligands, while the others (d_{xz}, d_{yz}, d_{xy}; t_{2g}) 'point between' the axes – it is thus not surprising that the d orbitals transform as two different I.R.s. The mean energy of the 5d orbitals is below that of the 6s in tungsten, and thus we arrive at the level of A.O.s shown in Fig. 3.5. We have put the level of the CH_3 σ orbitals slightly below that of the tungsten 5d, in accord with general views about their energy.

The construction of the molecular orbital diagram is now quite straightforward: the t_{2g} orbitals will be non-bonding, and thus their energy will change very little. The two sets of e_g orbitals will give one set of bonding and one of antibonding molecular orbitals. The a_{1g} orbitals will also give a bonding and an antibonding set, but the greater energy separation will, according to the principle established in Chap. 2, give a weaker bonding interaction, and thus the a_{1g} bonding orbital will lie above the e_g bonding orbital. A similar consideration will apply to the t_{1u} sets, which will interact even more feebly. We thus arrive at the molecular orbitals for $W(CH_3)_6$ (Fig. 3.5).

We may classify the orbitals in the following order:

e_g, a_{1g}	bonding
t_{1u}	weakly bonding
t_{2g}	non-bonding
e_g^*, a_{1g}^*	antibonding
t_{1u}^*	high energy, weakly antibonding.

Tungsten has 6 valence electrons, and the methyl groups will have one electron in each of their σ orbitals – we thus have 12 electrons to fill the molecular orbitals, which just fills the six bonding molecular orbitals e_g, a_{1g}, t_{1u}. We would thus predict that the compound would be stable as indeed it is (to dissociation – it is reactive simply because tungsten and methyl groups can form stronger bonds with other groups, for reasons such as improved overlap).

Let us now consider the nature of the first four sets of molecular orbitals. The e_g orbitals are composed of W e_g (5 d) orbitals and CH_3 e_g orbitals. Because the energies of these two sets are close, the M.O. will probably have roughly similar amounts of metal and ligand character i.e. the bonding is quite covalent. The a_{1g} orbital will be mainly methyl σ orbital (because of the higher tungsten 6s energy), and the t_{1u} prob-

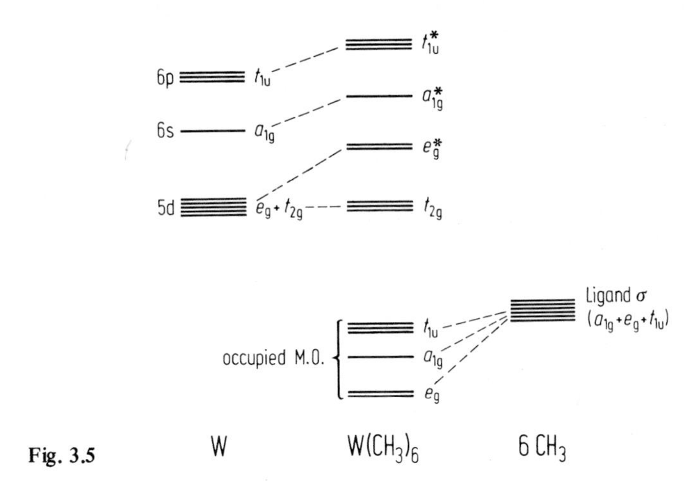

Fig. 3.5

ably almost all methyl σ orbital in character. The unoccupied t_{2g} orbitals will be almost unchanged tungsten 5d orbitals.

In sum the 6 occupied orbitals will have rather more methyl than metal character, and since each provides 6 electrons, there will be a slight loss of charge to the ligands. The photo-electron spectrum of this species has been measured and is quite in accord with this, also suggesting that the 6p orbitals play little part in the bonding i.e. that the t_{1u} bonding orbital is almost pure methyl σ in character.

In conclusion, the construction of the M.O. diagram has presented no special problems, and because of the high symmetry, has even been slightly easier than (for example) for NO_2. Note that the e_g and a_{1g} bonding orbitals will be highly *delocalised* over the metal and six ligands, and that it is not possible to draw suitable localised bonds.[1] Now let us turn to a complex which shows some of the features peculiar to transition metal compounds.

The electronic structure of $[Fe(NH_3)_6]^{2+}$

The ferrous ammonium ion is another octahedrally coordinated complex, and although the symmetry is not quite perfectly octahedral (see Chap. 4, the Jahn-Teller effect) we will assume an O_h microsymmetry. NH_3 has the same σ bonding orbital as CH_3, but it is now doubly occupied. It is normal to place the NH_3 σ orbitals well below those of the metal, from energy level and other considerations.

The molecular orbital diagram will not be substantially different from that for WMe_6, except that the lower energy of the ligand σ orbitals will imply that the σ bonding orbitals $a_{1g} + e_g + t_{1u}$ have less metal and more ligand character. The 3d orbitals of Fe^{2+} are much less diffuse than the W5d, and the overlap with the ligands is less – this leads to a smaller bonding effect, and the e_g^* is raised above the t_{2g} level much less than in WMe_6. In WMe_6 there were 12 valence electrons; in $[Fe(NH_3)_6]^{2+}$ there are 12 in each ligand σ orbital and 6 on the central ion (Fe^{2+} – $3d^6$). We will therefore need to fill more molecular orbitals, and 6 electrons must be placed in the

[1] The use of octahedral d^2sp^3 hybrids for the metal involves too much p orbital participation.

t_{2g} levels, or higher. The first three may be placed in the t_{2g} level with a clear conscience, but with the fourth electron we must realise that 1) to place it in the t_{2g} orbitals will involve pairing it with one of the three already there (Hund's maximum multiplicity rule requires that the first three have parallel spins), and this will involve a loss of spin pairing energy, and 2) the weak bonding interaction means that the e_g^* orbitals are only slightly higher than the t_{2g}. The lowest energy is obtained if the electrons are arranged to minimise the pairing of electron spins – this is achieved if two electrons are placed in the e_g^* levels. The final arrangement of electrons in the t_{2g} and e_g^* levels is shown in Fig. 3.6a. The energy difference between the two levels is generally referred to as the *ligand field splitting energy* and is denoted by Δ or 10Dq. Its value is typically around 15000 cm^{-1} (= 15 kiloKayser (kK) or 2eV, or 200 kJ/mole).[2] The term 'ligand field' derives from the idea that the d orbitals are split in energy by a potential field due to the ligands.

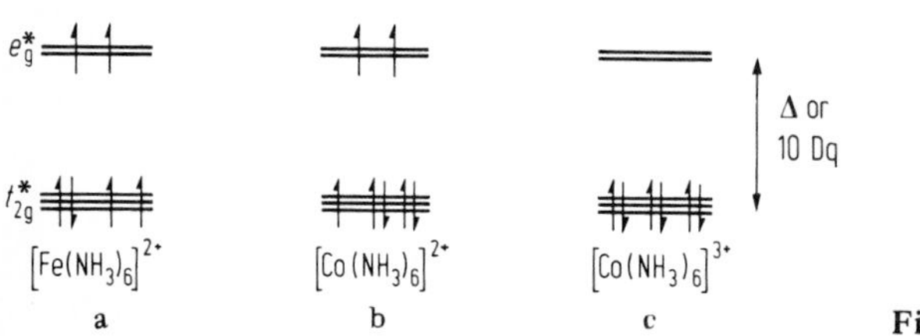

Fig. 3.6

The energy 15000 cm^{-1} falls in the visible region of the spectrum, and the visible and near ultra-violet spectra of transition metal complexes have been of incalculable importance in establishing their structure; rather than break this purely theoretical approach we discuss them separately (Chap. 4).

If we now move to the next element in the first transition series (cobalt) $[Co(NH_3)_6]^{2+}$ has seven electrons to be distributed in the e_g^* and t_{2g} levels. The ligand field stabilisation energy Δ is little changed, and the spin pairing energy is still sufficiently high for the population of the higher energy e_g^* to be favoured, and the electronic arrangement shown in Fig. 3.6b to be favoured. However, this complex may be oxidised readily to give $[Co(NH_3)_6]^{3+}$ with one less electron; the $e_g^* - t_{2g}$ gap Δ is now much larger (22.9 kK, compared with 10.1kK for $[Co(NH_3)_6]^{2+}$), and the complete occupation of the t_{2g} subset is favoured over partial occupation of both e_g^* and t_{2g}, since the gain in spin pairing energy for the occupation of the e_g^* is less than the 'promotion energy', Δ, that is required – the arrangement is shown in Fig. 3.6c.

Figures 3.6a and c represent two different ways of arranging 6 electrons in the system. Since both systems have 6 electrons in the t_{2g} and e_g^* orbitals which are predominantly d orbital in character, they are referred to as d^6 ions ($[Co(NH_3)_6]^{2+}$ is a d^7 ion). The arrangement in Fig. 3.6a where the electrons are arranged to give the maximum number of parallel spins (and lowest spin pairing energy) is described as *high spin* (as is Fig. 3.6b); that of Fig. 3.6c is a *low spin* arrangement. A low spin or high spin arrangement is chosen according to whether the ligand field stabilisation energy Δ is greater or less than the energy required to pair the spins. The reader should be able to show for himself that the ions d^1, d^2, d^3, d^8, and d^9 do not have available

---

2 For conversion between energy units, see Appendix II.

the two alternatives of high spin or low spin configurations in octahedral complexes.

So far, we have only considered ligands with one sigma bonding orbital; had we considered chlorine as a ligand we would have considered one 3p orbital as having σ symmetry, while the other two would have π symmetry about the metal chlorine bond axis; any bonding involving these orbitals is metal-ligand π bonding. Figure 3.7. shows how π bonding may take place in the xz plane with either the d_{xz} or p_x orbital of the metal; a similar diagram could of course be drawn for other axes. The ligand π orbitals transform as $t_{lu} + t_{lg} + t_{2g} + t_{2u}$; remembering that the p orbitals transfer as t_{lu} and the d orbitals as $e_g + t_{2g}$, we would expect there would be a bonding interaction between the ligand π orbitals and the metal.

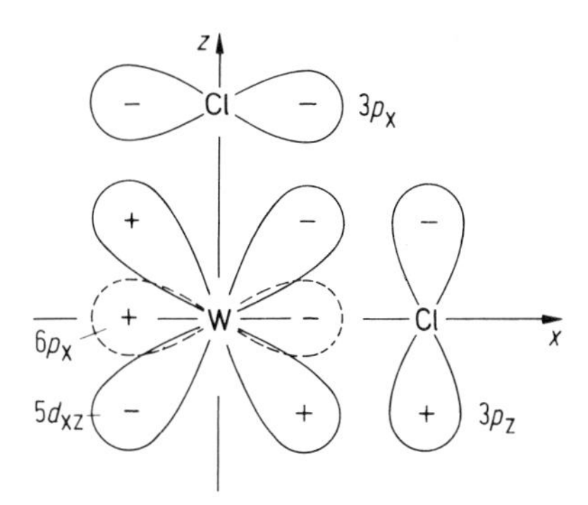

Fig. 3.7. Tungsten-chlorine π bonding in WCl_6

The nature of this interaction depends on the energy of the ligand π orbitals; for chlorine they will be strongly bound, while for ligand groups such as CO there are low lying antibonding orbitals of π symmetry. We will consider the two cases separately.

As an example of a chloride consider WCl_6 for which the molecular orbital diagram is drawn in Fig. 3.8a. The effect of the ligand π orbitals at low energy is to raise the energy of the predominantly metal t_{2g} and t_{lu} levels, now formally π antibonding; the $t_{lu}(p)$ orbitals were already weakly σ antibonding, and the π interaction would be expected to be even weaker. The π bonding t_{lu} have been put only just below the non-bonding t_{lg} and t_{2u} levels.

The t_{2g} π bonding levels will be slightly lowered as a result of their rather stronger interaction with the metal t_{2g} levels, previously non-bonding. The net effect of the π bonding is thus to lower Δ, by raising the energy of the t_{2g} metal orbitals, and to transfer an amount of charge from the occupied ligand π orbitals to the metal t_{2g} d orbitals as a result of the mixing.

WCl_6 is an octahedral compound of W(VI) similar to $W(CH_3)_6$, in which we predicted a transfer of charge to the methyl groups through the σ bonds. This will be accentuated in WCl_6, since Cl has a higher electron affinity than CH_3, but offset by the transfer of π electron density from the chlorine electrons to the metal t_{2g} d orbitals. This prevents the tungsten atom carrying a ridiculously high charge and is in accord with Pauling's *electro-neutrality principle,* which proposes it to be unlikely that any atom in a compound carries a charge of greater than ± 1 units.

Support for this M.O. diagram comes from the fact that the LFSE, Δ, is much less in complexes of ligands which are good π donors, such as chlorine, than those which

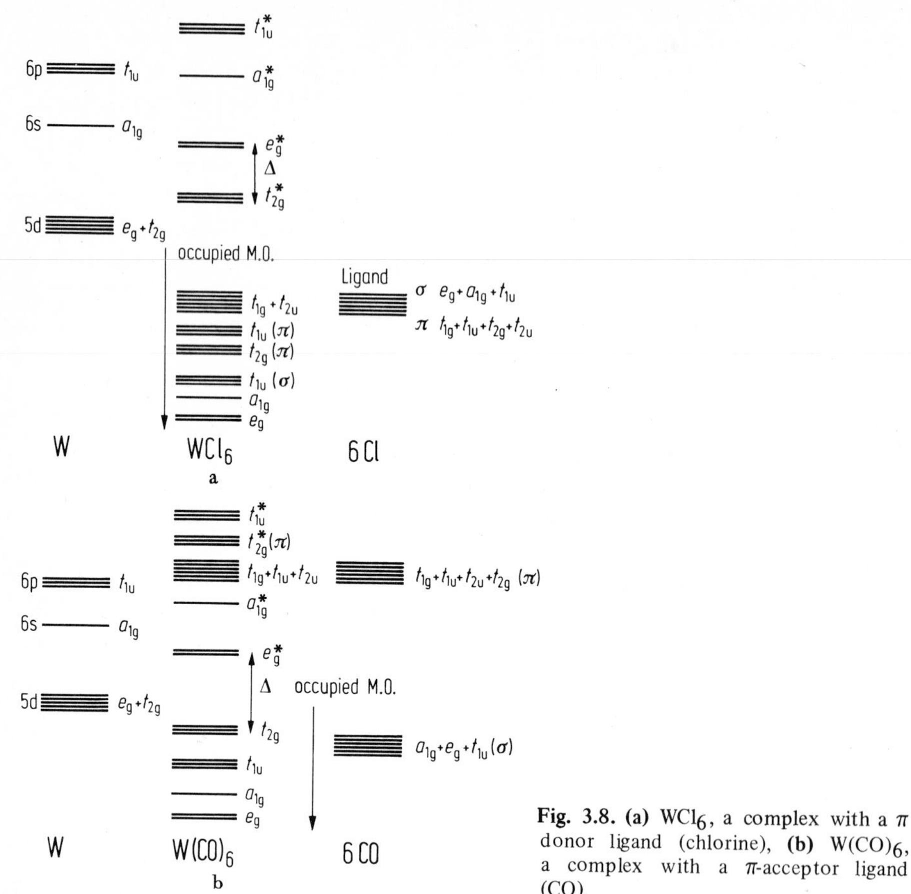

Fig. 3.8. (a) WCl_6, a complex with a π donor ligand (chlorine), **(b)** $W(CO)_6$, a complex with a π-acceptor ligand (CO)

are σ donors (such as ammonia). More graphically, the ion $[IrCl_6]^{2-}$ has five electrons in the weakly antibonding, mainly metal, t_{2g} shell. Clearly, one of these electrons must be unpaired, and if our arguments are correct, the small amount of chlorine p_π orbital mixed into this M.O. requires that this unpaired electron has a non-zero probability density on each chlorine atom. The E.P.R. spectrum of this ion shows a considerable coupling of this electron with the chlorine nuclear spins, confirming that the electron is delocalised over the ligand atoms.

The second case of metal-ligand π bonding is with ligands having empty π orbitals at a relatively low energy – a good example is the CO ligand which has low-lying unoccupied π^* antibonding orbitals. These have the same symmetry as the first case, but now, as both the t_{1u} (ligand) and t_{1u} (metal p) orbitals are unoccupied, we need not worry about their interaction. The metal t_{2g} orbitals will overlap with the ligand t_{2g}, and, since the metal orbitals are lower in energy, the bonding orbital resulting will be mainly metal in character, but will involve a slight delocalisation of d electron density onto the ligands. The metal t_{2g} orbitals are now bonding, and consequently Δ is increased by any π acceptance by the CO ligand. For the CO ligand, π acceptance corresponds to an occupation of the antibonding π^* orbitals, and thus a weakening of the

CO bond. This is shown by a fall in the C–O stretching frequency from 2189 cm^{-1} to about 2000 cm^{-1} in metal carbonyl complexes.

A good example of such a complex is tungsten hexacarbonyl $W(CO)_6$: the carbonyl group has a weakly bonding σ orbital mainly localised on the carbon atom which overlaps well with the metal σ orbitals. This orbital is doubly occupied, and consequently, there is a donation of charge from the 6 carbonyl orbitals to the metal, in the occupation of the 6 σ M.O.s ($e_g + a_{1g} + t_{1u}$). There remain six electrons (from the tungsten) to be placed in the next M.O., the t_{2g} bonding. This fills all the bonding M.O.s, giving a stable molecule, and also 'back donates' charge from the metal to the ligand – in total then, there is little transfer of charge between metal and ligands. This mode of bonding, in which charge is transfered in one direction by σ donation and in the opposite direction by π donation is known as *synergic bonding.*

π bonding is thus possible in the presence of either (i) a π *donor* ligand, which mixes some metal t_{2g} character into its occupied π orbitals, decreasing Δ, and leading to a partial transfer of charge to the metal, or (ii) a π *acceptor* ligand, which mixes some of its empty antibonding π orbitals with the metal t_{2g} orbitals, increasing Δ, and resulting in a partial charge transfer to the ligands. An example of the first case is WCl_6; of the second case, $W(CO)_6$.

Having dealt with the electronic structure of these complexes, we may now make some suggestions about their stability and reactivity. We mentioned that π accepting species increase Δ (as do strong σ donors), and consequently we would predict that complexes of such ligands would be low spin. Ligands which act as strong π donors (such as F^-) or which are weak σ donors will tend to give high spin complexes: thus $[CoF_6]^{3-}$ is a high spin $3d^6$ complex, while $[Co(NH_3)_6]^{3+}$ is low spin, NH_3 being a stronger σ donor, but a very weak π donor. Complexes with the t_{2g} orbitals occupied will be more stable if the ligands are π acceptors (the t_{2g} are then bonding) than with strong σ donors such as CH_3 (t_{2g} non-bonding) or π donors (t_{2g} antibonding). This leads to the result that π accepting ligands stabilise lower oxidation states. In complexes such as $[Co(NH_3)_6]^{2+}$ the weakest bound electrons are in weakly antibonding e_g^* levels; their weak binding enables them to be removed with no real loss in stability of the complex, and $[Co(NH_3)_6]^{2+}$ may be readily oxidised to $[Co(NH_3)_6]^{3+}$ without breaking any metal-ligand bond. The non-bonding or weakly bonding nature of the d electrons in many complexes accounts for the wide variety of oxidation states these species can show. In the case of manganese, it is even possible to remove electrons progressively from $[Mn(OH_2)_6]^{2+}$ (d^5) to $[MnO_4]^-$ (d^o), the oxygen becoming a progressively stronger π donor as the protons are removed. The effects of the ligands in such systems are considerable: for example, to remove an electron from $[WF_6]^-$. (d^1) to give WF_6 (with structure similar to WCl_6) requires only 5.2ev. The tungsten atom in such a complex definitely carries a slight positive charge, yet the energy to remove the lone electron from the t_{2g} orbitals is much less than the first ionisation energy of tungsten (7.98 eV).

Ligands may also affect redox equilibria in solution, since they change the energies of the d orbitals. Co^{3+} and Mn^{3+} are rapidly reduced to the +2 state in the absence of stabilising ligands, but $[Mn(CN)_6]^{3-}$ and $[Co(NH_3)_6]^{3+}$ are stable and are formed by atmospheric oxidation from the +2 oxidation state complexes.

The distinction between ionic and covalent bonding in transition metal complexes is not clear cut. The closeness of tungsten 5d and carbon σ orbital energies make it unrealistic to take an ionic model (W^{6+}, $6CH_3^-$) as a starting point for discussion of $W(CH_3)_6$. The greater orbital binding energy of the donor lone pair of NH_3, and the lower oxidation state of the metal makes the ionic model more plausible for $[Co(NH_3)_6]^{2+}$. Oxidation to Co(III) (i.e. going from $3d^7$ to $3d^6$) will cause a drop in 3d orbital energy which will result in greater mixing with the nitrogen σ donor orbitals. This explains why the e_g^* orbitals are raised much more with respect to the t_{2g} orbitals in $[Co(NH_3)_6]^{3+}$ than in $[Co(NH_3)_6]^{2+}$. As the metal atom is oxidised, the metal orbital energies drop, the covalent contribution to the bonding increases, and Δ rises.

Tetrahedral coordination

Tetrahedral coordination is the second most common coordination of transition metal complexes. In pure tetrahedral geometry the M.O. diagram may be obtained as for the octahedral case: the ligand σ orbitals transform as $a_1 + t_2$ (as for methane), and these can interact with the central atom s-orbitals (a_1) and d orbitals (t_2 set) and also (to a lesser extent because of their energy) the p orbitals (also t_2). The e orbitals are not bonding and thus we expect a splitting between e and t_2 levels, with the metal t_2 d orbitals slightly antibonding and at a higher energy. The order of the two subsets of the d subshell is thus inverted in comparison with the octahedral case. Both e and t_2-levels are available for π bonding.

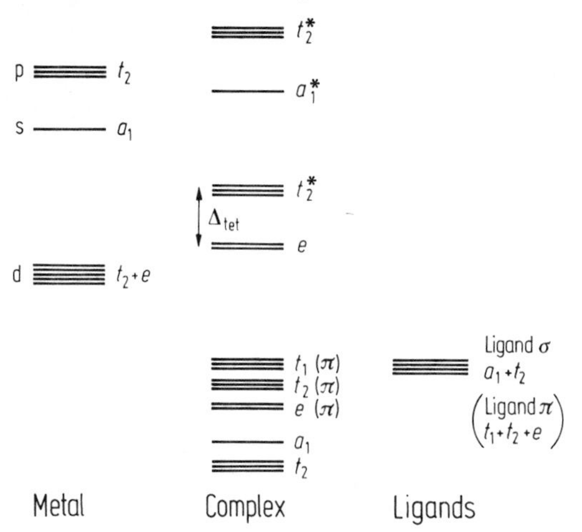

Fig. 3.9. Molecular orbital diagram for a tetrahedral complex

Calculations, based on either a pure ionic model, or an angular overlap model, and experiment show that the splitting between the e and t_2 d sub-shells is much smaller for tetrahedral than for octahedral complexes. The ionic model predicts that if metal ligand bond lengths remain the same, Δ tet $= \frac{4}{9}$ Δ oct and the experimental values found agree fairly well with this, although Δ tet is often slightly more than $\frac{4}{9}$ Δ oct as the metal-ligand bond lengths are often shorter for tetrahedral complexes.

The small value of Δ tet, and the fact that both sets of orbitals are equally affected by π bonding enable us to make certain predictions about tetrahedral complexes. Δ is smaller than for octahedral complexes while the electronic repulsion parameters are the same – thus all tetrahedral complexes are high spin. There are no particular numbers of d electrons for which the tetrahedral arrangement is very strongly favoured, and in general an octahedral coordination is more stable; however, with small central atoms, octahedral coordination may cause the ligands to overlap each other, and this generally produces an antibonding repulsive effect. The tetrahedral arrangement is common in such cases, e.g. $[CoCl_4]^=$, $[NiCl_4]^=$. Tetrahedral coordination is also found with d^{10} species such as $Ni(CO)_4$, $Ir(NO)(PPh_3)_3$ where the ligands may π-accept from both subsets of d orbitals. For d^0 systems tetrahedral coordination is also found e.g. in $[WS_4]^=$ and OsO_4 where there is appreciable π donation from the ligands giving a high double bond character to the metal ligand bond.

Finally we may note that the d–d absorption spectra of tetrahedral complexes are often more intense than their octahedral counterparts. Referring to the M.O. diagram (Fig. 3.9), we see that both metal d and p orbitals have t_2 symmetry and consequently the t_2^* M.O. may have some p character mixed in. The $e \rightarrow t_2^*$ transition thus has some $d \rightarrow p$ character, and is more intense since the $d \rightarrow p$ transition is allowed by the electric dipole mechanism.

Square planar complexes

There exist a considerable number of square planar four-coordinate complexes of symmetry D_{4h}. The relevant orbitals transform as shown in Fig. 3.10; if σ bonding only is considered then only the $d_{x^2-y^2}$ (b_{1g}) and d_{z^2} (a_{1g}) metal d orbitals will have suitable symmetry to overlap with the ligands. We would expect the $d_{x^2-y^2}$ to overlap much more strongly as it is wholly directed at the ligands, while only the toroid 'doughnut' part of the d_{z^2} can overlap with the ligands. We thus expect the $d_{x^2-y^2}$ b_{1g} orbital to be appreciably more antibonding. This explains the experimental observation that d^8 complexes which have the lowest four d orbitals occupied show a strong tendency to square planar geometry.

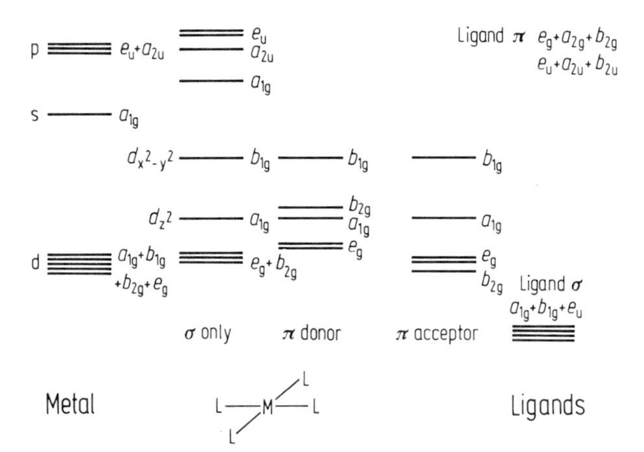

Fig. 3.10. Molecular orbital diagram for a square planar (or quadratic) complex with various types of ligand orbitals in the complex are mitted for clarity

The effect of π bonding are, as usual, smaller in magnitude. π acceptor ligands lower the energy of the σ non-bonding e_g and b_{2g} orbitals, and π donor ligands (e.g. Cl^-) raise their energy. The effect on the d_{xy} (b_{2g}) orbitals is more prononced as we would expect from the greater overlap with the ligands compared with the d_{xz}, and d_{yz} orbitals. The strongly antibonding nature of the b_{1g} ($d_{x^2-y^2}$) orbital is not affected by π bonding.

There are two useful comparisons to be made with the square planar M.O. diagram: firstly, if we consider an octahedral complex in which the two ligands on the z axis are progressively removed and the xy plane ligands compressed (an extreme case of a Jahn-Teller distortion – Chap. 4), then, if the ligand is only a weak π donor, the $d_{x^2-y^2}$ will rise in energy, the d_{z^2} will fall (the σ bonding effect), and the d_{xy} orbital will rise slightly as the xy plane ligands move in (the π effects) while the d_{xz} and d_{yz} will not be greatly changed. This is shown in moving from the left hand side to the centre of Fig. 3.11. Secondly, if we had considered squashing a tetrahedron to give a square of ligands, then the t_2 subset, which are antibonding and directed at the ligands for a tetrahedron will fall in energy, while the overlap with the $d_{x^2-y^2}$ will rise sharply, with an increase in energy. The d_{z^2} will not be greatly affected. These changes are shown in moving from the right hand side to the centre of Fig. 3.11.

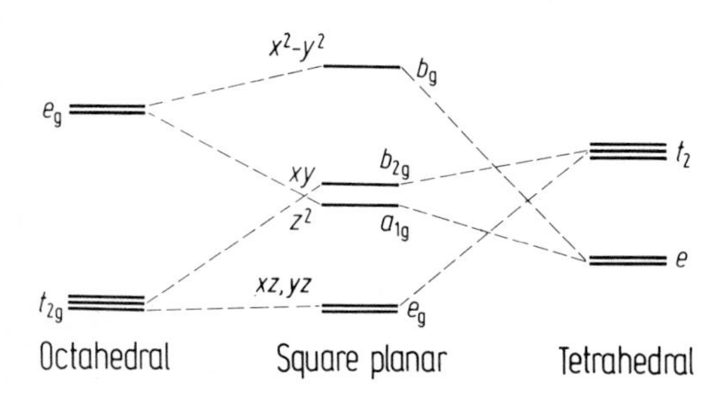

Fig. 3.11. The splitting of d-orbitals according to coordination

We have remarked that square planar geometry is favourable almost uniquely for d^8 species – we may now note that it will only be favourable if the ligand field splittings are sufficiently great to overcome the pairing of electrons necessary to leave the $d_{x^2-y^2}$ b_{1g} empty. Thus square planar complexes are typically strong field complexes. Thus Ni^{2+} ($3d^8$) forms octahedral $[Ni(H_2O)_6]^{2+}$ and tetrahedral ($[NiCl_4]^{2-}$) complexes with weak field ligands, and with stronger, more covalent ligands gives planar complexes ($[Ni(CN)_4]^{2-}$). Copper (II) ($3d^9$), which would be expected to show an appreciable Jahn-Teller distortion in octahedral geometry, also forms some planar complexes, but also many rather distorted octahedral and tetrahedral complexes corresponding to the whole range of Fig. 3.11.

There are a small number of nickel complexes of the type NiX_2L_2, where X is a halogen and L a phosphine or amine ligand, which exist in tetrahedral and square planar forms, or which change structure with temperature. This seems to be due (in some cases) to a balance of electronic effects (similar to spin crossover phenomena Chap. 4) and in others to the crowding and steric hindrance of bulky L ligands in the square planar form. Thus the complex shown in Fig. 3.12 is planar for R = CH_3, but tetrahedral for R = $(CH_3)_3C$, the bulkier t-butyl group.

Other coordination numbers

Although the three types of coordination described above are the most common encountered in the chemistry of the transition metals, other coordinations are by no means rare. With the larger second-and third row (4d and 5d) transition metals higher coordination numbers are found (as high as 9 for ReH_9^{2-}), and several 8-coordinate species may be found (e.g. $K_4Mo(CN)_8$); references to reviews of these coordination numbers are given in the bibliography.

Five-coordinate species are also found frequently, and are also believed to be formed as intermediates in many chemical reactions (e.g. after addition to a 4 coordinate, or elimination from a 6 coordinate complex). The two possible arrangements (Fig. 3.13) are the trigonal bi-pyramid (as in $Fe(CO)_5$) and the square based pyramid (as in one form of $[Ni(CN)_5]^{3-}$) with the metal atom slightly above the plane of the square towards the fifth ligand. Five coordinate species are often distorted, and, in cases where the five ligands are chemically identical (or even fairly similar) may show *fluxional* behaviour, in which molecular vibrations 'scramble' the different ligands. This is shown in Fig. 3.14; the axial L′ ligands in the left hand figure become equatorial in the right hand figure. This 'scrambling' occurs very quickly, and $Fe(CO)_5$ shows only one carbon n.m.r. signal, the average of the equatorial and axial sites. Similar behaviour is found for main group five-coordinate species such as PF_5.

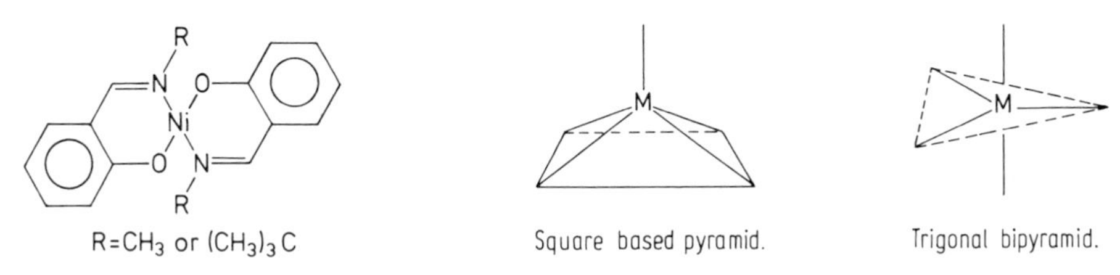

Fig. 3.12

Fig. 3.13. Penta-coordination

Coordination numbers of 2 and 3 are found for gold and silver complexes, but these complexes have full d shells, and the orbitals involved in the bonding are almost certainly the s and p orbitals above the d shell.

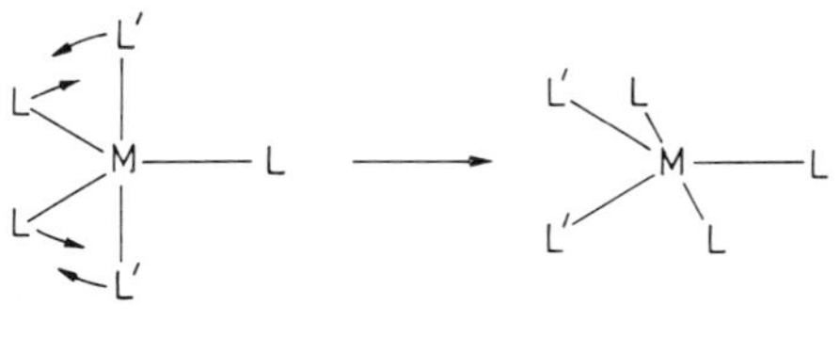

Fig. 3.14

All the discussions given for complexes involving d orbitals have assumed that all the ligands involved are identical. Most complexes do not have only one type of ligand bonded to the metal, so one might wonder why such detail for highly symmetrical complexes has been given. The answer is again connected with microsymmetry. For octahedrally coordinated species such as MA_xB_{6-x}, if A and B are not too different, while both the e_g and t_{2g} subshells are both split by the lower symmetry, the splittings within the e_g and t_{2g} subsets are much smaller than the energy difference between the e_g and t_{2g} levels, and so to a first approximation we may ignore them. This tacit assumption of pure octahedral microsymmetry is of course crude, but enables us to continue associating the e_g levels with σ bonding and the t_{2g} with π bonding; for qualitative discussion this is very useful. Care should be taken not to deduce too much from the electronic spectra of such systems, and the magnetic properties can only be understood by a thorough treatment of the electronic structure.

The participation of f orbitals in bonding

The lanthanides and actinides both have partially filled f subshells, and one might expect it to be necessary to derive molecular orbitals involving the f orbitals. In fact there is very little evidence for participation of the 4f orbitals in lanthanides in chemical bonding. Presumably the 4f orbitals are not sufficiently diffuse to give a good overlap with ligands. The lanthanides have an exclusively ionic (or metallic) chemistry and the 4f orbitals may be regarded as non-bonding. The 5f orbitals of the actinides might be expected to be more diffuse, and consequently more likely to overlap with the ligands. The early actinides do show rather more covalent character (witness the stability of the MO_2^{n+} species) but the bonding in their compounds is by no means well understood, and is complicated by the small energy differences between 5f, 6d, 7s and 7p levels. The most conclusive evidence comes from the study of the organo-metallic complexes formed by uranium, neptunium, and plutonium with cyclo-octatetraene, C_8H_8. A full M.O. description is given by Huheey. The lanthanides form no covalent organometallic compounds. Further discussion is given in Chaps. 4 and 7. M.O. treatment of lanthanides and actinides would be complicated by the high C.N. (e.g. 12 in $[Mg(H_2O)_6]_3 [Ce(NO_3)_6)_2]$. $6H_2O$.

C. Organometallic Compounds

In this section we shall discuss features of chemical bonding that are peculiar to organometallic chemistry, in particular that involving transition metals. In view of the huge growth that organometallic chemistry has undergone in the last 25 years, it is as well to emphasise that this section is concerned only with details of the bonding which have not been discussed before. For this purpose we shall consider organometallic chemistry to include species such as nitrosyls which are analogous to carbonyls. We have already discussed two organometallic compounds – $W(CO)_6$ and $W(CH_3)_6$, and have seen that their bonding involves no radical departure from general M.O. theory.

Organometallic compounds are almost invariably highly covalent, and the bonding electrons highly delocalised. Furthermore, the symmetry is generally much lower than in the complexes we have hitherto discussed: in few cases is it possible to draw an M.O. diagram for the whole molecule, and discussion of electronic structure tends to concentrate on the character of a particular metal ligand bond.

If it is impossible to draw M.O. diagrams for the molecule as a whole, it has been found possible to explain the stability of a large number of organometallic compounds by electron counting methods. It was noticed about 50 years ago that a large number of stable carbonyls had 18 electrons in the metal ligand bonding system. We have already seen an example of this with $W(CO)_6$ – there were 6 pairs of electrons in the (mainly ligand) σ orbitals and 6 electrons in the (mainly metal) π orbitals. Another example would be $Fe(CO)_5$ where each CO is considered to donate 2 electrons to the metal ligand bonding, and the iron atom has 8 electrons in its valence shell ($3d^6 4s^2$), again giving a total of 18. Further support comes from the observation that the unstable $V(CO)_6$ (17 electrons) readily takes up an electron to give $[V(CO)_6]^-$ – 18 electrons. Further examples will appear in this section, and we proceed immediately to a statement of the rule:

The sum of the electrons on the metal and the electrons donated by the ligands should equal the atomic number of the inert gas next after the metal in the periodic table or *the number of electrons in the valence shell of the metal plus the number contributed by the ligands should equal 18.*

Such compounds have a greater stability than equivalent compounds with a different number of electrons. This 'effective atomic number' or *'18 electron'* rule is essentially a rule of thumb, and is unreliable for ionic complexes, but covers a very large percentage of organometallic compounds of transition metals, irrespective of their symmetry.

One or two tentative explanations of the success of the rule have been proposed, of which the most reasonable is that the metal has nine atomic orbitals in its valence shell, (five d, one s and three p), and that the most stable bonding arrangement will be for it to form nine bonding and nine antibonding orbitals, with all the bonding orbitals filled – corresponding to 18 electrons in the valence shell. Some force is given to this argument by the fact that, for the first row elements, the rule works best in the middle of the 3d series where 3d, 4s and 4p orbitals are closest in energy (see also page 272).

The majority of the exceptions to the 18 electron rule are 16 electron compounds, many of them square planar d^8 complexes where, as has been mentioned, one of the d orbitals (the d_{z^2}) is non-bonding.

If the 18 electron rule is accepted (and it is by no means the only electron counting rule to work for no especial reason) it is clearly necessary to decide how many electrons each ligand can give. The numbers for a variety of common ligands are given in Table 3.3. The assignment will become clear as we proceed.

Table 3.3. The number of electrons donated by various ligands

	Number of e^-	Examples
Covalent bond	1	CH_3, H, Cl, bridging CO
Lone pair	2	terminal CO, NH_3, Cl^-, PPh_3
Olefin	2	C_2H_4, C_2F_4
Allyl group	3	C_3H_5
Nitrosyl	3	NO
Diene	4	C_4H_6
Aromatic ring	6	C_6H_6, $C_5H_5^-$

Carbonyls

There are a very large number of carbonyls, and substituted carbonyls and we note here only a few of interest. $Mn(CO)_5$ is a 17 electron system and might be expected to be unstable; in fact it is unknown, preferring to form a Mn-Mn single bond to give a dimer $Mn_2(CO)_{10}$, each Mn atom donating one electron to the valence shell of each other (i.e. $Mn(CO)_5$ acts as a 1 electron ligand) to give two electron systems (Fig. 3.15). $[Mn(CO)_5]^-$ can however, be prepared, as may $[Mn(CO)_6]^+$. Alternatively, a series of one electron donors may occupy the sixth coordination site of the $Mn(CO)_5$ species to give an 18 electron system e.g. $CH_3Mn(CO)_5$; $HMn(CO)_5$; $ClMn(CO)_5$.

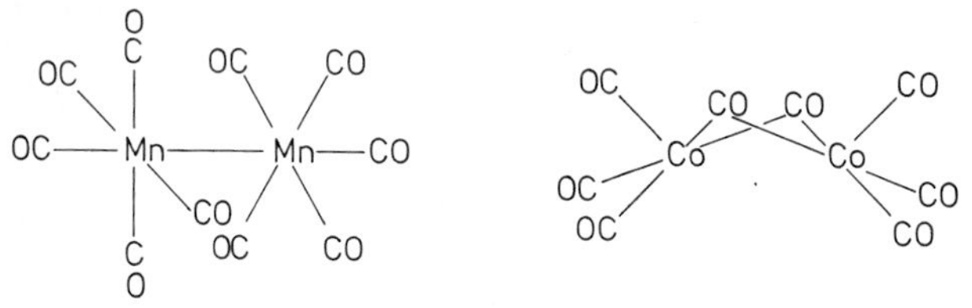

Fig. 3.15

$Co(CO)_4$ is again a $17e^-$ system and is unknown – however the dimer $Co_2(CO)_8$ exists, and has the solid state structure shown in Fig. 3.15. Rather than form a simple $(OC)_4Co\text{-}Co(CO)_4$ structure it adopts a structure with two bridging CO groups and a metal-metal bond (see Fig. 3.15). The electron count is now (for each cobalt atom):

Cobalt	9
3 terminal CO	6
2 bridging CO (1 e^- each)	2
1 from the other Co (metal metal bond)	1
	18

The X-ray crystal structure shows that the CO bond length is slightly greater for the bridging carbonyl groups than for the terminal groups. This implies that the CO bond is slightly weaker in this case, and this is confirmed by the infra-red stretching frequency. Terminal carbonyl groups generally absorb near 2050 cm^{-1} (free CO is 2143 cm^{-1}) while bridging groups absorb typically at 1850 cm^{-1} – this is in fact not too far from an organic ketone absorption showing the strong covalent nature of the M-CO-M linkage. Triply bridging carbonyls are known (e.g. $Rh_6(CO)_{16}$), and absorb at lower energy (1750 cm^{-1}).

The weakening of the CO bond results (as was mentioned earlier) from the withdrawal of electrons from σ bonding or non-bonding orbitals and the donation of electrons into the π antibonding orbitals. Thus the metal-ligand bond is strengthened at the expense of the internal bonding of the ligand. This is extremely common in organometallic chemistry.

It may be noted that when the CO groups in a carbonyl are substituted by a weaker or non-π acceptor (e.g. pyridine (C_5H_5N), ethylendiamine) the infra-red frequency falls further, indicating that the CO now π accepts a greater charge. CO frequencies have been used extensively as an indication of the bonding nature of other ligands in the complex.

Analogues of carbonyls

There are several ligands which have a electronic structure similar to carbon monoxide, notably NO^+, N_2, $C{\equiv}CH^-$, CN^- all iso-electronic with CO, iso-cyanides (RN=C) and the CS molecule, unknown except as a ligand. All are assumed to bond in a roughly similar way, acting as σ donors and π acceptors.

Nitrosyl complexes form the greatest number of analogues to carbonyls, and incidentally also give us further information on the nature of the bonding of these groups. If we take a simple example, $(Ph_3P)_2Ir(NO)(CO)$ the NO stretching vibrations absorb at 1645 and 1660 cm^{-1}. Free NO is paramagnetic with one electron in the π^* orbitals and absorbs at 1860 cm^{-1}. The lower frequence in the complex suggests that there is further π acceptance into the π^*.

Nitrosonium salts (e.g. NO^+ BF_4^-) exist and absorb at about 2200 cm^{-1}. The iridium complex is diamagnetic showing that the unpaired electron is sufficiently involved in bonding to pair with the unpaired electron on the iridium. The electron count may be carried out in three ways:

2 PPh_3	4	2 PPh_3	4	2 PPh_3	4
1 CO	2	1 CO	2	1 CO	2
Ir	9	Ir^+	8	Ir^-	10
NO	3	NO^-	4	NO^+	2
	18		18		18

which one is used essentially a matter of choice, although the third is often used to stress the analogy with CO. Bridging NO groups may also be found, and as with CO, absorb in the infra-red at a lower frequency.

A feature of NO complexes is the variation in the metal N-O bond angle. In most nitrosyls this is close to 180°, but a small number have been found where the bond angle is close to 120°. In such cases (e.g. $[Ir(CO)Cl(PPh_3)_2(NO)]^+$) the NO stretching frequency is generally rather lower than for complexes with linear metal-NO groups. The metal-nitrogen bond in such cases is rather longer than usual, and it has been suggested that the N atom is sp^2 hybridised (giving the bond angle of 120°) and rather closer to NO^-. This makes the complexes obey a 16 electron rule rather than an eighteen electron rule, but as they generally involve a d^8 metal (Ir(I), Ru(O) which gives many 16 electron complexes this is perhaps not surprising.

Molecular nitrogen forms a fairly large number of complexes in which the linear mode of bonding is similar to CO. The stretching frequency is reduced from 2331 cm^{-1} to between 1900 and 2150 cm^{-1} and bond length increase; for example the bond length in $Co(PPh_3)_3(N_2)H$ is 1.112 A compared with 1.098 A in free N_2. It appears to be a rather weak σ donor, but a strong π acceptor, and molecular nitrogen in such complexes can sometimes be reduced to ammonia or amines under relatively mild conditions. There has been a great of research carried out on these compounds as possible catalysts for the fixation of nitrogen.

Finally we may mention that the bonding of cyanide ion and isocyanides is very similar to that of carbon monoxide as the similarity of their electronic structure would suggest. Several other simple molecules, in particular phosphines and arsines (such as $P(C_6H_5)_3$, $AsEt_3$) may be substituted for CO in metal carbonyls. Evidence suggests that they also act as σ donors and π acceptors, but the π electrons are generally held to be accepted into the empty d orbitals of the ligating atom (i.e. the 3d in phosphorus, the 4d in arsenic, see also page 260).

Olefin complexes

The olefin complex $K^+[PtCl_3(C_2H_4)]^-$ prepared by Zeise has been known for nearly 150 years, but a satisfactory bonding description of the complex was only given in the early 1950's. The structure of the ion is shown in Fig. 3.16. The platinum shows square planar coordination typical of a third row transition metal d^8 complex. The sketch of the π orbitals of the ethylene molecule shows that the π bonding orbital ϕ_1 has π symmetry about the Pt-olefin (z) axis, and the π antibonding orbital ϕ_2 has π symmetry with respect to this axis. Thus we have again a ligand with a full σ and empty π orbitals.[3] The model for the bonding proposed by Dewar and Chatt is that there is a transfer of charge from the ϕ_1 orbital to the platinum accompanied by back donation from the platinum d_{xz} orbital into the orbital ϕ_2 – in other words the familiar synergic bonding found in carbonyls. The Dewar-Chatt model is generally accepted as the most satisfactory explanation of the bonding of olefins. For the purposes of electron counting, the ethylene ligand is regarded as a 2 electron donor.

The effective transfer of charge from the π bonding orbital ϕ_1 to the π antibonding orbital ϕ_2 would be expected to weaken the C=C double bond – this may be shown by the drop in the C–C stretching frequency (1551 cm^{-1} in Zeise's salt, 1623 in ethylene) and a lengthening of the C–C bond (by up to 0.1 Å). Electron withdrawing substituents (such as fluorine or cyanide) on the ethylene would be expected to increase

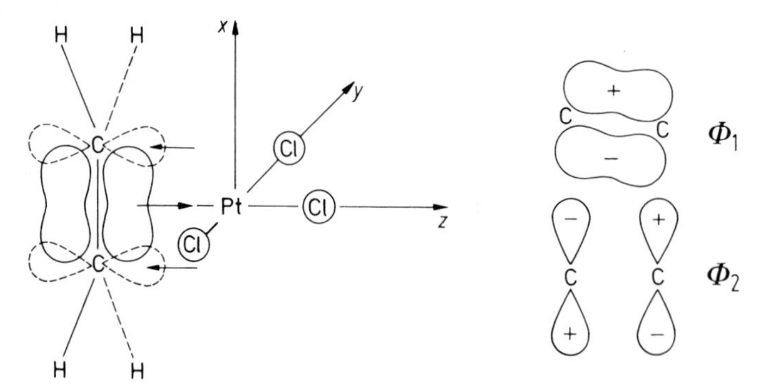

Fig 3.16. The structure of $[PtCl_3(C_2H_4)]^-$. The C_2H_4 (xy) plane lies perpendicular to the plane of the paper. Simplified representations of the π bonding ϕ (continuous line) and π antibonding ϕ_2 (dotted line) ethylene orbitals are given

---

3 These symmetries refer only to the metal-ligand bond.

the π acceptance by ϕ_2, and weaken still further the C=C bond; in fact $C_2(CN)_4$ complexes show an 'olefinic' carbon-carbon bond length little shorter than a carbon-carbon single bond. Alkenes coordinated to transition metals are much more susceptible to hydrogenation; this is the basis of many catalysts.

From a molecular orbital point of view, the transfer of charge from ϕ_1 and to ϕ_2 is similar to a photochemical activation involving the jump of an electron from π to π^*. The excited states of ethylene and similar molecules are frequently found to be bent, and it is found that the C_2X_4 species is no longer planar, and that the two CX_2 fragments are bent away from the metal atom. This has led to the suggestion that coordinated olefins are similar to cyclopropanes, and are essentially σ bonded:

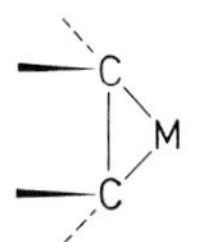

One of the more remarkable features of coordinated ethylene is the rotation of the C_2H_4 molecule about the metal-C_2H_4 bond axis (the z axis in Fig. 3.16). The nuclear magnetic resonance spectrum of the compound $(C_5H_5)Rh(C_2H_4)_2$ shows that the ethylene rotates about the Rh-alkene bond axis with a rotational barrier of about 60 kJ/mole. Presumably the destruction of π bonding with the d_{xz} orbital on rotation is compensated by π bonding with the d_{yz} orbital.

Ferrocene

Much of the interest in organometallic chemistry was stimulated by the discovery in 1951 of the extremely stable compound ferrocene $(C_5H_5)_2Fe$.

The structure is shown in Fig. 3.17. Only the π molecular orbitals of the ligand are assumed to be involved in bonding with the metal and their form is shown in the figure. Viewed down the 5 fold axis of symmetry (the z axis) the a_1 clearly has σ symmetry and can overlap with the $3d_{z^2}$, 4s, or $3p_z$ orbitals of the iron; the doubly

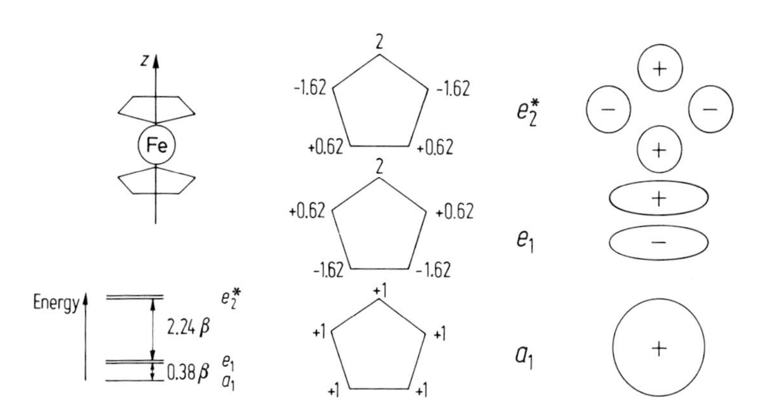

Fig. 3.17. The structure of ferrocene and the Hückel π molecular orbitals of $C_5H_5^-$

degenerate e_1 orbitals have π symmetry and may overlap with $3d_{xz}$, $3d_{yz}$, $4p_x$ or $4p_y$ orbitals; the e_2 orbitals have δ symmetry, and may overlap with the $3d_{x^2-y^2}$ and $3d_{xy}$.

If the normal assignment of $Fe^{2+}(3d^6)2C_5H_5^-$ is taken, then the a_1 and e_1 orbitals of the cyclopentadiene rings are occupied, and act as σ and π donors to the metal respectively, while the $3d_{z^2}$ is non-bonding and doubly occupied, and the $3d_{x^2-y^2}$ and $3d_{xy}$ orbitals are occupied and stabilised by donation of charge to the empty δ e_2^* orbitals of the ligand. This leads to the order of orbitals shown in Fig. 3.18. The energy difference between the a_{1g} (non-bonding $3d_{z^2}$) and e_{2g} (weakly δ bonding) is not very great and seems to cross over in the vanadium analogue $(C_5H_5)_2V$. In ferrocene, they are both occupied, and their order is thus irrelevant.

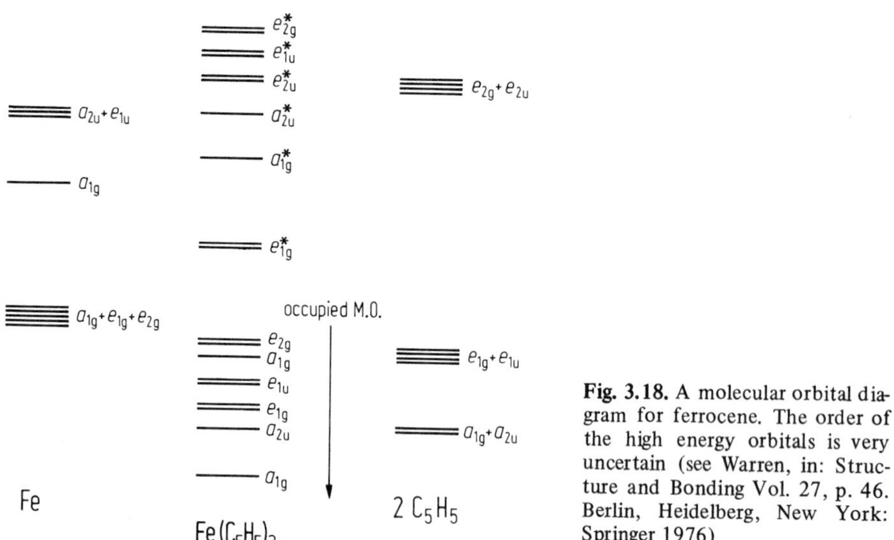

Fig. 3.18. A molecular orbital diagram for ferrocene. The order of the high energy orbitals is very uncertain (see Warren, in: Structure and Bonding Vol. 27, p. 46. Berlin, Heidelberg, New York: Springer 1976)

Ferrocene and its analogues with other transition metals are unusual for organometallic compounds in having a sufficiently high symmetry for a full M.O. diagram to be drawn. Extensive spectroscopic and magnetic measurements have been made. As far as chemical reactivity is concerned, the formulation as $Fe^{2+}(C_5H_5^-)_2$ or, in the M.O. scheme, the full occupation of the a_1 and e_1 levels of the $C_5H_5^-$ species is confirmed by the typical reactions of an aromatic ring (e.g. Friedel Crafts reactions) that ferrocene undergoes. According to Hückel M.O. theorie $C_5H_5^-$ is expected to show aromatic behaviour.

Ferrocene has the two cyclopentadiene rings staggered in the solid phase, but adopts the eclipsed conformation in the vapour phase, with a rotation barrier of only 3.8 kJ/mole. Many other metallocenes adopt the eclipsed conformation in the solid state. C_5H_5 is always considered as a five electron donor, or, as $C_5H_5^-$, a six electron donor. Such is the stability of the M.O. framework that many metallocenes do not obey the 18 electron rule and some are paramagnetic.

Other π electron systems

Many other organic π electron systems may act as ligands, but they present no real difficulties in the general description of their bonding. Thus the allyl group C_3H_5 acts as a 3 electron donor: it may be regarded as a combination of an alkene ($2e^-$) and a 1 electron σ donor, but the bonding appears to be more delocalised and the second figure in Fig. 3.19 is a better representation. Dienes may act as 4 electron donors, as might be expected from 2 separate alkenes in compounds such as butadiene iron tricarbonyl (iii).

An interesting example of the capacity of metal-alkene bonding to stabilise unusual species is the compound with tetramethyl-cyclobutadiene (iv), cyclobutadienes being otherwise unknown.

Acetylene is unusual in being able to act as a 'bidentate' ligand, forming two alkyne metal bonds using the different sets of π and π^* orbitals at right angles to each other as in Fig. 3.19 (v).

By analogy with $C_5H_5^-$, benzene can act as a 6 electron donor as in $(C_6H_6)Cr(CO)_3$.

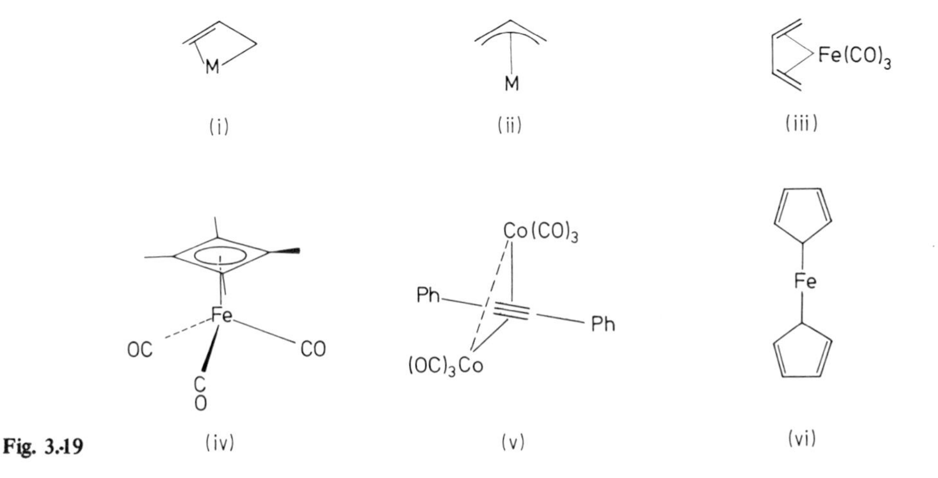

Fig. 3.19

Oxygen complexes

These have attracted some interest in view of their biological importance. The basic scheme of bonding is very similar to that of ethylene, except that it is necessary to pair the two electrons in the degenerate π_g^* orbitals. This gives one full π^* orbital and one empty. The empty π^* orbital and the full π bonding orbital coplanar with it may then act as π acceptors and σ donors to a metal exactly as for ethylene (Fig. 3.20). This mode of bonding is generally held to occur in $(PPh_3)_2PtO_2$ and $Ir(CO)Cl(PPh_3)_2O_2$ where the O–O bond length is lengthened (as might be expected).

It should be added that the dioxygen molecule does have other bonding schemes available: in complexes of iron and cobalt, it is found bonded as a 'bent' group analogous to "NO^-", with only one oxygen atom bonded to the metal, and a M–O–O angle close to 120°. Dioxygen is also found to act as a bridging bidentate ligand with a M–O–O–M linkage; again the M–O–O angles are close to 120°. An interesting review of dioxygen complexes is given by Vaska.

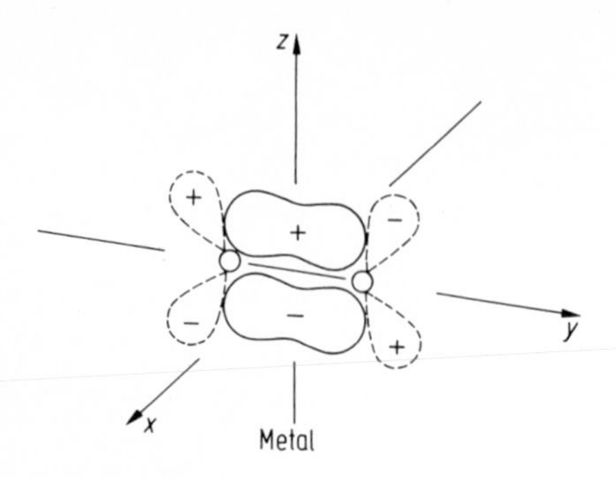

Fig. 3.20. The π and π^* orbitals in the xy plane are both occupied. In the yz plane the π bonding orbital is occupied, and the π^* antibonding orbital empty. See Fig. 3.16 for the platinum-ethylene bond

Nomenclature

There are frequently ambiguities in the naming of organometallic compounds; for example the species C_5H_5 may bond as in ferrocene, or by a straightforward σ bond – as in the structure originally proposed for ferrocene, (Fig. 3.19 (vi)).

It would be useful to be able to distinguish these two modes of bonding, and a simple scheme proposed by Cotton enables this. A count is made of the number of atoms of the ligand actually involved in the metal-ligand bond. For ferrocene this number is 5, for the hypothetical structure only 1. The two structures are distinguished as *pentahapto* or *monohapto*. In a similar way an allyl group would be *trihapto* and a benzene ligand *hexahapto*. The system for writing this is not finally agreed, but the most common method is bis-η^5-cyclopentadienyl iron for ferrocene (as opposed to bis-η^1-cyclopentadienyl iron for the *monohapto* species) – it is however still common to find the *pentahapto* species described as 'π cyclopentadienyl' and the *monohapto* as 'σ cyclopentadienyl'. A good discussion is given by Huheey.

In passing we may remark that bridging ligands are indicated by the prefix μ. Thus the bridged dioxygen species are μ-dioxo compounds, and $Co_2(CO)_8$ (Fig. 3.15) is described as di-μ-carbonyl-bis(cobalt-tricarbonyl).

Fluxional molecules

Organometallic chemistry furnishes many examples of fluxional molecules, where the molecular framework is not rigid. We have already mentioned $Fe(CO)_5$, and here give two examples of these remarkable effects. The compound $(C_8H_8)Fe(CO)_3$ has a structure in which two of the double bonds act as a 4 electron donor (Fig. 3.21), and the structure shown would certainly lead one to predict more than one ^1H nmr. signal. However, the n.m.r. spectrum at 25°C shows only one line. It has been established that this is due to a movement of $Fe(CO)_3$ species round the ring in a time fast by comparison with the n.m.r. observation time[4] – even cooling to 175°K gives no change. However the Ru analogue $(C_8H_8)Ru(CO)_3$ can be 'frozen out' to give the expected spectrum.

The compound $Ti(C_5H_5)_4$ (Fig. 3.21) has C_5H_5 groups bonded in two different modes σ (or *monohapto*) and π (or *pentahapto*). Surprisingly enough the proton n.m.r. signal shows only one signal at room temperature or above – this is due to three distinct fluxional processes:

[4] See page 297 for discussion of observation times.

1) A very fast 1-5 shift of the metal carbon σ bond around the σ-C_5H_5 ring (sometimes referred to as 'ring whizzing')
2) A rotation of the π bonded (*pentahapto*) ring about the metal – C_5H_5 axis
3) A scrambling of σ (*monohapto*) and π (*pentahapto*) bonding rings.

These remarkable rearrangements result in the n.m.r. spectrum showing only a time average of all the various proton resonances, it is however possible to 'freeze out' the resonances (i.e. slow down the exchange by lowering the temperature) and to see more than one signal. A fuller treatment is given by Jesson and Muetterties.

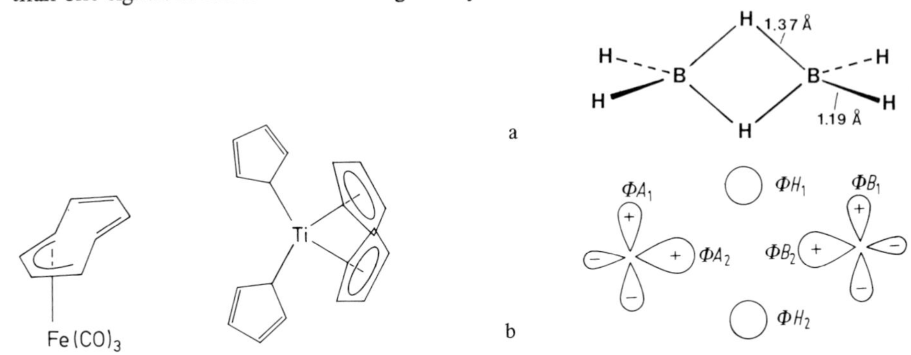

Fig. 3.21

Fig. 3.22 a and b. The structure and atomic orbitals of diborane

D. Electron Deficiency

The earliest descriptions of chemical bonding were based on the principle of a localised 2 electron bond between two atoms; this is essential for simple valence bond theory, and convenient for simple M.O. theory, and even in compounds where delocalisation was known to occur (such as benzene), formal electron pair bonding schemes could be drawn for benzene (the Kekule structures). However, as synthetic chemistry progressed, and as studies of the structures of compounds became more common, it became clear that there were several compounds for which such simple bonding schemes could not be proposed, as there were not enough electron pairs to give the correct number of bonds. Such compounds were dubbed *electron deficient* – a little unfairly as the deficiency lies with the bonding theory, and the name tends to imply that such compounds would be good electron acceptors (or oxidising agents), which is by no means always the case. The accepted definition of an electron deficient compound is one in which m atoms are held together by less than $2(m\text{-}1)$ electrons, *or* one for which no two centre electron pair bonding scheme can be drawn. From the point of view of the more general molecular orbital theory we have been using, there is nothing *extraordinary* about electron deficient compounds, but they exhibit a sufficient number of rather unusual properties to justify a separate discussion.

The classic example of electron deficiency is diborane (B_2H_6) whose structure is shown in Fig. 3.22 a; the simple molecule BH_3 is unknown. Chemically, the terminal BH bonds show no peculiarities, and we may simplify our discussion without any real loss by assuming them to be localised electron pair bonds. Furthermore, the terminal

H–B–H angle of 122° suggests that we may regard the terminal H as bonding with sp^2 hybrids. For the bonding involving the other bridging hydrogens we have thus to consider the sp^2 hybrids not involved in terminal BH bonding (these are labelled ϕA_2 and ϕB_2 in Fig. 3.22 b), p orbitals not involved in the sp^2 hybrids (ϕA_1 and ϕB_1), and the two hydrogen orbitals (ϕH_1 and ϕH_2). We thus have 6 orbitals; these may be split up by their symmetry properties (the point group is D_{2h}) but we can take a short cut, and notice that ϕA_2, ϕB_2 and the combination ($\phi H_1 + \phi H_2$) are symmetric with respect to rotation about the B–B axis, while ϕA_1, ϕB_1 and ($\phi H_1 - \phi H_2$) are antisymmetric. These two sets will thus be mutually orthogonal. Within the sets, the three orbitals will give bonding, non-bonding and antibonding molecular orbitals: the form of these MO for the set ϕA_2, ϕB_2 and ($\phi H_1 + \phi H_2$) is shown in Fig. 2.23.

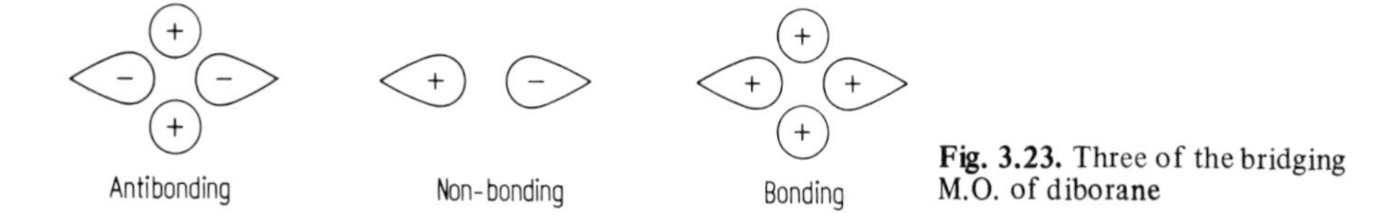

Fig. 3.23. Three of the bridging M.O. of diborane

Boron has three electrons in its valence shell, and hydrogen has only one; the total number of valence electrons in the molecule is thus 12. Eight electrons are taken by the simple B–H terminal bonds, leaving four for the bridging groups. These four electrons will fill the two bonding orbitals of the bridging M.O. described above, and the description of the molecule involves only bonding orbitals being filled – consequently the molecule is stable.

The bridging hydrogen system of the molecule is delocalised over 4 atoms (2 hydrogen and 2 borons). A suitable combination of the two occupied bonding orbitals gives 'localised' bonds having maxima of electron density along the two B–H–B curves. These localised bond are *three centre, two electron* bonds (Fig. 3.24.) The importance of the three centre two electron bond is that it enables a stable structure with one electron pair forming two 'bonds' to be proposed; this is impossible according to classical electron pair bonding theory. If a full M.O. treatment of the molecule is undertaken one arrives at exactly the same picture of the bonding, although one of the unoccupied 'non-bonding' M.O. of the bridge system is appreciably antibonding. A full discussion is given by Wade.

There is nothing fundamentally startling in the M.O. description of the bridge bonding system. However, one electron pair delocalised over three centres and providing

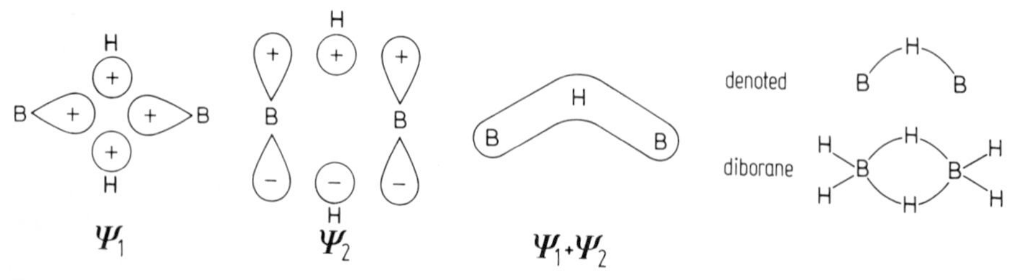

Fig. 3.24. The boron-hydrogen bonding orbitals, and their localised representation

two formal bonds produces a weaker bond than two 2 centre 2 electron bonds. Wade gives the estimated energy of the terminal B–H bonds as 375 kJ/mole and of a B–H–B bond as 450 kJ/mole – much less than twice the 2 single B–H bond value. Similarly, Fig. 3.22a shows the B–H distance is longer for the bridging than for the terminal bonds.

The B–H–B bridging system is easily cleaved by a Lewis base which has a pair of electrons which can overlap well with an acceptor molecule and which is not too strongly bound. Thus the reaction

$$B_2H_6 + 2Me_3N \rightarrow 2Me_3NBH_3$$

is very favourable, involving the interaction of 2 lone pairs and two 3 centre 2 electron bonds to give four 2 centre 2 electron bonds. All boron hydrides act as Lewis acids, and are decomposed by bases. The instability of these compounds is further shown by their positive heats of formation showing them to be unstable with respect to decomposition into boron and hydrogen. (e.g. B_2H_6 ΔH_f^o = +32.5 kJ/mole, BCl_3 ΔH_f^o = –395 kJ/mole, B_2O_3 ΔH_f^o = –1281 kJ/mole)

The results of the extremely difficult preparative and structural studies of the boron hydrides undertaken by many chemists (notably Stock and Lipscomb) showed that many more complex boron hydrides existed, frequently sharing the same type of B–H–B bridged bonds, and forming cages and rings with boron-boron bonded skeletons. It was also found that many of these compounds showed boron-boron bonds which could not be explained according to classical electron pair bond theory. Two types of three centre two electron bonds for boron polyhedra were proposed: one in which three 'σ type' boron orbitals overlapped to give bonding, non-bonding, and antibonding molecular orbitals, and the second in which two boron σ type and one boron p orbital overlapped to give 3 M.O. (see Fig. 3.25). Using these three types of three centre bond it was possible to draw a bonding scheme for many boron hydrides.

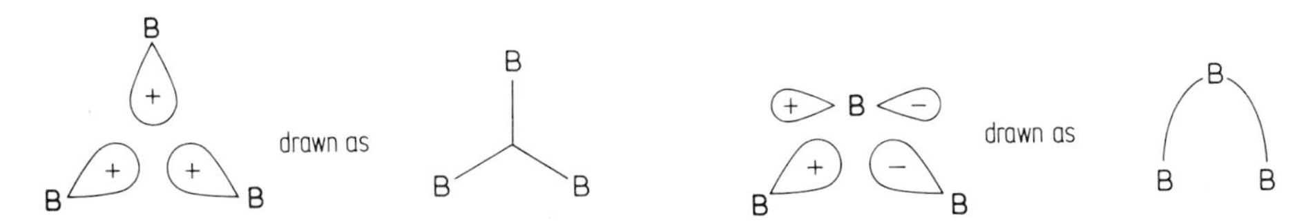

Fig. 3.25. Boron-boron-boron bonding molecular orbitals

Hexa-borane(10) B_6H_{10} is shown in Fig. 3.26 – its structure consists essentially of a pentagon of boron atoms with a BH group above the centre of the pentagonal plane.

This approach has the disadvantage that sometimes several bonding schemes can be drawn for the same compound and that the molecule often has a higher symmetry than the bonding pattern would predict, as for example in the anions $B_6H_6^{2-}$ (octahedral) and $B_{12}H_{12}^{2-}$ (icosahedral). This problem can be resolved by a full M.O. treatment allowing for delocalisation throughout the molecule. Several examples of this are given in the excellent review by Longuet Higgins and we shall only give a simple example here.

$B_6H_6^{2-}$ has an octahedral structure with a B–H unit at each apex of the octahedron. In total there are 18+6+2 = 26 valence electrons of which 12 will be involved in

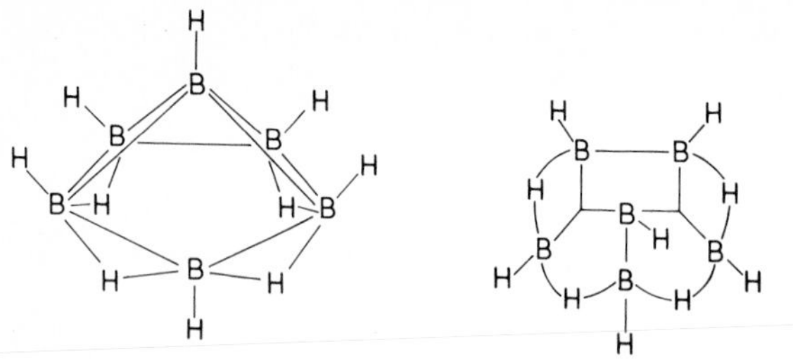

Fig. 3.26. The structure of hexaborane (10) B_6H_{10}

simple BH bonds, leaving 14 to hold the octahedron together. The boron hydrogen bond is assumed to involve a sp hybrid, leaving one sp hybrid directed towards the centre of the octahedron and two p orbitals in the plane perpendicular to the sp hybrid axis (Fig. 3.27). The sp hybrids directed to the centre of the octahedron will form one strongly bonding M.O. (of symmetry a_{1g}) of the form $(\phi_1 + \phi_2 + \phi_3 + \phi_4 + \phi_5 + \phi_6)$ where ϕ_i is the sp hybrid of the i-th boron atom. The other M.O. formed by these orbitals will be antibonding. There are 12 p orbitals perpendicular to these sp hybrids which will produce 6 antibonding and 6 bonding M.O. The total of bonding M.O.s is thus 7 – corresponding to 14 electrons, the observed number of *skeletal bonding electrons* (i.e. the number required to hold the boron cage together). A similar treatment for B_4Cl_4 (a tetrahedron of B–Cl units) yields an equally satisfactory picture of the M.O. diagram if it is assumed that the Cl atom can act as a donor of π electron density (as was assumed for transition metal complexes).

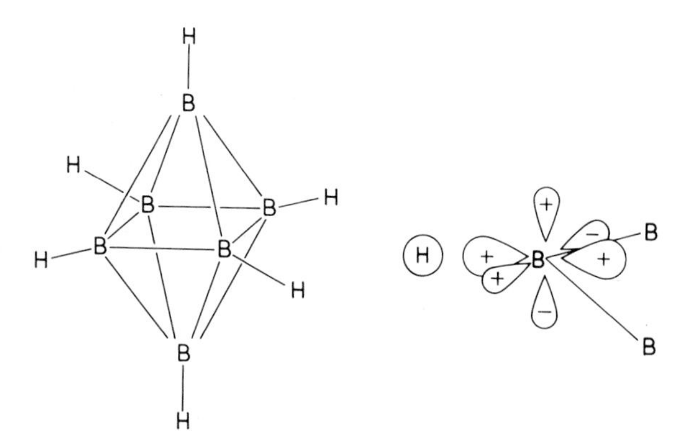

Fig. 3.27. The $B_6H_6^{2-}$ anion

These full M.O. treatments are satisfactory for species with high symmetry, but much more difficult for those species with lower symmetry. As more preparative and structural work appeared, certain general features of the boron skeleton became apparent. The boron skeleton always formed either the most spherical triangular faced polyhedron, or a fragment of such a polyhedron (for example: a square based pyramid is an octahedron with one vertex missing; a square planar structure is an octahedron with two *trans* vertices missing). The form of this polyhedron was governed by the number of skeletal bonding electrons. A species with n boron atoms in the cluster would form a closed polyhedron with n vertices if it had $2n + 2$ skeletal bonding electrons: if it had $2n + 4$ skeletal electrons it formed a polyhedron with $(n + 1)$ vertices with one vertex missing; if it had $2n + 6$ electrons it formed a polyhedron with $(n + 2)$ vertices, two of which were missing.

To give a few examples:

No. of skeletal atoms	No. of skeletal electrons	Structure
6	$14 = 2n + 2$	Octahedron
5	$14 = 2n + 4$	Octahedron, one vertex missing = Sq. pyramid
4	$14 = 2n + 6$	Octahedron, two vertices missing = Sq. plane
7	$16 = 2n + 2$	Pentagonal bipyramid

The compounds with $2n + 2$ skeletal electrons are described as *closo-*, those with $2n + 4$ as *nido-* (nest-like), and those with $2n + 6$ as *arachno-* (web-like). It will be seen that the fundamental structural pattern of the molecule is determined by the number of skeletal electron pairs, not by the number of boron atoms which determines only how many vertices are left 'empty'. The theory may be justified (if not proved) by the following argument, very similar to that used for $B_6H_6^{2-}$. At each vertex there will be a boron atom with one hydrogen atom bonded to a boron sp hybrid pointing radially away from the polyhedron; there will be one sp hybrid pointing radially into the polyhedron, and there will be 2 p orbitals tangential to the surface of the polyhedron at the vertex. The sp hybrids pointing inwards will form one strongly bonding M.O. If there are n vertices, then the $2n$ tangential p orbitals will form n bonding and n antibonding orbitals. The total number of bonding orbitals is then $n + 1$, corresponding to $2n + 2$ bonding electrons for a stable structure for *closo* (i.e. complete polyhedric) compounds. For the *nido* and *arachno* species it is found that the removal of one or two BH units from a *closo* species does not change the number of skeletal bonding M.O., and consequently for a stable configuration (all bonding M.O. filled) the structure is derived from the *closo* species with the same number of skeletal electrons.

This theory is extremely successful in predicting structures of complex boranes and the related, rather more stable carboranes, where a CH species may be substituted for a BH^- unit, to give, for example, $C_2B_4H_6$ from $B_6H_6^{2-}$. The effectiveness of the theory is dramatically illustrated by the changes in structure accompanying reduction:

$$\textit{closo} \underset{-2\,e^-}{\overset{+2\,e^-}{\rightleftarrows}} \textit{nido} \underset{-2\,e^-}{\overset{+2\,e^-}{\rightleftarrows}} \textit{arachno}$$

as e.g.

$$\underset{\textit{closo}}{C_2B_{10}H_{12}} \xrightarrow[\text{napthalene}]{\text{Na}} \underset{\textit{nido}}{C_2B_{10}H_{12}^{2-}}$$

This theory has not considered the possibility of bridging hydrogens – these are very common in these structures and may be found bridging either two borons (edge-bridging) or three (face bridging). They may even be found in *nido* and *arachno* species 'bridging' between boron atoms and unoccupied vertices (*endo* hydrogens). R.E. Williams has recently reviewed this subject and proposed several rules for prediction of bridging hydrogen behaviour.

It may seem strange to have spent so much space discussing a series of unstable compounds of one specific element. However it is becoming increasingly clear that the bonding patterns found in boron hydride (or borane) and carborane chemistry have a very much wider application. Elements to the left of group IV in the main groups of the periodic table have valence shells with more atomic orbitals than valence electrons. If their chemistry is sufficiently covalent one might expect the occurrence of three centre two electron bonds. This is indeed found, notably in organometallic and hydride chemistry. Aluminium forms a borohydride salt $Al(BH_4)_3$ with six B–H–Al bonds. In organometallic chemistry the methyl group may act as a bridging group as in Al_2Me_6 (Fig. 3.28) and $Mg(AlMe_4)_2$. It also seems that many 'non-classical' carbonium ions studied in organic chemistry involve 'electron deficient' bonding corresponding to a delocalisation of the positive charge over many centres. As a simple example the structure of $CH_5{}^+$ is shown in Fig. 3.29. The structure may be regarded as $CH_3{}^+$ group bridging two hydrogen atoms to give a two electron three centre H–C–H bond.

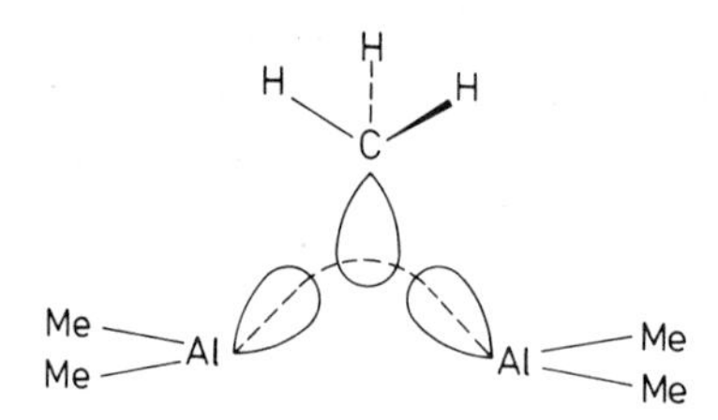

Fig. 3.28. The three-centre bonding in Al_2Me_6

It is however particularly interesting to examine some of the ring, cage, and cluster compounds found in other fields of inorganic chemistry. In a general review, Mingos classified such compounds as (i) *electron deficient,* having too few electrons to draw a 2 centre 2 electron bond structure for each bond in the molecule (e.g. the boranes) (ii) *electron precise,* having exactly the number necessary (e.g. P_4, a regular tetrahedron with 6 P–P bonds along the edges, has 12 electrons) or (iii) *electron rich,* having more electrons than is necessary for such a structure. We will consider these classes separately.

Electron precise compounds

In principle these present little problem since 2 centre 2 electron bonds may be drawn for every bond between two atoms. There are two possible drawbacks to such a simple scheme:

1) The electrons counted in the valence shell may in fact be too tightly bound to take part readily in chemical bonding. For example does the valence shell of phosphorus ($[Ne]3s^23p^3$) contain 3 or 5 electrons?

2) A simple edge-bonding scheme may be unfavourable with respect to a more open structure containing double bonds. The simplest example is prismane which could equally be found as Dewar benzene or normal benzene – the most stable form (Fig. 3.30).

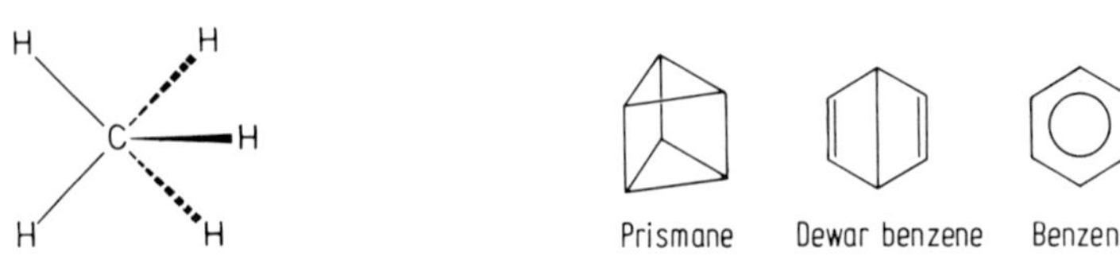

Fig. 3.29. The structure of CH_5^+

Fig. 3.30

Electron rich compounds

In these compounds the excess of electrons must either go into non-bonding orbitals or antibonding orbitals. If they go into antibonding orbitals then the cluster will 'open up' as shown by Mingos for the series S_4N_4 (or As_4 S_4), $S_8{}^{2+}$, S_8 (Fig. 3.31). If they remain in non-bonding atomic orbitals, then, in terms of the number of electrons involved in holding the molecule together, the skeletal bonding may actually be electron deficient. A simple example is the trigonal bipyramidal cation $Bi_5{}^{3+}$. If the 6s orbital is considered as accessible for bonding then there are (5 x 5) – 3 = 22 electrons, too many for edge bonding of the trigonal bipyramid structure. If the 6s electrons are inert chemically, there are only 12 electrons in the bonding scheme, whilst there are 15 atomic orbitals available, and, from a 2 centre 2 electron bond point of view, 9 molecular orbitals to form. It is most probable that the 6s orbital is non-bonding, and consequently we might expect the skeletal bonding pattern to show similarities to electron deficient compounds.

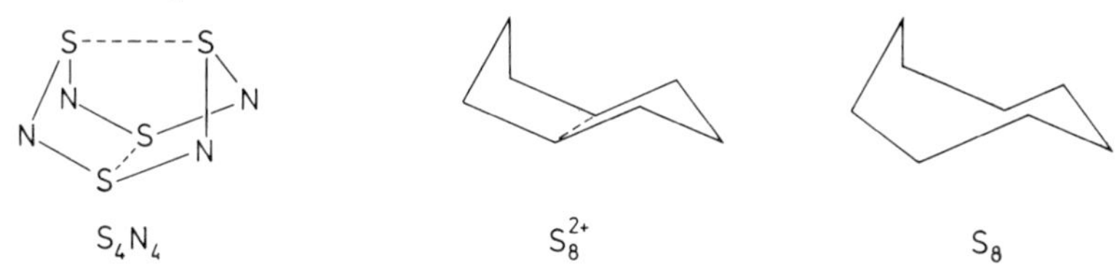

Fig. 3.31

Electron deficient clusters

Having shown that apparently electron rich or electron precise species can in fact be electron deficient as far as skeletal bonding is concerned, it is now necessary to generalise the description of skeletal bonding.

We are in fact approaching chemical bonding in its most *general* sense, using some of the ideas gained from the discussion of the boranes as a starting point. The very broad application of these ideas is the justification for the relatively detailed discussion of the bonding in boranes.

As before it is assumed that three orbitals are provided by each vertex for cluster bonding, one directed to the centre of the polyhedron, and two tangential to the surface. For a *closo* polyhedron with n vertices this produces n + 1 bonding M.O. as before. The assigment of structure (*closo, arachno, nido*) according to the number of electrons involved in skeletal bonding is exactly the same as before. Thus for Bi_5^{3+}, if it is assumed that the 6s atomic orbitals play no part in the bonding, each bismuth atom has three orbitals (the three 6p orbitals) for cluster bonding. There are (5 x 3) – 3 = 12 electrons available requiring the 'parent' polyhedron to have 5 vertices (i.e. a trigonal bipyramid), and so we predict Bi_5^{3+} to be a *closo* trigonal bipyramid – as is observed.

The only problem is the assignment of electrons to non-bonding orbitals with no part in the skeletal bonding. Wade has proposed the following scheme for a species ML_n acting as a vertex of a cluster compound:

1) The valence shell of M contains 4 A.O. for a main group element (one s, three p) and 9 A.O. for a transition metal element (one s, three p, five d)

2) Three of these orbitals are used for the skeletal bonding and n orbitals for bonding the ligands. Those atomic orbitals remaining are non-bonding and are filled.

3) Electrons not occupying the ligand bonding or metal non-bonding orbitals are involved in cluster bonding.

As an example let us consider the octahedral anion $Ru_6(CO)_{18}^{2-}$. Ruthenium has eight valence electrons, and the structure is formed from 6 $RU(CO)_3$ groups. Thus three A.O.s are used to bond the carbonyl groups, and three for the cluster bonding leave 9 – 3 – 3 = 3 non-bonding orbitals. Each carbonyl donates one electron pair to the metal, filling the carbonyl bonding M.O.s, and six electrons go into the non-bonding M.O.s, leaving 2 for the cluster bonding. The total number of skeletal bonding electrons is thus (6 x 2) + 2 = 14 (the extra two electrons come from the charge). Fourteen is the number of electrons required to stabilise an octahedron.

Bridging species

Just as hydrogen atoms may act as bridging species for the boranes, so they, and other groups, may bridge in these cluster compounds. For electron counting purposes the electrons donated by such groups are counted as skeletal bonding electrons: thus H donates one electron (making $H_2Ru_6(CO)_{18}$ isostructural with $Ru_6(CO)_{18}^{2-}$) bridging CO groups 2 and so on. Even metal carbonyl groups may act as bridging species. $Os_7(CO)_{21}$ consists of 7 $Os(CO)_3$ species, giving 14 skeletal electrons. Its structure is an octahedron with one $Os(CO)_3$ group bridging a triangular face – a capped octahedral structure. $Fe_5(CO)_{15}C$ has a bridging carbon atom, and 5 $Fe(CO)_3$ groups – the total number of skeletal electrons is 14 (the carbon atom gives 4) so the structure is basically octahedral with one vertex missing (a *nido* species) and the carbon approximetely in the Fe_4 plane (see Fig. 3.32 a).

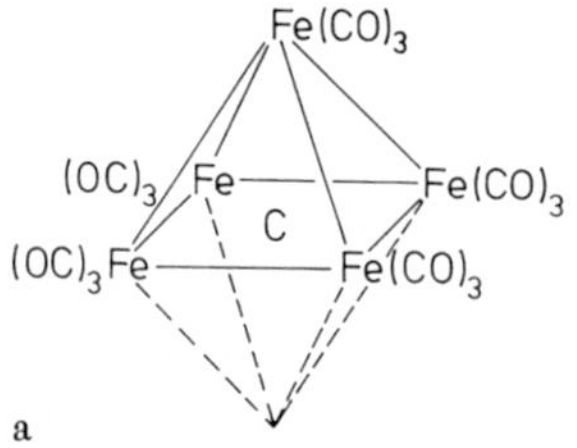

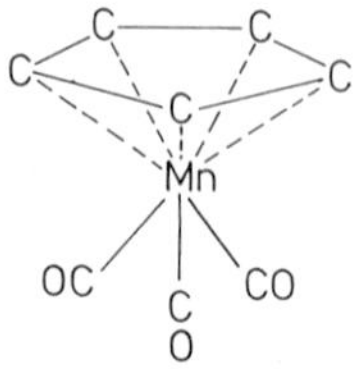

Fig. 3.32 a and b

Organometallic compounds

This theory may equally be applied to organometallic compounds. Consider for example the compound *pentahapto*C_5H_5)Mn$(CO)_3$ (Fig. 3.32 b). Manganese has 7 electrons, 3 orbitals bonding with the carbonyls, 3 orbitals bonding with the C_5 ring (according to cluster theory) and thus 3 non-bonding orbitals. Therefore there is $7 - (3 \times 2) = 1$ electron available for the cluster bonding. Each carbon has one CH bond (taking one electron pair) and thus has 3 electrons left for skeletal bonding. The total number of electrons for skeletal bonding is now $(5 \times 3) + 1 = 16$, corresponding to a pentagonal bipyramid. There are only six vertices (5 carbons and one Mn) so the structure is *nido*. Similar reasoning leads one to describe the planar $C_5H_5^-$ anion as an *arachno* structure. A very large number of organometallic compounds may have their structures predicted (or rationalised) using these rules initially developed for very different compounds.

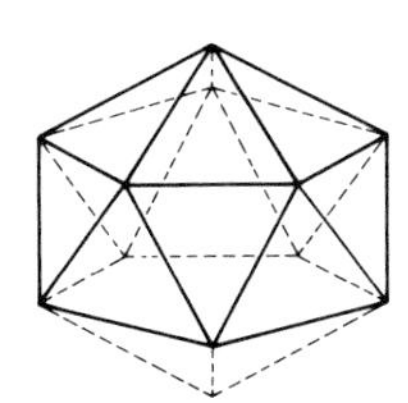

Fig. 3.33. $C_2B_9H_{11}{}^{2-}$

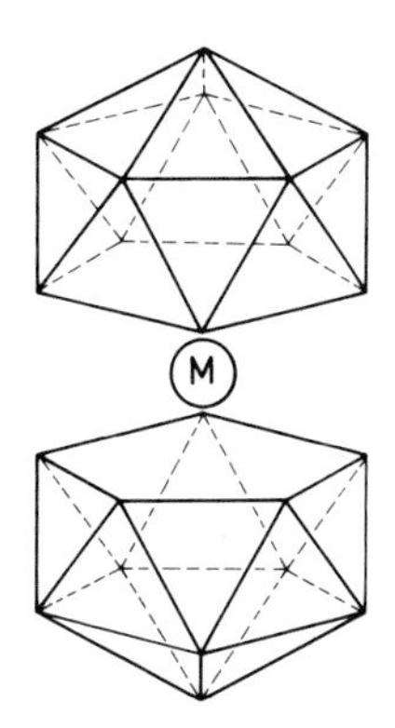

Fig. 3.34. A bis(discarbollyl) complex

We might regard the compound (η^5 C_5H_5)Mn$(CO)_3$ as the addition of a Mn$(CO)_3{}^+$ group (with three empty A.O. available for cluster bonding) to an *arachno* species ($C_5H_5{}^-$) to give a *nido* species. Is it possible to add similar species to *nido* complexes to give *closo* species? The anion $C_2B_9H_{11}{}^{2-}$ is a *nido* species with the structure of an icosahedron with one vertex missing (Fig. 3.33). It is possible to place Mn$(CO)_3{}^+$ in this missing vertex to give the complex $(C_2B_9H_{11}Mn(CO)_3)^-$. Similarly the species E^{2+} (E = Ge, Sn, Pb) where E^{2+} has the structure ns^2 (i.e. 3 p orbitals available for skeletal bonding) may be inserted to give *closo* $C_2B_9H_{11}E$. A final example is the ferrocene analogue $Fe(C_2B_9H_{11})_2{}^{2-}$ which may be regarded as a complex of Fe^{2+} with $(C_2B_9H_{11})^{2-}$, giving 3 empty A.O.s to each cluster and keeping 3 non-bonding A.O.s This bis(dicarbollyl) complex is actually more stable thermally than ferrocene (Fig. 3.34).

The success of this model of cluster bonding in so many applications is remarkable. However there are many cases where it is not entirely satisfactory and gives false predictions. This is particularly true for cases where π bonding between ligands and skeletal M.O.s is important (e.g. B_4Cl_4) or where some of the assumptions about non-bonding electrons are dubious. The clusters $[Mo_6Cl_8]^{4+}$ (see Fig. 7.19) and $[Nb_6Cl_{12}]^{2+}$ have been known for a long time, and were initially thought to be simple chlorides in low oxidation states. The structure of $[Mo_6Cl_8]^{4+}$ can be explained by simple edge bonding of the Mo_6 octahedron but for a really satisfactory explanation of their structure it is necessary to undertake a complete M.O. description allowing for the part played in the bonding by all the chlorine electrons.

In triangular clusters with three metal atoms, it seems that the two highest occupied M.O. are actually antibonding with respect to the metal-metal bonds – this is supported by metal-metal bond length data as given in the recent review by Mason and Mingos. There is a fair amount of evidence that in some transition metal complexes some antibonding orbitals are occupied. In the compounds where a full M.O. treatment is possible, a good explanation of the structure is possible; however the electron counting theory given often offers a short cut, especially for compounds of low symmetry, since it is based on a simple M.O. model. It is of course quite possible that the molecular orbitals adapt themselves to a structure imposed by other effects: in a recent paper Johnson derives the structures of carbonyls from the packing possibilities of the CO groups.

Electron deficiency – conclusions

Electron deficient bonding is not inherently different from other M.O. systems, although it is generally more delocalised. If the discussion of MgF_2 is recalled from Chap. 2 it will be recalled that the most strongly bonding M.O. (σ_g) is a three centre two electron bond and that the σ_u bonding M.O. is only bonding if the $3p_z$ orbital of Mg is available. It is only because there are 4 electrons present that it is possible to draw two localised bonds. It may well be more reasonable to draw a three centre bond than involve an empty, high energy atomic orbital in a hybrid orbital. We have already seen examples where it was more realistic to regard some electrons as non-bonding and the others as involved in electron deficient bonds. The delocalisation of M.O.s necessary in electron deficient bonding has encouraged the development of approximate M.O. methods such as the skeletal electron counting scheme. This seems to the author to be a field where advances and improvements are possible in the near future which is one reason why it has been described in some detail here.

E. The Electronic Structure of Solids

Hitherto our discussion has been concerned with isolated molecules consisting of a definite number of molecules joined together to give a discrete unit. It is a matter of common observation that most chemical compounds, when pure, are solids and, indeed, some compounds exist only in the solid state. It may therefore seem that our previous discussions have been something of a digression unless our conclusions for molecular species can equally be applied to solids, and it is this problem that we discuss here. In passing we may note that although solids are very far from being unreactive, most chemical reactions take place in solution or in the vapour phase where the mixing of the reactants can take place more readily, so a discussion of discrete molecules may have more relevance for chemical reactivity. By contrast the physical properties of solids (such as physical strength or conductivity) have a much greater importance.

Almost all solids are crystalline and have a regular and extended disposition of their constituent atoms. They frequently have high density, and they retain their shape unless subjected to harsh physical treatment. From this we may deduce that the interactions of the constituent particles can be quite strong, and that discussing these extended structures in terms of the sum of separate units may overlook the symmetry requirements of the crystal structure. We may therefore expect delocalised bonding schemes.

Classification of solids

We will classify solids into four groups to simplify discussion. As usual, there are many borderline cases between groups, each group corresponding to the dominance of one particular effect, but the following separation may be made:

1) Molecular solids (and solidified noble gases)
2) Ionic solids
3) Covalent solids
4) Metallic or band structure solids.

It should be emphasised that this classification is artificial, and that a solid is not obliged to belong to only one of these classes, or to show only one type of bonding. For example, most silicate minerals are composed of covalently bonded lattices, sheets or chains of SiO_4 tetrahedra carrying negative charges, close-packed with cations to give a tightly bound solid.

Molecular solids have structures consisting of a well defined molecular unit joined to other similar units by non-bonding interactions (see later in this chapter) such as the hydrogen bond or Van der Waals forces. The interactions between the molecules are relatively weak and the solid has relatively low physical strength, and low heats of fusion (~400 J/mole). Most neutral organic compounds crystallise to give molecular solids. They may be regarded essentially as aggregates of free molecules, and their electronic structure is essentially the same as that of the free molecule.

The other three classes may not be regarded as aggregations of free molecules as they involve regular, extended groups giving either *chains* (extended in 1 dimension), *sheets* (extended in 2 dimensions) or *lattices* (extended in three dimensions). Any attempt to draw an isolated molecule involves an arbitrary cutting up of the structure. Structures with sheets or chains may have the sheets or chains bonded together by the same set of interactions as molecular solids. These solids generally have a higher physical strength and much higher heats of fusion than molecular solids. They may be subdivided into:

i) Ionic solids, constructed from at least two different species carrying an electrical charge corresponding to an exact multiple of the electronic charge. These *ions* are held together by forces of electrostatic attraction. Ionic compounds will normally be found in condensed phases (i.e. solid or liquid state) since the electrostatic energy will be lowest when each ion is closely surrounded by oppositely charged ions. It is often possible to identify the ions present in these solids as those found in aqueous solution or in other different crystals. Mn^{2+} in the ionic solid MnF_2 has a similar optical spectrum to $MnSO_4$ in solution, for example.

ii) Covalent solids may be represented as networks of simple 2 electron 2 centre bonds. Diamond, for example, has a 'giant molecule' structure in which each carbon atom forms four 2 electron 2 centre bonds to the four carbon atoms surrounding it – this is a particularly strong structure (Fig. 3.35 a). Arsenic, antimony, and bismuth form layer structures with the atoms alternately just above or just below the plane of the layer, and linked to three other atoms. The bond angles suggest that the atom uses its three half-filled p orbitals to form these bonds, with slight participation by the s orbital. The layers are held together by Van der Waals attractions (Fig. 3.35 b).

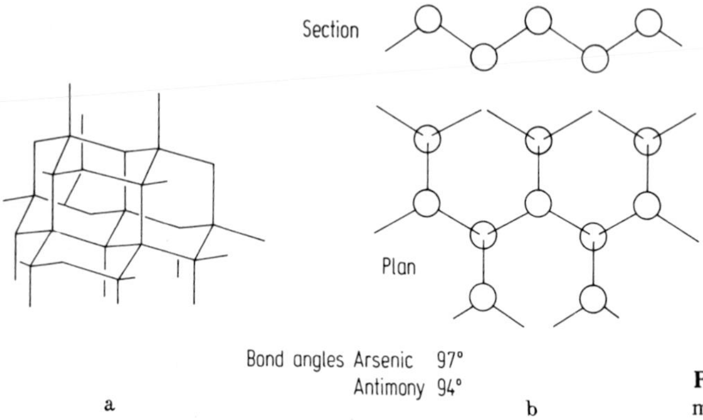

Section

Plan

Bond angles Arsenic 97°
Antimony 94°

a b

Fig. 3.35. (a) Diamond, **(b)** metallic arsenic or antimony

iii) Metallic or band structure solids essentially comprise those solids which cannot be described satisfactorily by the first three models, and include all metals. Many compounds which are not metals fall into this group, so the 'metallic' title is perhaps confusing and we shall refer to them as band structure compounds, although it may equally be argued that this title is rather too general. In fact, the band model of bonding may equally be used to describe ionic and covalent solids, and is often more useful for the discussion of physical properties. Band structure solids frequently have high C.N.s and close packing of the constituent atoms, giving high densities.

The band model

We may usefully approach the band model by considering the consequences for M.O. theory of the close packing and consequent good overlap of atoms in a solid. A crystal contains many millions of millions of atoms packed together closely so that all the atoms have the same environment, or one of a limited number of environments (except near the surface). In such conditions we would expect the molecular orbitals to be highly delocalised. If we bring two atoms close together, we expect their atomic orbitals to overlap to give bonding and antibonding M.O.; if we bring up another atom, its atomic orbitals will interact with these molecular orbitals and a new pattern will result. In a crystal with high coordination numbers for its constituent atoms we would expect a great deal of overlap and consequently a spreading of the molecular orbitals. In a crystal with say 6×10^{23} atoms (Avogadro's number) it will no longer be possible to distinguish the individual orbitals and we will appear to have a continuous *band* of energy levels (Fig. 3.36 a). The number of energy levels will be equal to the number of atomic orbitals involved in the bonding, and consequently proportional to the number of atoms in the crystal. The spread will be proportional to the degree of interaction, and consequently will increase with overlap. This is confirmed by the increase in band

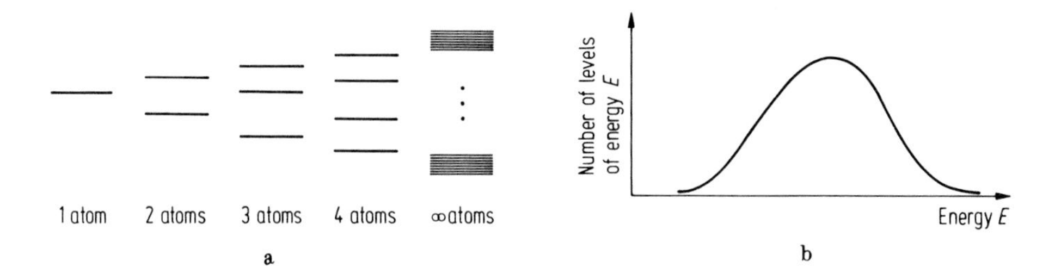

Fig. 3.36 a and b

widths observed on compressing solids. The disposition of levels within a band as a function of energy may be shown by a diagram plotting the number of states of energy E against E (Fig. 3.36 b) (a density of states diagram).

In this discussion there are two factors we have ignored: the symmetry of the system, and the bonding or antibonding properties of the bands.

The symmetry of the system, in particular the packing of the constituent atoms, has a considerable effect on the energy levels, and limits the possible energies of the bands to certain regions. These regions are judged to be bonding or antibonding according to their energy relative to the initial energy of the atomic orbitals.

This will perhaps be made clearer by refence to a diagram (Fig. 3.37). The solid element E has three atomic orbitals, the core orbital a and two valence orbitals b and c.

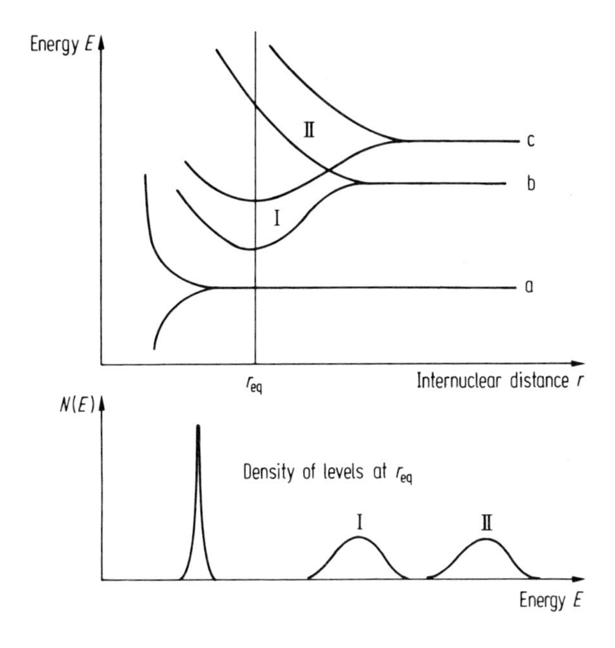

Fig. 3.37

At high internuclear distance none of these will interact. As the internuclear distance is decreased, the valence orbitals overlap and there is a spreading of orbitals *b* and *c* to give bands. At a certain value of the internuclear distance these will overlap in energy. There now develop two bands I and II separated by a *band gap* where there are no allowed energy levels. Band I is clearly bonding and band II is antibonding. The equilibrium internuclear distance is at the minimum energy for Band I. If the element is compressed beyond this point, the energy of the bands rises, and finally the non-bonding core orbital *a* begins to overlap to give a band.

The separation of the bands depends on the crystal structure; it is quite possible for bands to overlap, and this is quite common for the transition metals. The highest energy valence orbital, the s orbital, overlaps with its neighbours to give a broad band. The rather more tightly bound and more spatially compact d orbitals interact less strongly to give a narrow band with an energy 'inside' the s band (see Fig. 3.38).

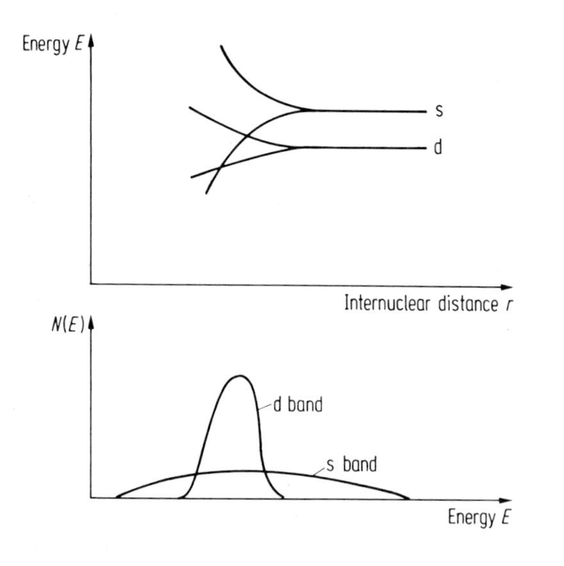

Fig. 3.38

The filling of the bands with the valence electrons proceeds as for other molecular orbitals and we again have the problem of whether to pair electrons or leave them unpaired and place them in orbitals of different energies. The problem of electron repulsion in band structures is complicated, and we will content ourselves with remarking that paramagnetism of metals arises, as for complexes, from the presence of unpaired electrons.

Another method of deriving the bands in metals that is used frequently is the 'free electron' or electron gas model. In this approach the metal is assumed to be ionised, and the electron to be free to move around in the crystal. Its wave function is then given by the 'particle in a box' treatment, with the size of the box equal to the size of the crystal. This gives a continuous series of energy levels. If we now add the potential of the positive ions in the crystal as a perturbation, this continuous series is split up into separate bands as we found for the molecular orbital approach. The splitting up of the bands depends on the potential, and consequently on the way the ions are packed in the crystal. The end result is exactly the same as if we had started from the atomic orbitals and constructed molecular orbital type bands: the so-called 'tight binding approximation' that was given first.

The physical properties of solids

We may now begin to discuss the pyhsical properties of solids in terms of the electronic structure. Firstly we see that a condition of strong overlap for the maximum interaction favours high coordination numbers. Next we may discuss the conductivity properties: if an electric field is applied to a solid where all the bands are full, then electrons moving against the field will be raised in energy, those moving with it will be lowered in energy. If the band is full and there is no other empty band near it, it is not possible for the electrons to rearrange themselves so that they can move with the field rather than against it as there are no vacant energy levels. Consequently there will be no change in electron disposition, as many electrons will travel against as with the field, and the solid will be an insulator. If, however the band is only partially full, then the field will cause the electrons to enter levels where they travel with rather than against the field. There will then be a majority of electrons travelling with the field, and consequently a net transfer of charge through the solid. This is *metallic* conduction. It therefore follows that metallic conductors have partly filled bands. This accounts for the excellent conductioning properties of the alkali metals – even if the valence s orbital does not mix with the p orbitals above, the band formed by the one valence orbital will be exactly half full, and can consequently conduct electricity. The highest partially filled band in a solid is called the conduction band. Metallic conductivity is decreased by increasing temperature as the increased vibrations of the atoms scatter the conduction electrons. To summarise, a solid with all its bands full will be an insulator, a solid with a partially filled band will be a conductor.

Semiconductors

Consider a solid with its highest energy band full, and involved in bonding; we will call this the valence band. The energy level above this will be empty unless it is occupied as a result of excitation of the solid; the excitation necessary will depend on the band gap, the distance between the top of the valence band and the bottom of the empty band. If this band gap is small it will be possible to excite an electron thermally into this empty band which will now be partially occupied, and can act as a conduction band. At the same time, the valence band is now partially occupied and can conduct electricity. The upper, generally empty, band is referred to as the conduction band. It is clear that this type of conduction will increase with increasing temperature as there will be more electrons in the conduction band. A solid such as we have described is called a *semiconductor,* and its method of conduction as semiconduction. The conduction depends strongly on the band gap which may be studied by spectroscopic methods as described in the next section.

Spectroscopic studies of solids

Some of the best evidence for the band theory comes from the spectroscopic studies of solids, in particular their X-ray spectra. An X-ray beam has a sufficiently high energy to eject an electron from a tightly bound core orbital; this ejected electron can either enter one of the empty or partially occupied bands of the solid, or it can be completely expelled from the vicinity of its parent atoms. We will consider the two cases separately.

In the first case, X-rays will be absorbed by the solid when their energy corresponds to the difference between the core level and an empty band level. For isolated atoms (e.g. in the gas phase) sharp transitions (between two well defined levels) are seen, but in solids, a wide region of absorption is seen, implying that the electron can enter one of several closely spaced levels, in agreement with band theory.

If the core electron is completely ejected, then an excited ion is produced. This species can reduce its energy if an electron drops from one of the higher occupied levels to fill the 'hole' in the very low energy core orbital. This relaxation is accompanied by the emission of an X-ray photon with an energy equal to the difference between the core and higher occupied orbital energies. The energy of the X-ray emitted thus gives an indication of the energy of the occupied upper orbitals relative to the core orbital. A band of energies is seen in accord with the 'spreading out' of the upper enerby levels due to formation of bands. In summary, X-ray absorption measures the energy between the core levels and the empty levels of bands, and X-ray emission measures the difference in energy between core levels and occupied energy levels (see Fig. 3.39).

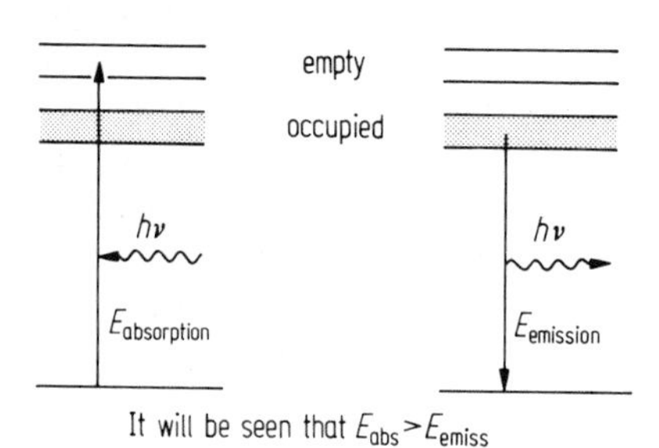

Fig. 3.39

If we now turn to the three specific cases of insulators, semiconductors and metals, we may predict their X-ray spectra. Insulators will show X-ray absorption corresponding to excitation to the empty band and X-ray emission from electrons leaving the valence band. Since there is a considerable band gap, the absorption and emission curves will be at different energies (see Fig. 3.40). As the band gap decreases the material becomes semiconducting and the absorption and emission regions move gradually closer.

For metals only part of the conduction band is occupied; at absolute zero, there will be no thermal electronic energy and only the lowest energy levels will be occupied – the emission line will show a sharp cut-off when the highest occupied level is reached and the absorption line will start just after this energy (the Fermi level), equally sharply (see Fig. 3.40). At higher temperatures, the electrons will have some thermal energy, and some of the upper part of the band will be occupied and some of the lower part empty – the emission and absorption lines will then overlap slightly. This is observed experimentally. The observation of these broad bands (with widths of approximately 5 eV) is one of the most positive proofs of the existence of band structure in solids.

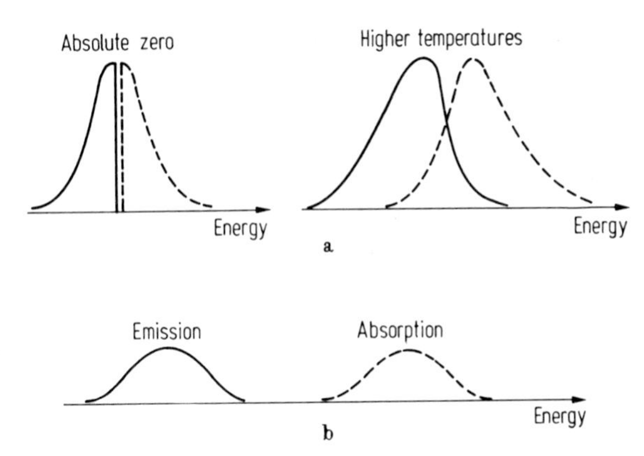

Fig. 3.40 a and b. X-ray spectra for solids. **(a)** metal: solid line – emission; dotted line – absorption. **(b)** an insulator or semiconductor

The band theory also explains the interaction of radiation of lower energy with solids. Metals have partly filled bands, and consequently have a large number of empty bands just above the highest occupied levels. Radiation of almost any energy (less than the width of the band) will thus be absorbed and can then be re-emitted by the electron dropping back – this explains metallic lustre and their high reflectivity even to infra-red and radio-frequency photons. Semiconductors have a definite band gap and photons with energy lower than the band gap will not be able to excite an electronic transition, and will not be absorbed or reflected – there is now a definite absorption edge, above which excitation is possible. For germanium the gap is about 5000 cm^{-1} and there is an absorption edge in the infra-red. Since the conduction band must be thermally accessible for semiconductors, the absorption edge is frequently in the infra-red and high energy radiation (such as visible light) is able to excite transitions to the conduction band. Consequently, many semiconductors have metallic lustre, and they frequently exhibit photo-conductivity, acting as conductors on irradiation. Insulators, with much larger band gaps (~5 eV) are frequently colourless; the exceptions generally involve electronic transitions other than interband jumps.

Having established the band model of describing bonding in solids we may now re-examine the three classes of solids in which it is important. Its importance for molecular solids is negligible, since there is relatively little interaction between the component units.

Ionic solids

In these solids one component is presumed to carry a positive charge and the other a negative charge, and the stability of the solid comes from the electrostatic attractions between oppositely charged ions stacked in the crystal lattice. If the ions are assumed to carry charges which are exact multiples of the electronic charge, then the energetic stability of the solid may easily be calculated from standard thermodynamic properties of the elements involved (at least for simple species). This approach enables thermodynamic predictions of considerable use to be made, and is discussed further in Chap. 5.

The disadvantge of the approach is that the assumption of charge distribution made is never accurate. This defect may be circumvented by a band model which assumes that the full valence band is localised almost entirely on the anions, while the empty conduction band is composed mainly of the empty orbitals of the positively charged cation.

Ionic solids have large band gaps and are insulators, although in the liquid state they conduct electricity as a result of the displacement of their constituent ions. Those without other chromophoric (i.e. light absorbing) centres (such as internal d-d transitions in transition metal compounds) are colourless.

Covalent solids

It is possible to represent the structure of these compounds by localised two electron bonds; alternatively, however, it is possible to describe two bands, one full and one empty, which correspond to the bonding and antibonding M.O. The band gap depends on the strength of the covalent bond – in diamond, where the bonding may be regarded as resulting from the good overlap between the tetrahedral sp^3 hybrids of carbon, the band gap is 6.0 eV. In other cases such as arsenic (see Fig. 3.35) the overlap is weaker, and the band gap only 1.2 eV. The strong directional nature of the bonding in structures such as diamond and SiO_2 leads to fairly low coordination number and consequently rather 'open' lattices. The breakup of these rigid lattices to produce liquids requires a great deal of energy, and consequently the melting points of strongly bonded covalent solids are often very high – boron 2300°, diamond ~3550°, silicon dioxide (quartz) 1610°, while most ionic solids melt below 1000°. The more weakly bonded species such as tin often have an alternative, denser, metallic (i.e. band structure) form available which is found either as an allotrope in the solid state or in a slightly disordered form in the liquid phase. Thus tin has a diamond structure (grey tin) at low temperatures but above 13.2 ^{o}C the stable form is a metallic structure (white tin) with distorted eightfold coordination and a much higher density (grey – 5.75 g/cc, white 7.28 g/cc). Similarly, germanium normally has the diamond structure and is a semiconductor, but on compression it gives a high density form with a very high conductivity. Bismuth has a structure similar to arsenic (Fig. 3.35) but on fusion, its density increases by 3 percent implying a closer packing of the atoms.

Band structure solids

This group includes not only those whose bonding *cannot* be described in terms of one of the other three models, but also those where another model can be used, but does not satisfactorily describe the physical and chemical properties of the solid. Thus $TaSe_2$ might be described as an ionic solid composed of Ta^{4+} and Se^{2-} ions; however this does not explain its metallic lustre and high conductivity. This compound has a layered structure with trigonal prismatic coordination of the tantalum between layers of Se atoms (Fig. 3.41). The crystals have a flaky appearance with a well developed cleavage parallel to the layers. The conductivity is greater parallel rather than perpendicular to the layers.

The proposed band structure is given in Fig. 3.42. The 'bonding' bands marked σ are fully occupied, and there is one electron per tantalum atom in the non-bonding band

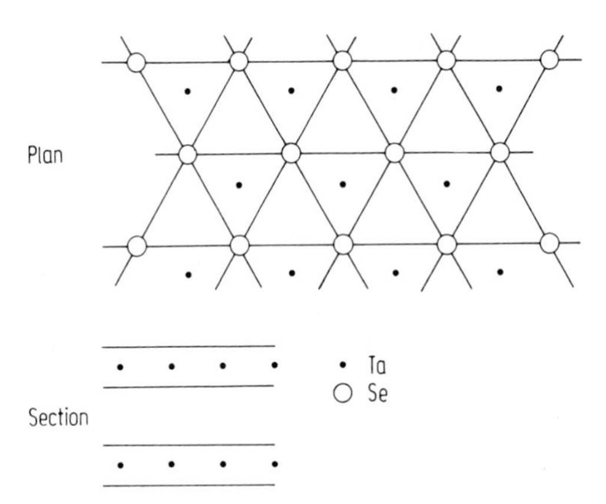

Fig. 3.41. The structure of $TaSe_2$

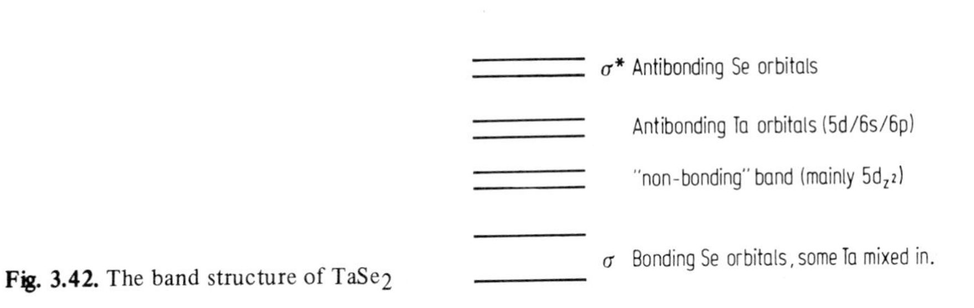

Fig. 3.42. The band structure of $TaSe_2$

composed principally of the $5d_z{}^2$. This partially occupied band gives the metallic conduction. In the analogous compound WSe_2, there is one more valence electron in the non-bonding band, effectively filling it, and WSe_2 is a semiconductor. Mixtures of $TaSe_2$ and WSe_2 show metallic conduction. The electronic structure and behaviour of these systems is difficult to explain on an ionic basis, but fits in well with a 'band structure' mode of bonding.

Having established the usefulness of the band structure approach we may now ask if we can predict when a compound will have a particularly stable band structure. It was first noticed experimentally that a very large number of compounds and alloys with the same value of the ratio (no. of atoms: no. of bonding electrons) had the same structure. This is the basis of Hume-Rothery's rule for predicting the existence of stable phases: a very large number of stable phases with electron: atom rations of 3:2, 21:13 and 7:4 are known – for example CuZn (3:2), Cu_9Al_4 (21:13) $CuZn_3$, Cu_3Sn (7:4). Theoretical analysis shows that these ratios correspond to systems where the bands are full or very nearly full.

We may predict that a band structure description involving considerable delocalisation of bands will be the best description of systems where there is 1) good overlap of adjacent atoms and 2) a fairly close match in the orbital energies of the atoms involved. These two conditions are of course the same as we deduced to be necessary for a stable bonding M.O. to be formed in homonuclear diatomic molecules. In the solid state these conditions will be met by:

1) High coordination number and close packing (good overlap)

2) Fairly diffuse wave functions for the atomic orbitals involved – this favours *a*) elements towards the left of the periodic table (there is a progressive shrinkage of orbitals on crossing a given period); *b*) heavier rather than lighter elements (the wave functions become more diffuse as the number of radial nodes (Chap. 1) increases).

3) In solids with more than one element present, a similarity of energy for those orbitals involved in bonding. If the orbital energies are too far apart, the ionic model may be more suitable, as the bonding bands will be strongly localised on the anions – the consequent positive charge on the cations leads to very poor overlap as a result of the 'shrinking' of atomic orbitals on ionisation.

Examples

1) The alkali metals themselves are excellent metallic conductors since their weakly bound ns valence electrons overlap well. With halogens they effectively lose their electrons to the tightly bound p orbitals of the halogens to give ionic halides.

2) On traversing a period of transition metals, the d orbitals become increasingly strongly bound and less radially diffuse – they are thus less involved in bonding, and the heats of atomisation of the metals fall off from the middle to the end of the period.

Transition metal oxides are invariably much more ionic than the sulphides or selenides since the latter have valence orbitals closer in energy to those of the metal, and with greater radial diffuseness.

3) Elements on the right of the periodic table with a relatively large number of valence electrons which are quite tightly bound tend to form small discrete molecules with themselves and each other. The large number of valence electrons would involve some occupation of antibonding levels. Thus diamond, with 4 valence electrons per atom has the low energy bonding band full – addition of one more electron per atom (to give a Group VB element) would involve occupation of the high energy antibonding band. This problem is overcome by reducing the coordination number (i.e. breaking some of the bonds) to give a structure where all electrons are in bonding or non-bonding orbitals (thus N_2, C.N. = 1, P_4, C.N. = 3, S_8, C.N. = 2).

Interstitial compounds

The transition metals, the lanthanides and the actinides have the capacity to form compounds of frequently ill-defined stoichiometry with hydrogen, boron, carbon and nitrogen. X-ray structure analysis shows the light atom in these compounds to be incorporated in the interstices (or gaps) between the heavier metal atoms which are packed together closely. The incorporation of the light atom may often change the structure of the metal and usually increases the distance between metal atoms. Metallic conductivity is generally retained, although the absolute value changes slightly. The carbides and borides frequently have immense physical strength, and constitute some of the hardest substances known. The melting points of these solids are also amongst the highest known.

The properties of these solids clearly suggest a band structure model of bonding, and it seems clear that the presence of the light element produces a considerable stabilisation of the structure. Their properties are very different from the carbides, nitrides, etc. of the main group metals whose structures are clearly much more ionic.

Defects and non-stoichiometry in solids

Hitherto we have considered all crystalline materials to be perfectly crystalline. This corresponds to the minimum energy for the structures; however, the stability of a species is determined by its *free* energy, and consequently we should consider the possibility of a non-perfect structure being stabilised by its increased entropy. It is equally possible that in the process of formation of the solid the most regular structures is not adopted in the whole crystal. In fact all crystals contain a concentration of defects due to these two factors, thermodynamic and kinetic.

Three major classes of *point defect* have been recognised, and we discuss them here with reference to ionic solids, although they may also be found in other solids.

1) The Schottky defect. A Schottky defect arises from a cation and an anion vacancy in the solid; it may be created by removing one anion and one cation from randomly chosen points in the bulk of the crystal, and putting them on the surface of the crystal. The structure is thus 'opened out' and is now more disordered, so there is a positive entropy gain, offset by a slight fall in the energy of packing the crystal together. This type of defect is common in the alkali halides.

2) The Frenkel defect. This defect occurs when an ion is removed from its lattice site, and placed in an interstitial position in the structure, leaving a vacancy. In general, the cation, which is normally smaller than the anion, is the ion which moves into the interstitial site. This type of defect is common when there is a considerable difference between the sizes of the anion and cation: a good example is AgBr.

3) The F-centre or colour centre. This involves an excess of cations in the crystal, resulting in a number of anion sites with no anions to fill them; local charge balance is restored by trapped electrons in these anion vacancies. Thus potassium will diffuse into KCl to give an excess of K^+ ions in the cation sites, balanced by electrons trapped in the anion vacancies. The trapped electron often has a strong optical absorption spectrum (with energy levels similar to the particle in a box problem) and consequently the crystals are coloured.

Simple diagrams of all these defects are given in Fig. 3.43 (see page 114).

Impurities

Most solids have fairly tightly packed structures, and so it is not easy to introduce *any* foreign atom into the lattice, but if atomic sizes are roughly equal, substitution at low levels is possible, and in some practical cases, almost impossible to prevent. It is easy, for example, to dope germanium with gallium or arsenic, the elements on either side of germanium in the periodic table. It is possible to diffuse small atoms into interstitial sites in solids composed of large atoms. A final type of impurity we may consider is the valence impurity in which one of the elements present has a small fraction of its atoms present in a different valence state – thus the solid NiO has a certain amount of nickel present as Ni^{3+} rather than Ni^{2+}.

The facility with which a solid can accommodate defects and impurities determines its tendency to non-stoichiometry. For a simple binary compound $A_{1-x}B$, a stoichiometric compound will be formed it the free energy of the system shows at least one sharp minimum as a function of x. For an alkali halide for example, the energy will be at a minimum for $x = 0$. A slight excess of alkali metal will produce F-centre defects, and the loss of lattice energy will be partially compensated by the entropy increase of the defect crystal, but as the crystal can only accommodate a small number of colour centres, the free energy of the crystal will rise very rapidly for more than a minute excess of metal. For an excess of halide, the energy will rise even faster since the only way of accommodating the excess in the crystal is as halogen atoms, which is unfavourable, or by doubly ionising the alkali metal to give M^{2+} which is equally unfavourable. The alkali halides have very sharp minima of free energy against composition plots, and are thus easy to prepare as stoichoimetric compounds. By contrast, compounds such as nickel oxide NiO, or ZnO, are almost always found to have a non-stoichiometric composition such as $Ni_{0.995}O$. For nickel oxide, some of the metal is present as Ni^{3+} ions, while for zinc oxide excess metal, present as interstitial Zn^{2+} ions with associated trapped electrons is common. For such compounds the free energy versus composition curves have shallower minima, and the possible entropy gains from disorder and non-stoichiometry are important.

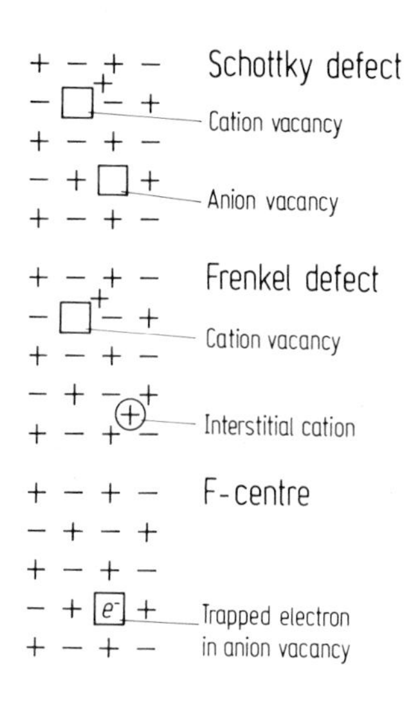

Fig. 3.43. Common crystal defects

The electronic structures of solids containing defects and impurities

The band structure of a regular solid is of course changed by the presence of defects and impurities. Two effects may be distinguished:

1) The appearance of an excess or deficiency of electrons in the band structure of the perfect solid

2) The appearance of new electronic energy levels as a result of the impurity or defect. These new levels are frequently localised close to the impurity or defect, and are particularly important when their energies fall in the band gaps of the solid.

Some examples will illustrate this:

1) A trapped electron in an F-centre has an energy between that of the valence band and the empty conduction band. An anion vacancy can act as an electron trap i.e. it provides an empty level above the valence band and below the conduction band. This level will be localised on the anion vacancy.

2) A Ni^{3+} ion in "NiO" will be a much stronger electron trap than the Ni^{2+} ions whose orbitals form the empty conduction band. The valence impurity Ni^{3+} thus produces an empty electronic level below the conduction band and above the valence band. This level will of course be localised on the Ni^{3+} ion.

3) Germanium doped with gallium (which has one valence electron less) will have an empty gallium energy level just above the valence band; by contrast, if the germanium was doped with arsenic, which has one more valence electron than Ge, the extra electron will be in an energy level localised on the arsenic, just below the empty conduction band.

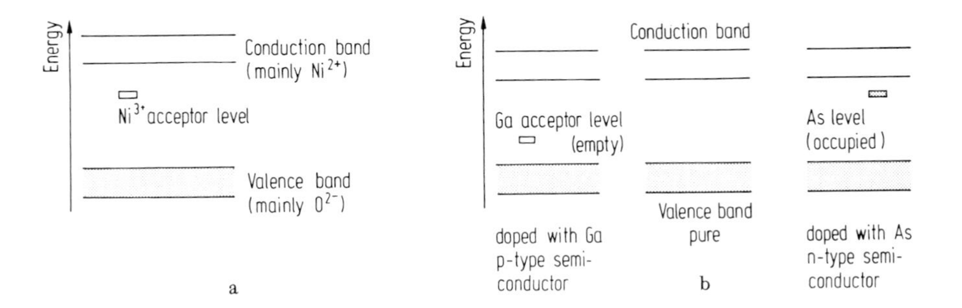

Fig. 3.44. (a) NiO, **(b)** Germanium

The most pronounced effect of these new levels is seen in the conductivity properties. Electrons in levels just below the conduction band (e.g. As in Ge) can be thermally excited into the conduction band more easily than the electrons in the more tightly bound valence band – the net effect is to reduce the band gap for conduction. Such conduction arises from the motion of the negative electrons in the conduction band, and is called n-type conduction.

Solids with empty levels below the conduction band can be excited more easily than those where the electrons have to be excited into the valence band. The excitation leaves a 'hole' in the valence band, which, now being partially occupied, can act as a conductor. In this case, the conduction occurs as a result of the motion of the 'hole' through the valence band – since the hole is positively charged with respect to the valence band this is referred to as p-type conduction. An example is germanium doped with gallium. The difference between n and p-type conduction is of immense importance in the construction of semiconductor devices.

The presence of impurities and point defects in crystals can thus fundamentally affect their properties, electronic structure, and stoichiometry. We may also note that *dislocations,* that is, errors or non-regularities in the stacking of the crystal planes can also have a considerable effect; for example, impurities may concentrate in such regions where the crystal structure is already distorted. The important effects of defects and impurities on reactivity are discussed in Chap. 6.

Packing of atoms in solids

Since we have emphasised the importance of close packing in solid structures, we should discuss the possible arrangements. Three structures are particularly common in the solid state, and many others may be derived from them. They are:

1) *Cubic close packed,* or face centred cubic. There is an atom at each corner of a cube, and one at the centre of each face. Each atom has 12 equidistant nearest neighbours (C.N. = 12)

2) *Hexagonal close packed.* The unit cell in Fig. 3.45 shows the structure geometrically, but it is easier to regard the structure as close packed layers stacked on top of each other so that the third layer is vertically above the first, the fourth above the second, and so on (one may regard cubic close packing as a similar structure with a repeat distance of three rather than two units). Each atom has 12 equidistant nearest neighbours.

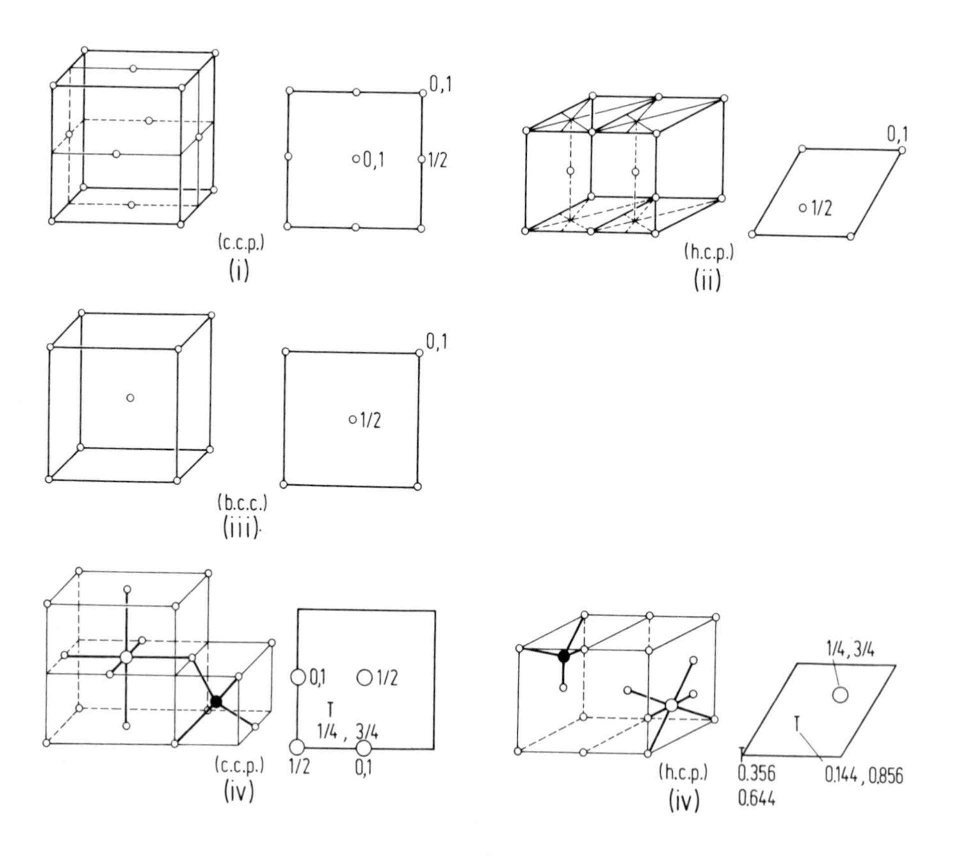

Fig. 3.45. (i) Face centred cubic (f.c.c.) or cubic close packed (c.c.p.), (ii) hexagonal close packed (h.c.p.), (iii) body centred cubic (b.c.c.), (iv) octahedral and tetrahedral holes in c.c.p. and h.c.p. lattices. Open circles – octahedral holes; full circles or T – tetrahedral holes. The figures indicate the vertical coordinate of the hole or holes. The position of the other holes may be deduced from the symmetry of the system

3) *Body centred cubic.* There is an atom at the centre of a cube, surrounded by atoms at each corner of the cube. Each atom has 8 nearest neighbours (C.N. = 8) but there are 6 other atoms only slightly further away.

Cubic and hexagonal close packing are the two most efficient methods of packing identical spheres with 74 percent of the available space filled while body centred cubic structures have 68 percent of the available space filled.

All these structures have 'holes' where other spheres may be introduced – thus each cubic close packed (c.c.p.) unit cell has 4 atoms, but 4 holes with octahedral symmetry and 8 holes with tetrahedral symmetry. Similarly the unit cell for a hexagonal close packed (h.c.p.) structure contains 2 atoms and has 2 octahedral holes and 4 tetrahedral holes. The body centred cubic (b.c.c.) structure has octahedral holes but these are much smaller than their equivalents in the close packed structures and are rarely occupied.

It is now possible to derive a very large number of inorganic structures either by filling the holes in the lattice, or by substituting some of the atoms in the close packed array. Some examples are:

1) Sodium chloride – all the octahedral holes are occupied by sodium ions in a cubic close packed array of chlorine ions. The packing of the sodium ions is also c.c.p.

2) Spinels have the general formula AB_2O_4 where A and B are bivalent and trivalent ions respectively. The oxygen atoms have a cubic close packed structure, with the metal ions occupying both octahedral and tetrahedral holes. A quarter of the available holes are occupied, giving the correct stoichiometry.

3) Caesium chloride has the body centred cubic structure with a caesium ion at the centre of the cube, and chloride ions at the vertices (or vice versa).

4) Zinc sulphide (zincblende) has half the tetrahedral holes in a c.c.p. lattice of sulphide ions occupied by zinc ions – this gives a CN of 4 for both Zn and S. The arrangement of the zinc ions is also cubic close packed. If the zinc and sulphur atoms are replaced by carbon atoms, the diamond structure is obtained.

These are just a few examples taken from a huge subject: for further examples and a complete critical discussion, the reader is referred to the excellent book by Wells.

Prediction of crystal structure

The prediction of the structure of solids has attracted great attention since the invention of X-ray crystallography, but has proved to be a formidably complicated subject. This section is intended only to introduce the subject, and the reader should consult the bibliography for more detailed studies.

One of the earliest approaches was the radius-ratio rule used for ionic solids. An atom may be regarded as being surrounded by a coordination polyhedron which may be characterised by its coordination number: 6 for NaCl, 8 for CsCl, 4 for zincblende, and so on. The most efficient packing arrangement will be for the atoms in the coordination polyhedron to be just in contact with each other; if the central atom is too large, the coordinating spheres will be forced apart, if it is too small, it will be favourable to go to a lower coordination number to give a closer coordination of the central atom (see Fig. 3.46 a). If simple geometry is used to calculate the ratios (of cation and anion radius) at which the ions just touch, and the cation does not 'rattle' in between the anions, the following result is obtained:

Ratio (r_c:r_a)	> 0.732	> 0.414	> 0.225
Favoured C.N.	8	6	4

The problem of determining ionic radii is discussed in Chap. 5, but whichever scheme is used, the radius-ratio rule is not very satisfactory. Although compounds with high r_c:r_a values show higher C.N. than those with low r_c:r_a values, many compounds predicted to have C.N. 8 show sixfold coordination. According to the pure electrostatic model, the gain in energy on going from the NaCl structure (C.N. = 6) to the CsCl structure (C.N. = 8) is under 1 percent so any slight non-ionic effect in the bonding may dominate the choice of crystal structure, and invalidate the radius ratio rule. It is indeed the slight covalency present in the bonding of even the most ionic systems which is responsible for this failure. Since covalent bonding depends on overlap, the stereochemistry which gives the best overlap will be strongly favoured. The presence of covalency can thus stabilise a particular structure.

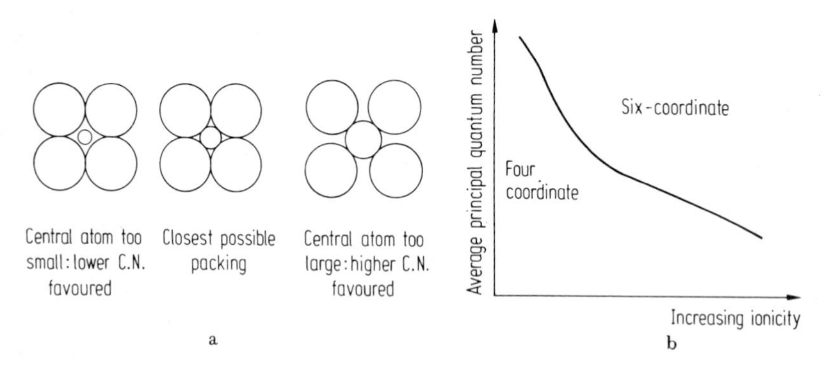

Fig. 3.46 a and b. A Mooser-Pearson plot, showing the division between four and six coordinate structures

Experimentally it is found that high C.N. are most common for compounds of the heavy elements, and for highly ionic compounds. Mooser and Pearson first showed this by examining a series of AB compounds where A and B are elements with s and p electrons only in their valence shells, and the sum of the valence electrons of A and B is 8 (e.g. NaCl, CaS, GaAs, elemental Ge). They plotted the average principal quantum number of the valence shells of the elements A and B (i.e. 3 for NaCl, 3.5 for CaS, 4 for Ge) against an estimate of the ionicity (measured by electronegativity difference – Chap. 5) and found that a sharp dividing line could be drawn between species with C.N. 4 or 6 (Fig. 3.46 b). A more theoretical treatment by Phillips and van Vechten gives a even better separation of the two structure types.

The Mooser-Pearson approach has been applied to a wide variety of compounds and the plots of average principal quantum number against ionicity always give a good separation of compounds of different structure. Mooser and Pearson point out that the more ionic compounds have close-packed arrays of cations and dense structures, while the more covalent compounds show close packing of the anions, and have more open structures.

It is possible to explain these observations qualitatively. High ionicity will favour closer packing and higher C.N.; although the $NaCl \rightarrow CsCl$ transformation increases the lattice energy by only 0.9 percent, the C.N. 4 → C.N. 6 change increases the stability of a purely ionic lattice by 6–7 percent, and the covalent interaction needed to stabilise C.N. 4 with respect to C.N. 6 will be greater than that needed to stabilise C.N. 6 with respect to C.N. 8. Increasing covalence will increase the importance of overlap and favour bond lengths and a coordination which maximise overlap. The geometry chosen will depend on the orbitals available, and the orbital energies must thus be considered. Elemental silicon has both s and p orbitals available for bonding, and a tetrahedral sp^3 hybridised structure will be favoured. For the isoelectronic salt NaCl, the orbitals involved in any covalent bonding will be the empty, spherical 3s orbital of Na^+, and the occupied 3p orbitals of Cl^- which will have greatest overlap if coordinated octahedrally. The 3s orbital of chlorine is very tightly bound, and will not be stabilised by any covalent interaction; for NaCl therefore, any covalent interaction will stabilise octahedral rather than tetrahedral coordination. The tendency to high C.N. for the heavier elements shown by the Mooser-Pearson diagrams is a reflection not only of the greater diffuseness of the valence orbitals of heavy elements (leading to a less directional character in the bonding), but also of the disinclination of the tightly bound s orbitals of heavy elements to take part in bonding. This effect is sometimes referred to as dehybridisation.

The bonding of elemental metals must involve covalency, and the structure adopted will therefore depend on overlap, the number of electrons available, and the orbital energies. If packing considerations alone were important, we would not expect any b.c.c. metals, since the packing is less efficient than c.c.p. or h.c.p., but b.c.c. metals are commonly found. Two theories are frequently used to explain the variations in metal crystal structure, that of Brewer and Engel, and that of Altmann, Coulson and Hume-Rothery. The Brewer-Engel approach assumes that the structure adopted will be dependent on the number of electrons available for bonding in s and p valence orbitals; electrons in d orbitals contribute to the cohesive energy, but will not affect the structure adopted. Brewer estimates that with less than 1.5 s or p electrons available per atom, a b.c.c. structure will be adopted, with between 1.7 and 2.1, a h.c.p. structure, and with between 2.5 and 3 electrons a c.c.p. structure. For transition metals a more stable structure *may* be obtained if electrons are excited from non-bonding d orbitals into s and p orbitals. The numbers of s and p electrons used for bonding, as deduced from the structure adopted, correlates well with the energy needed to excite the electrons determined from atomic spectroscopy.

The Altmann, Coulson, and Hume-Rothery approach considers the nature of the hybrids giving the greatest overlap with surrounding atoms, and includes the sterochemical preferences of the d orbitals. Both theories explain the progression from b.c.c. structures at the beginning of the transition series to h.c.p. and c.c.p. structures at the end (see Adams and Brewer). The structure of alloys may also be understood with this approach if the average number of electrons per atom is calculated.

In conclusion, the prediction of crystal structure is a difficult business as it is necessary to consider every bonding effect, even if its contribution to the total bonding energy is very small. If it is impossible to generalise, it is nonetheless possible to understand the effects involved using the general bonding ideas we have discussed in this chapter.

Liquids, glasses and amorphous solids

So far we have discussed only crystalline species, allowing for the fact that there may be some defects or dislocations. This covers the vast majority of solids however non-crystalline they may appear. A substance which appears to be completely randomly packed, even to X-ray diffraction, is described as amorphous; however, there is usually local ordering (over a few atoms or so) and it is extremely difficult to prepare a completely random solid.

Glasses generally have a very random array of small molecular units, so that the environment of each atom is relatively constant but the linkages between them are not. This is a metastable state, and crystals can separate out of glasses. There is a sufficient degree of 3-dimensional linkage to withstand stress, but no cleavage planes are found, in contrast to crystalline solids. The randomness of the linkage and consequent variation in strength of the linkages gives a glass a very ill-defined melting point, and it is usually plastic over a considerable temperature before melting. Impurity atoms are often much more soluble in glasses than regular closely packed solids, a fact of some use to manufactures of stained glass.

Liquids represent yet another step towards randomness, yet even here it is possible to distinguish most probable atom-atom distances, and it seems that the coordination number of each atom is roughly constant, although there is much more relative motion than for solids. In solutions of ions, it is often possible to show that the optical spectrum and other properties of the ions are unchanged on dissolution, but isotope exchange studies, for example, often show very rapid exchange between those molecules solvating a dissolved ion, and those effectively free. A sufficiently long time average of liquids shows rather more order than the instantaneous picture.

F. Metal-Metal Bonds

Metal atoms may interact with each other in many ways. We have already discussed the bonding in metallic solids, and simple metal-metal covalent bonds as in $[Hg_2]^{2+}$, the metal complexes of $[SnCl_3]^-$, and $Mn_2(CO)_{10}$ may be treated quite normally. Certain other metal-metal interactions display either unusual strength or unusual weakness, and merit separate discussion.

Strong metal-metal interactions

Although compounds with very strong metal-metal bonds were first prepared over 100 years ago, it was only with the determination of the crystal structure of the compound $KReCl_4$ $2H_2O$ in 1964 that the existence of this special type of interaction was recognised. $KReCl_4 \cdot 2H_2O$ was found to contain discrete $Re_2Cl_8{}^{2-}$ anions with the structure shown in Fig. 3.47. Particularly interesting was the very short rhenium-rhenium distance, shorter than that found in rhenium metal (2.75Å), and the fact that the two Cl_4 squares were eclipsed – that is, they were in the position where chlo-

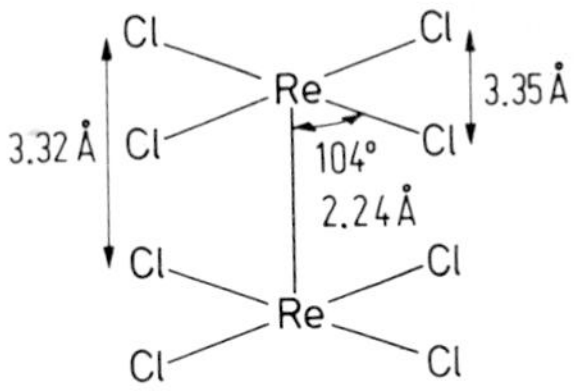

Fig. 3.47. $Re_2Cl_8^{2-}$

rine-chlorine repulsion would be maximised. We may assume that the apparently unfavourable arrangement of the chlorine ligands is compensated by the increased strength of the metal-metal bond. This may be explained by reference to the molecular orbital diagram proposed for this compound (Fig. 3.48).

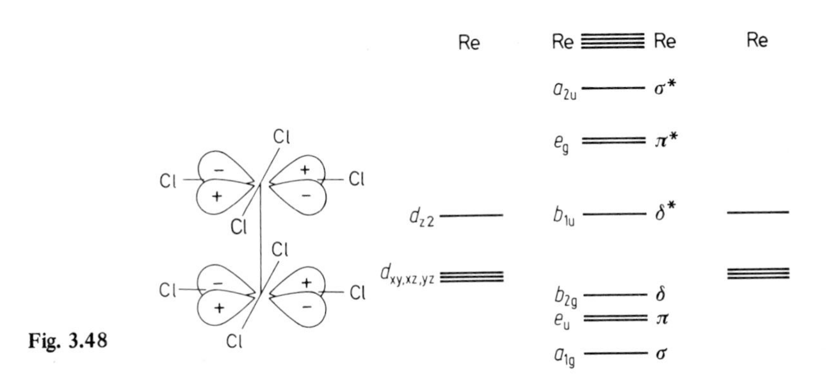

Fig. 3.48

The coordination of the rhenium atoms by the chloride ligands is approximately square planar, and, following the discussion of square planar complexes, we may assume that the $d_{x^2-y^2}$ orbital is strongly antibonding, and that the other d orbitals are available for interaction with the neighbouring rhenium atom. Viewed along the Re-Re axis, the d_{z^2} orbital has σ symmetry, the d_{xz} and d_{yz} have π symmetry, and the d_{xy} has δ symmetry. Fig. 3.48 shows that if the chlorine atoms are eclipsed, the d_{xy} orbitals on each rhenium atom will overlap, and consequently will give a bonding and an antibonding orbital, but if one of the rhenium planes is rotated by 45° the overlap will be zero – the fall in repulsion energy of the chloride ligands is offset by the destruction of the δ bond between the two rhenium atoms. This appears to be unfavourable, so the eclipsed rather than the staggered forms are adopted.

The molecular orbital diagram may be constructed relatively easily: if we ignore the filled ligand-metal σ bonding orbitals (having too low an energy), then the orbitals involved in the metal-metal bond are the four d orbitals discussed above. The d_{z^2} orbitals overlap strongly, the d_{xz} and d_{yz} less so, and the d_{xy} weakest of all. We thus arrive at the orbital scheme shown in Fig. 3.48. Rhenium has 7 electrons in its valence shell ($5d^5\ 6s^2$), and each chlorine supplies one electron. $Re_2Cl_8^{2-}$ thus has (2 x 7) + (8 x 1) + 2 = 24 electrons in the valence orbitals. 16 will be involved in the metal-chlorine bonds, leaving 8 to fill the metal-metal bonding orbitals. This corresponds to filling the σ, two π and the δ bonding molecular orbitals, giving four filled bonding M.O., or a *quadruple bond.* This high bond order accounts very satisfactorily for the short metal-metal distance.

Given the stability of this structure it seemed very probable that other compounds of this type could be prepared, and after the initial explanation of the structure of $Re_2Cl_8{}^{2-}$ several other examples were found, notably with carboxylic acids such as CH_3CO_2H. In $Mo_2(CH_3CO_2)_4$ the CO_2 group bridges the two metal atoms, one oxygen ligating each metal. It has even been possible to prepare organometallic quadruply bonded species, notably $Mo_2(C_3H_5)_4$. Cotton has recently reviewed quadruply bridged species. The proposed M.O. scheme is supported by the preparation of $Mo_2(SO_4)_4{}^{3-}$, with only one electron in the presumably weakly bonding δ orbital. The E.P.R. spectrum shows this unpaired electron to be delocalised over both molybdenum atoms. A recent molecular orbital calculation supports the assigned order of orbitals, and suggests that the metal-metal bonding orbitals are mixed very little with the metal ligand bonding orbitals, as we assumed initially.

Similar strong metal-metal interactions are found in species such as $W_2Cl_9{}^{3-}$ which has the structure shown in Fig. 3.49. The structure is distorted, with the tungsten atoms closer together than would be expected for perfect octahedral coordination of the tungsten. The short bond length, the diamagnetism, and the Mössbauer spectrum of this compound all suggest a multiple bond character in the tungsten-tungsten bond. By contrast, the chromium analogue, in $Cs_3Cr_2Cl_9$, has a distorted structure with the chromium atoms further away from each other than expected, implying a repulsion between the Cr atoms, which appear from magnetic measurements to be completely independent of each other.

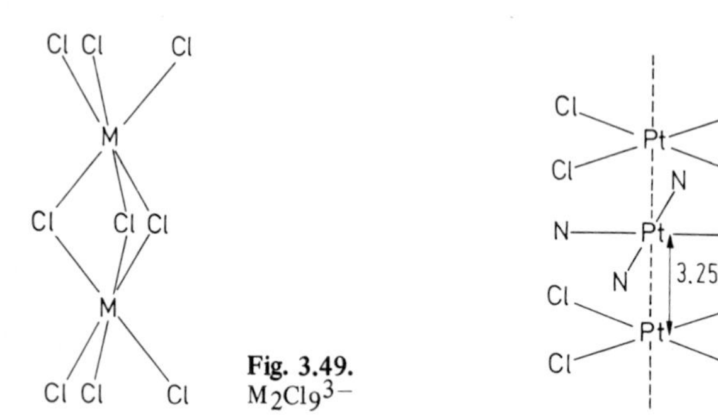

Fig. 3.49. $M_2Cl_9{}^{3-}$

Fig. 3.50. Metal-metal interactions in $Pt(NH_3)_4PtCl_4$

Weak interactions between metals

The crystal structures of certain metal complexes suggest a weak interaction between metal atoms which is, however, too weak to be formally identified as a bond. This is particularly common for complexes with a relatively 'open' structure (e.g. square planar or linearly coordinated complexes). Such compounds often have a completely different optical spectrum in the solid state to that found in solution: for example $K_2Pt(CN)_4 3H_2O$ is a yellow solid giving a colourless solution. Even more spectacular is Magnus's green salt, a green solid formed from the colourless $[Pt(NH_3)_4]^{2+}$ ion and the red $[PtCl_4]^{2-}$. The structure of this salt is shown in Fig. 3.50. These compounds all have structures with stacked planar molecules, the metal atoms vertically above each other and sufficiently close to lead one to expect some sort of an interaction. Their physical properties are markedly anisotropic, with strong absorption of light parallel to the metal-metal axis, and unusual conductivity properties. It is possible to diffuse a small amount of bromine into crystals of $K_2Pt(CN)_4 3H_2O$, to give $K_2Pt(CN)_4 Br_{0.3} 3H_2O$. This compound has much shorter Pt-Pt distances, only a little greater than in Pt metal. The crystal show metallic conduction parallel to the metal chains.

The electronic structure of these compounds is not well understood. The stacking geometry would suggest that the interaction between the metals involves the $d_z{}^2$ orbital, and (to a lesser extent) the d_{xz} and d_{yz}. In a pure d^8 complex, all these orbitals are occupied and one would initially expect a repulsion, with any bands formed being completely occupied. The behaviour of the bromine doped species is reasonable if it is assumed that the bromine acts as an electron acceptor impurity, forming Br^- and producing holes in the bands formed by the interaction of the $d_z{}^2$ orbitals. The one-dimensional metal behaviour of these compounds is then explained by conduction through the $d_z{}^2$ bands. The electronic structure of these compounds is discussed by Yoffe. Those readers with a liking for exotica may like to notice that $K_2Pt(CN)_4 Br_{0.3} 3H_2O$ shows an almost unbelievable variety of bond 'types': ionic (between K^+ ions and the anions present); covalent (the Pt-C-N linkage); band structure (along the metal chain); impurities (the bromine); and (presumably) some type of intermolecular attraction involving the water of crystallisation.

Although all the compounds discussed above involve platinum, other elements show similar properties: the square planar complexes $[Rh(CO)_2Cl]$ (a dimer with two bridging chlorine ligands) and $[Ir(CO)_2$ acac] (acac-acetylacetone a chelating β-diketone) show the same metal-metal chains. A recent crystal structure determination of $[C_5H_{11}NAuCL](C_5H_{11}N$ (piperidine) is a monodentate nitrogen ligand) shows a linear complex with four molecules fairly close together giving Au_4 squares.

G. Intermolecular Attractions

All the interactions between atoms that we have discussed so far (with the exception of the weak metal-metal interactions) have involved fairly strong bonds between atoms with energies greater than ~100 kJ/mole. It is usually necessary to change the chemical nature of the species to break these bonds. There exist however weaker interactions with an energy lower (often much lower) than 100 kJ/mole which are much easier to break up. It is these interactions which provide the cohesive forces for molecular solids and liquids – their relative weakness is shown by the volatility of many of these species. Thus it requires only 5 kJ/mole to dissociate N_2 molecules from a liquid to a gas, but 940 kJ/mole to dissociate the molecule into two nitrogen atoms. It is the origin and effect of intermolecular attractions and repulsions that we discuss in this section.

Intermolecular potentials

Intermolecular attractions will cause an association between atoms or molecules, but this association will be controlled by the repulsion between the nuclei and the closed shells of electrons of two adjacent atoms; it was mentioned in Chap. 2 that the noble gases do not form diatomic molecules, effectively for this reason. Several intermolecular potentials have been proposed to express the balance between attraction and repulsion in mathematical terms (see Maitland and Smith) – their general properties are (i) an attractive potential which falls with the internuclear or intermolecular distance r (e.g. $\frac{A}{r^6}$), and (ii) a repulsive term which rises very steeply at small r (e.g. $\frac{B}{r^{12}}$). The general form is shown in Fig. 3.51; despite the similarity of this figure to potential diagrams for classical chemical bonds, it should be realised that the equilibrium distance r_{eq} is greater and the energy dip E_{eq} is much smaller than for classical bonds.

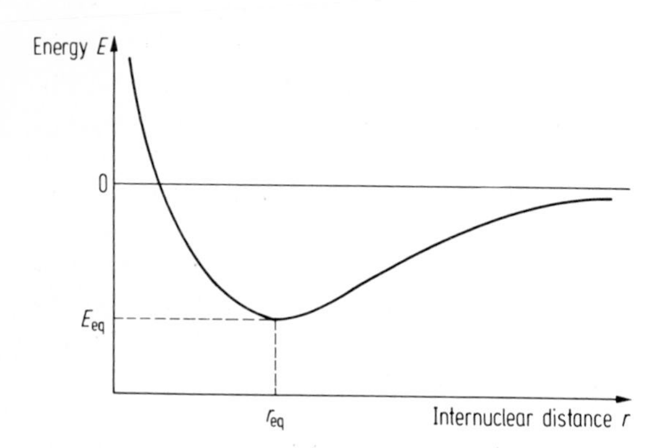

Fig. 3.51

It should be noticed that intramolecular repulsions similar to the intermolecular repulsions discussed above can affect the stability of molecules – this is often found in the chemistry of cycloalkanes, where the most stable arrangement of the carbon skeleton can produce strong repulsions between the hydrogens attached to the skeleton. Thus in cyclodecane $C_{10}H_{20}$ the carbon-carbon skeleton is slightly distorted, and there are several short H–H distances; the heat of combustion of this compound is 50 kJ/mole higher than that predicted for an unstrained molecule.

Bearing in mind that repulsions will counterbalance the effects of the attractions, we will devote the rest of this section to the origins and effects of the attractions. The origins may be classed under three headings:

1) Van der Waals forces
2) Charge transfer
3) Hydrogen bonding.

Van der Waals forces

The existence of these forces was proposed by the Dutch physicist Van der Waals as a result of studies of the non-ideal behaviour of gases. Their quantum mechanical description splits them into three classes.

1) Dipole-dipole interactions
2) Dipole-induced dipole attractions
3) Instantaneous dipole-induced dipole

Dipole-dipole and dipole-induced dipole moments are relatively easy to understand.

Two molecules with dipole moments can arrange themselves so that the positive end of the dipole of one molecule is close to the negative end of the dipole of the other molecule (and vice versa). There will consequently be an attraction between the molecules. This sort of attraction is important in gaseous hydrogen fluoride. Dipole-induced dipole attractions arise from the polarisation of a non-polar molecule in the electronic field of a dipolar molecule, leading to an attraction between the two species. Both these types of attraction are important for small polar molecules and in solutions of polar solvents.

The instantaneous dipole-induced dipole interaction is more subtle in its origin, but is more important. Hitherto we have regarded the electron density in an atom or molecule as being constant except for when, with a twinge of conscience, we discussed correlation effects. In fact there is a rapid, but continuous fluctuation of electron density, and this fluctuation produces instantaneous dipole moments which may attract other molecules by a dipole-induced dipole mechanism. This attraction is the origin of the cohesive forces between atoms of the noble gases, and causes them to liquefy and solidify at low temperatures. It becomes more important as the number of electrons in the particles considered increases, and as the particles become more polarisable. This explains the increase in boiling point with atomic number for the noble gases – actual weight has little effect on the boiling point of a substance, H_2, D_2 and T_2 having similar boiling points although T_2 (with two tritium nuclei) is three times heavier than H_2. The attraction is variously known as the *dispersion* force or the *London dispersion* force (after the physicist London who derived an expression for the attraction – see Atkins). It is a relatively short range effect, falling off as the inverse sixth power of the internuclear distance.

Dispersion forces are the main source of attraction between non-polar or weakly polar chemical species. Their magnitude is variable – the 'lattice energy' of the noble gases when solidified is only a few percent of an ionic solid such as NaCl. However, in a supposedly ionic solid such as TlI it has been estimated that the dispersion attraction between the two large, polarisable ions accounts for 20 percent of the lattice energy.

The London dispersion energy will be maximised by the interaction of two very polarisable groups – this fact is used by some enzymes with hydrophobic pockets, clefts in the protein structure lined with non-polar, but polarisable groups. This 'pocket' will have an attraction for non-polar polarisable groups (such as a carbon-carbon skeleton) but not for polar molecules such as water. It is proposed that these pockets 'hold' the non-reactive part of the enzyme substrate in a suitable conformation for the reactive part to be attacked.

From the crystal structures of molecular solids, it has been possible to extract a series of Van der Waals radii, the Van der Waals radius being the mean effective radius of an atom bonded to its neighbours by Van der Waals forces only. If in a crystal structure two apparently non-bonded atoms are found to be closer than the sum of their Van der Waals radii, one has good reason to expect some rather stronger interaction between these two atoms. This method is used for the detection of weak metal-metal bonds (see above) and is also popular for the detection of hydrogen bonds.

Charge transfer complexes

There are several examples known where two different molecules show weak attractions between each other (stronger than Van der Waals attractions, but weaker than 'normal' chemical bonds), and the association between these two molecules is marked by the appearance of a new absorption band in the electronic spectrum of the complex. These are charge transfer complexes, where there is a slight transfer of charge from one species to another. This corresponds to the formation of a very weakly bonding M.O. in which a small amount of the acceptor orbital is mixed into the wavefunction of the donor orbital. The absorption band is due to an excitation into the corresponding antibonding M.O. It has been shown that the energy of the charge transfer band follows quite closely difference between the ionisation potential of the donor, and the electron affinity of the acceptor. Charge transfer complexes are best regarded as a very weak form of 'normal' chemical bonding. The interactions are often so weak (molecular iodine dissolved in benzene for example) that the pure complex cannot be isolated.

The hydrogen bond

The crystal structures of many compounds have shown that hydrogen is often close to two coordinate, being bonded to one atom but having a second atom much closer than the sum of the Van der Waals radii would predict (see Fig. 3.52). Thermodynamic data suggest that this is a fairly strong interaction. The hydrogen atom appears always to be bonded strongly to one atom A, yet interacting with a second atom B – the bond lengths H–A and H–B are (with perhaps a few exceptions such as F–H–F) always asymmetric. There appears to be no particularly strong requirement that the A–H. . .B grouping be linear or otherwise. This rather unusual interaction is the *hydrogen bond*. It is found when A and B are chosen from C, N, O, P, F, S, Cl, Se, Br, or I, that is, elements which are usually considered to be more electronegative than hydrogen (see Chap. 5). Its asymmetric form shows it to be very different from the B–H–B bridges of borane chemistry.

Four factors are thought to be important in establishing a hydrogen bond:

1) Electrostatic attraction: When the atom A is more electronegative than hydrogen, the A–H bond will be polarised so that the atom A carries a negative charge and the hydrogen atom a positive charge. If the electronegative atom B carries a negative charge there will be an electrostatic attraction between H and B.

2) Electrostatic repulsion: There will be very little repulsion between the hydrogen atom and B as the hydrogen atom has no core electrons, but there will be a repulsion between the atoms A and B, both of which may be carrying negative charges.

3) Charge transfer: The atom B can act as a 'donor' to the A–H group to form a charge transfer type complex. This corresponds to forming a weak covalent bond between B and H. For a strong symmetric bond such as in HF_2^- we may even imagine a three-centre bond.

4) London dispersion forces: These will also give a slight positive contribution to the A–H. . .B bond strength.

It seems that these effects are about equally important. For some time it was thought that the electrostatic effect was the most important, but calculations have shown that this is due to a coincidental cancelling out of the other effects. Coulson calculated the following values for ice:

Electrostatic	−25 kJ/mole
Repulsion	+ 33 kJ/mole
Charge transfer	−33 kJ/mole
Dispersion	−13 kJ/mole
Total	−38 kJ/mole

The experimental value is −25 kJ/mole. The electrostatic model alone is not capable of explaining the spectroscopic evidence of the transfer of charge: the A–H stretching vibration in the infra-red increases appreciably in intensity (suggesting a re-distribution of charge) and the nuclear quadrupole coupling data for ice shows an environment for the oxygen atom very different from that found for H_2O vapour.

The hydrogen bond is quite strong by the standards of intermolecular attractions, and in HF_2^- the bond energy has been estimated as over 100 kJ/mole, although values below 30 kJ/mole are more common. This gives rise to some quite important effects in chemistry, and a few are discussed below. For further details of the vast amount of literature this subject has produced, the reader is referred to the reviews cited in the Bibliography.

Effects of hydrogen bonding

The most obvious effect of the hydrogen bond is to increase the association, and hence the boiling point of compounds in which there is strong hydrogen bonding. The most famous example is the high boiling point of water (373°K) compared with H_2S (211°K). Normally one would expect a smaller difference, with H_2S boiling at a higher temperature due to the increased London dispersion energy (in fact H_2Se boils at 230°K in accordance with this). The deviating value for water, (also found for NH_3 compared with PH_3, HF compared with HCl) is due to the very strong hydrogen bonding in water. Unexpectedly high boiling points of liquids are often caused by hydrogen bonding. Even in the vapour phase, hydrogen bonding can encourage the formation of dimers, or polymers.

In solids, hydrogen bonding can often radically change structures. In ice, the strong hydrogen bonding causes each hydrogen atom to be effectively two coordinate, and each oxygen effectively four coordinate. The structure has some similarities to SiO_2, and many of the polymorphs found for SiO_2 are also found for ice. In ammonium fluoride, one would expect six coordination of anion and cation but a wurtzite structure with both ions four coordinate is adopted – this enables the fluoride ions to form hydrogen bonds with the four hydrogen atoms attached to the nitrogen. Acid salts (such as $NaHCO_3$), solid acids, hydroxides, and salts with water of crystallisation frequently show hydrogen bonding – some examples are given in Fig. 3.52 (see page 128).

Hydrogen bonding is particularly important in determining secondary and tertiary structure – the conformations and arrangements adopted once the basic bonds and bond angles have been settled. The full structure of protein chains are very considerably influenced by the hydrogen bonding interactions between the peptide and amide links and between side chain residues. Hydrogen bonding is also useful for holding a molecule with a suitable orientation on the surface of a catalyst or enzyme so that it may be attacked easily. The hydrogen bond is sufficiently weak for the product of the reaction to be released afterwards without great difficulty.

Fig. 3.52. The forms of simple hydrogen bonds

Bibliography

Many of the subjects treated in this chapter are discussed in the general texts cited at the end of Chapter 2. The references below give more detailed discussion of specific topics.

Sect. A

Atkins, P.W., Symons, M.C.R.: The Structure of inorganic radicals. Amsterdam, London, New York: Elsevier 1966

Burdett, J.K.: 'Angular overlap' approach to Walsh diagrams. Structure and Bonding *31,* 67 (1976)

Coulson, C.A., Neilson, A.H.: Discussion of Walsh's rules. Disc. Faraday Soc. *35,* 71 (1963)

Herzberg, G.: Molecular spectra and molecular structure, vol. 3: Electronic spectra and structure of polyatomic molecules. New York: Van Nostrand-Rheinhold 1966

Walsh, A.D.: The derivation of Walsh diagrams. J. Chem. Soc. *1953,* 2260 et seq.

Sect. B

Galyer, L., Wilkinson, G., Lloyd, D.R.: Spectrum of $W(CH_3)_6$. Chem. Comm. *1975,* 497

Lippard, S.J.: Eight-coordinate complexes. Prog. Inorg. Chem. *8,* 109 (1967)

Muettterties, E.L.: Schunn, R.A.: Penta-coordination. Q. Rev *20,* 245 (1966)

Muetterties, E.L., Wright, C.M.: Molecular polyhedra of high coordination number. Q. Rev. *21,* 109 (1967)

Wood, J.S.: Penta-coordination. Prog. Inorg. Chem. *16,* 227 (1962)

Sect. C

Chatt, J.: The bonding of olefins to metals. J. Chem. Soc. *1953,* 2939, and Dewar, M.J.S.: Bull. Chim. Soc. France *18,* C71 (1951)

Coates, G.E., Green, M.L.H., Powell, P., Wade, K.: Principles of Organometallic Chemistry. London: Methuen 1971

Cotton, F.A.: *Hapto* nomenclature. J. Amer. Chem. Soc. *90,* 6230 (1968)

Eisenberg, R., Meyer, C.D.: Complexes of NO. Acc. Chem. Res. *8,* 26 (1975)

Huheey, J.E.: Inorganic Chemistry. London, New York: Harper and Row 1972

Mason, R.: Valence in transition metal compounds. Chem. Soc. Rev. *1,* 431 (1972)

Mitchell, P.R., Parish, R.V.: Review of the 18 electron rule. J. Chem. Educ. *46,* 811 (1969)

Muetterties, E.L.: Stereochemical non-rigidity. Acc. Chem. Res. *3,* 266 (1970)

Muetterties, E.L., Jesson, J.P.: Dynamic molecular processes in inorganic and organometallic compounds. In: Dynamic nuclear magnetic resonance spectroscopy. Jackman, L., Cotton, F.A. (eds.). New York: Academic Press 1975

Sidgwick, N.V., Bailey, R.W.: The 18 electron rule. Proc. Roy. Soc. A *144,* 521 (1934)

Vaska, L.: Complexes of molecular oxygen. Acc. Chem. Res. *9,* 175 (1976)

Vrieze, K., van Leeuween, P.W.N.M.: Fluxionality studied by n.m.r. Prog. Inorg. Chem. *14,* 1 (1971)

Warren, K.D.: M.O. theory of ferrocenes. Structure and Bonding *27,* 46 (1976)

Sect. D

Johnson, B.F.G.: Chem. Comm. *1976,* 211
King, R.B.: Transition metal cluster compounds. Prog. Inorg. Chem. *15,* 287 (1972)
Longuet-Higgins, H.C.: Electron deficient bonding. Q. Rev. *11,* 121 (1957)
Mason, R., Mingos, D.M.P.: Chemical Crystallography (MTP) International review of science, Physical Chemistry, Series 2, Vol. 11. J.M. Robertson (ed.). London: Butterworths 1975
Mingos, D.M.P.: Nature, Phys. Sci. *236,* 99 (1972)
Wade, K.: Electron deficient compounds. London: Nelson 1971
Wade, K.: Adv. Inorg. Chem. and Radiochem. *17,* 1 (1976)
Williams, R.E.: Adv. Inorg. Chem. and Radiochem. *17,* 67 (1976)

Sect. E

Adams, D.M.: Inorganic Solids. London, New York, Sydney: J. Wiley 1974
Brewer, L.: The structure of metals. Science *161,* 115 (1968)
Evans, R.C.: An introduction to crystal chemistry. 2nd Ed. Cambridge: Cambridge University Press 1966
Galwey, A.K.: The chemistry of solids. London: Methuen 1967
Pauling, L.: The nature of the chemical bond, 3rd Ed. Ithaca, N.Y.: Cornell University Press 1960
Wells, A.F.: Structural inorganic chemistry, 4th Edition. An extremely comprehensive source of structural data for all inorganic compounds. London: Oxford University Press 1975

Sect. F

Cotton, F.A.: Metal-metal multiple bonds. Chem. Soc. Revs. *4,* 27 (1975)
Yoffe, A.D.: Chem. Soc. Revs. *5,* 51 (1976)

Sect. G

Allen, L.C., Kollman, P.A.: Theory of hydrogen bonding. Chem. Rev. *72,* 283 (1972)
Hamilton, W.C., Ibers, J.A.: Hydrogen bonding in solids. New York: W.A. Benjamin 1968
Maitland, G.C., Smith, E.B.: Forces between simple molecules. Chem. Soc. Revs. *2,* 181 (1973)
Mulliken, R.S., Person, W.B.: Molecular complexes. London, New York, Sydney: J. Wiley 1969
Pimentel, G.C., McClellan, A.D.: The hydrogen bond. San Francisco: W.H. Freeman 1960, *and* Ann. Rev. Phys. Chem. *22,* 347 (1971)
Viogradov, S.N., Linnel, R.H.: Hydrogen bonding. New York: Van Nostrand-Rheinhold 1971

Problems

1. Draw a Walsh diagram correlating the M.O. of pyramidal and planar ML_3 molecules.

2. What will be the molecular structure of $[(CH_3)_3S]^+$? Would you expect $[(C_3H_7)(C_2H_5)(CH_3)S]^+$ to have the same stereochemistry about the sulphur atom?

3. Comment on the variation of bond angle in the series $CF_2 : SiF_2 : GeF_2$

4. Would you expect the metal-ligand interaction in $[Co(NH_3)_6]^{3+}$ (low spin) to be stronger or weaker than in $[Co(NH_3)_6]^{2+}$? Why?

5. Consider the weakest bound molecular orbitals of H_2O and NH_3 (Table 3.1). Which of these two molecules will interact most strongly with a metal cation? Draw a M.O. diagram for $[Fe(H_2O)_6]^{2+}$ and predict whether it will be high or low spin.

6. Tetra-glycine can function as a polydentate ligand with four nitrogen atoms binding to the metal. The complex initially formed with Ni^{2+} is blue and paramagnetic, but loses protons in neutral solution to give a yellow diamagnetic species. Can you explain this?

7. $HCo(CO)_4$ is a strong acid – what will be the structure of the anion produced by the dissociation? Draw a M.O. diagram for this anion.

8. Comment on the change in carbonyl stretching frequency in the series:

	$[V(CO)_6]^-$	$[Cr(CO)_6]$	$[Mn(CO)_6]^+$
ν_{CO} cm^{-1}	1860	2000	2090

9. It has been known for a very long time that solutions of silver salts will absorb unsaturated organic species such as alkenes and aromatic compounds. Propose an explanation for this.

10. (a) Coordinated η^3 allyl groups (as in $[(\eta^3\text{–}C_3H_5)Mn(CO)_4]$) frequently show one methene (CH) n.m.r. signal, and two methylene (CH_2) signals. At higher temperatures, the two methylene resonances merge to give one signal only. Discuss, in general terms, the significance of this observation.
(b) $[Fe(C_5H_5)_2(CO)_2]$ shows a temperature dependent n.m.r. spectrum. At room temperature two equal singlets are observed at $\delta = 4.4$ and 5.7 ppm. On cooling to -100 °C, a singlet (5 protons) at $\delta = 4.4$ ppm and a single proton at $\delta = 3.5$ ppm, coupled with four inequivalent protons at 6.3 ppm are observed. Suggest an explanation for this behaviour.

11. Nickel is generally rather difficult to oxidise to the +III oxidation state, but nickelocene (*bis* η^5-cyclopentadienyl nickel (II), $Ni(C_5H_5)_2$) is readily oxidised. Assuming the molecular orbitals to be similar to those of ferrocene (Fig. 3.18), try to explain this.

12. Suggest possible structures for the following complexes:

	$[Ir(OH)(NO)(PPh_3)_2]^+$	$[IrCl(NO)(CO)(PPh_3)_2]^+$	$[Ru(NO)_2Cl(PPh_3)_2]^+$
ν_{NO} cm^{-1}	1860	1680	1687 and 1845

13. Use skeletal electron counting methods to predict the structure of the boron skeleton of pentaborane (11) B_5H_{11}, pentaborane (9) B_5H_9, and tetraborane (10) B_4H_{10}.

14. The structure of CaB_6 consists of Ca^{2+} ions surrounded by $B_6{}^{2-}$ octahedra, linked by apices to six other octahedra. Is this structure in accord with electron counting schemes?

15. Consider the structures of diamond and metallic arsenic (Fig. 3.35), and classify them as electron deficient, precise, or rich. Does this classification enable the two structures to be related?

16. Show that filling the octahedral holes of a c.c.p. lattice gives a second c.c.p. lattice.

17. Why do compounds such as pyrite (FeS_2) have a metallic appearance?

18. Figure 3.38 shows that the band formed by the s orbitals is much broader than that formed by the d orbitals of a transition metal. Can you explain this?

19. The group IVB elements crystallise with the diamond structure, with the exception of lead which adopts a c.c.p. structure. Comment.

20. Molybdenum hexacarbonyl reacts with acetic acid to give a yellow compound with a molecular weight close to 430. The infra-red spectrum shows no sign of coordinated carbon monoxide, or of free OH groups. Propose a structure for this product.

21. WF_6 boils at 17.5 °C, WCl_6 at 346.7 °C. Why are these temperatures so different?

22. Maleic acid and fumaric acids have the structures shown below:

Maleic acid: H(HOOC)C=C(COOH)H (both COOH groups on the same side)

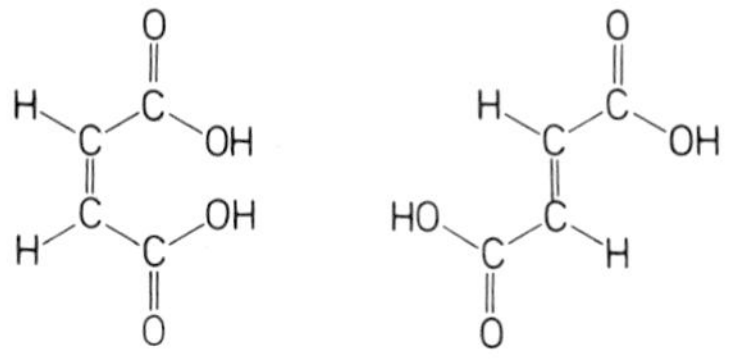

Fumaric acid: H(HOOC)C=C(H)COOH (COOH groups on opposite sides)

Both are dibasic acids, but while fumaric acid loses both protons with approximately equal facility, maleic acid loses its first proton with greater, and its second proton with less, ease than fumaric acid. Propose an explanation.

23. Draw molecular orbital diagrams for:

(i) $PbCl_6^{2-}$

(ii) $AuCl_2^-$

24. $Re_2Cl_8^{2-}$ was presented as the first example of δ bonding, but δ bonding orbitals were mentioned earlier for another compound. Which?

4. Electronic Spectra and Magnetic Properties of Inorganic Compounds

The investigation of the electronic absorption spectra and the magnetic properties of inorganic compounds has yielded an enormous amount of information, and for the transition metals has dominated the development of bonding theories; it is useful to devote a chapter to relating these observations to the theories we have proposed.

A. Electronic Spectra

It is usual to study spectra in the range 200 nm (50 kK) to 1000 nm (10 kK), the near ultra-violet to the near infra-red. These energies correspond to electronic transitions between weakly bonding and weakly antibonding, bonding and non-bonding, and non-bonding and antibonding levels. At high energies almost all compounds are opaque (as a result of strongly bonding – strongly antibonding transitions, or of photo-ionisation). At lower energies, molecular vibrations complicate the spectra. The spectra of transition metal compounds are discussed first, followed by sections on the lanthanides and actinides, and main group elements. We may split up the electronic spectra of transition metal compounds into d-d transitions and charge transfer spectra.

a) d-d Transitions

Many transition metal complexes show absorptions at relatively low energies for electronic transitions (in the visible part of the spectrum). These are associated with transitions between d orbitals whose degeneracy has been removed by an environment of lower than spherical symmetry. For example, in a octahedral complex, one may see transitions involving electron jumps from the t_{2g} orbitals to the e_g orbitals. As mentioned in Chap. 3, thes orbitals are often only weakly bonding or antibonding – their energy separation is small. We may use the absorption energies to measure the ligand field splitting energies Δ, but it should be noted that only in a few cases does the absorption energy correspond exactly to Δ.

Intensity of d-d transitions

As mentioned in Chap. 1, the intensity of a transition is proportional to the transition probability $|\langle\psi_{\text{excited}}|\hat{\mu}|\psi_{\text{ground}}\rangle|^2$, and the symmetries of ψ_{ex}, ψ_{gd}, and $\hat{\mu}$ may therefore indicate if the transition will have zero intensity. Electronic transitions are almost invariably caused by interaction with the oscillating electric dipole moment of electromagnetic radiation; there are perhaps half a dozen exceptions where electric dipole forbidden transitions acquire visible intensity by other mechanisms. The electric dipole moment operator $\hat{\mu}$ transforms as the quantities x, y, and z, corresponding to the t_{1u} I.R. in octahedral symmetry, and the p I.R. in spherical symmetry.

If we consider a transition of the kind $t_{2g}^1 \rightarrow e_g^1$, then ψ_{ex} and ψ_{gd} are both g type; $< \psi_{ex} \mid \hat{\mu} \mid \psi_{gd} >$ has g x u x g = u type symmetry and consequently is zero (Chap. 1). Thus the transition is forbidden, yet $3d^1$ complexes such as $[Ti(OH_2)_6]^{3+}$ are quite strongly coloured. This is caused by the excitation of vibrations in the complex by the radiation at the same time as the electronic excitation: if we simultaneously excite a vibration from its g type groundstate to a u-type excited state, the total wave function ψ_{ex} will now be u-type and the product will have u x u x g = g type symmetry, and the transition moment will not vanish. This co-excitation of electronic and vibrational levels leading to an apparent breakdown of a selection rule is known as a *vibronic* transition. It requires that the orbitals involved in the transition should be 'coupled' with, or affected by the molecular vibrations, i.e. that they should not be completely non-bonding, and results in the absorption spectra being broadened by the vibrational energy taken up or released. These spectra are consequently much broader than the sharp single lines of atomic spectra where the transitions are pure electronic and rigorously obey the selection rule $\Delta l = \pm 1$. In compounds of the lanthanide elements, the 4f orbitals are almost completely non-bonding, and spectra resulting from f-f transitions are much weaker and narrower than their d-d counterparts.

The selection rule $\Delta S = 0$ is much more rigorously obeyed, as it is not affected by any vibronic coupling. Thus complexes with a d^5 high spin configuration ($t_{2g}^3\ e_g^2$), which cannot undergo a $t_{2g} \rightarrow e_g$ transition without pairing an electron spin in the e_g levels, and consequently a change in S, have very weak d-d absorption spectra – for example the very pale pink manganous Mn^{2+} salts. The reason that they are seen at all is the result of a spin-orbit coupling linking the pure atomic quantum numbers L and S – if S is no longer a pure quantum number (i.e. the Russell-Saunders coupling scheme is not 100 percent accurate) then the $\Delta S = 0$ rule may be broken.

The secondary effects which allow transitions to be seen, are not however powerful enough to give the absorption bands great intensity. Thus if we measure the strength of the transition by the molar extinction coefficient, typical values for the various transitions mechanisms are:

Fully allowed electric dipole	10^5
Vibronically allowed	10^2
$\Delta S \neq 0$, but spin-orbit allowed	<1

In tetrahedral symmetry, three of the d orbitals belong to the same I.R. as the p orbitals (t_2), and some d-p mixing can occur. This results in a considerable increase in intensity compared with the pure d-d transitions of octahedral complexes: the d-d transitions of tetrahedral $[CoCl_4]^{2-}$ have extinction coefficients nearly a hundred times higher than octahedral $[Co(OH_2)_6]^{2+}$.

Electron repulsion effects

In the discussion of atomic spectra it was emphasised that a partially filled shell had different electronic energies depending on the relative arrangements of the electrons, and that these energy states could be described by Russell-Saunders terms if the spin-orbit coupling was not too great. In the interpretation of the spectra of complexes with partially filled d shells we would expect such electron repulsion effects to be important. We may discuss these effects from one of two extreme positions – 1) the electron repulsion effects much greater than the ligand field effects (the weak field case); 2) the ligand field effects much stronger than the electron repulsion effects (the strong field case).

The weak field case

This is treated by taking the Russell-Saunders terms and applying the ligand field splitting as a small perturbation. For the d^1 case there are no electronic repulsion effects, but for d^2, the Russell-Saunders terms are ^3F, ^3P (spins parallel); ^1S, ^1D, ^1G (spins paired). Hund's rules would lead us to predict that ^3F is the lowest term. The energy differences between terms of the same multiplicity may be expressed in terms of Racah's parameters B and C (page 29). In this case, the difference between the groundstate and the ^3P state is 15B. If we now consider the effects of lowering the symmetry to octahedral, the ^3F term in spherical symmetry transforms as $^3A_{2g} + {}^3T_{2g} + {}^3T_{1g}$, while the ^3P transforms as $^3T_{1g}$. Note that this is very similar to the I.R.s spanned by f orbitals ($a_{2u} + t_{1u} + t_{2u}$) and p orbitals (t_{1u}) respectively but the terms are g-type since they arise from the g-type d orbitals. As with perturbations in general, the octahedral ligand field causes a splitting of the 21-fold degenerate ^3F term, in this case into three different terms. It can be shown readily for the ^3F terms, that as Δ increases $^3A_{2g}$ rises in energy, $^3T_{1g}$ falls, and $^3T_{2g}$ rises slowly. The ^3P term, now $^3T_{1g}$ (written $^3T_{1g}$ (P) to distinguish it from the other $^3T_{1g}$ (F)), rises slowly in energy. We may plot the variation in energy of the terms, against increasing Δ as in Fig. 4.1. We would thus expect three absorptions arising from transitions from the $^3T_{1g}$ (F) ground state to the

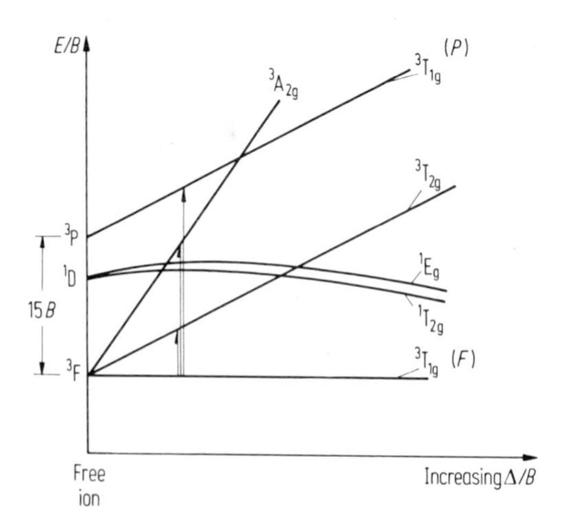

Fig. 4.1. Tanabe Sugano diagram showing the splitting of Russell-Saunders terms relative to the groundstate $^3T_{1g}(F)$ with increasing ligand field (for a d^2 system). Spin allowed transitions are marked.

$^3T_{2g}$, $^3A_{2g}$, and $^3T_{1g}$ (P) levels. If we consider an example $[V(H_2O)_6]^{3+}$, only the first two transitions can be observed, the third high energy band being masked by a more intense charge transfer band (q.v.). The problem of masking by other bands is quite common. There is a ^1D level lying between the ^3F and ^3P levels, but any transition to this level (^1E_g and $^1T_{2g}$ in the ligand field) is spin forbidden, and the transitions are masked by the spin allowed transitions, some of which are, in turn, masked by the dipole moment allowed charge transfer bands. It should be noted that the strength of the interelectronic repulsion is such that one cannot generally assign a precise number of electrons to the e_g or t_{2g}, (i.e. an $e_g{}^a t_{2g}{}^b$ configuration) but must consider the ligand field as changing the energy and lifting the degeneracy of well-defined Russell-Saunders terms.

Strong field case

In this case it is assumed that the ligand field effects are much greater than the electronic repulsion effects, as for example in a complex with a ligand which is a strong σ donor and π acceptor, giving a large Δ. In such a case one begins with a definite occupation of the e_g and t_{2g} levels (for example, for d^2, $e_g{}^2$, $e_g{}^1$ $t_{2g}{}^1$, or $t_{2g}{}^2$), and introduces the electron repulsion as a perturbation. This is best done by the group theoretical method described in Chap. 1: $e_g{}^2$ gives rise to the terms of the direct product $e_g \oplus e_g = A_{1g} + A_{2g} + E_g$. Inspection shows that only the A_{2g} term is a triplet, the others being singlets. In this case the effect of the electron repulsion perturbation is to mix slightly the well-defined $e_g{}^2$, $e_g{}^1 t_{2g}{}^1$, and $t_{2g}{}^2$ configurations, and lower their degeneracy. In the weak field case, it is the ligand field which mixes the well-defined Russell-Saunders terms.

Derivation of parameters from the spectra

In the weak field case, the ligand field is applied as a perturbation. The calculations are relatively straightforward, and are detailed in several books (Sutton; Murrell, Kettle and Tedder; Jørgensen (1962)). The effects are i) to lift the degeneracy of the Russell-Saunders terms, e.g. ^3F becomes $^3T_{1g} + {}^3T_{2g} + {}^3A_{2g}$ – this splitting is dependent only on the magnitude of Δ, and ii) as Δ increases, to cause mixing of levels of the same symmetry but originating in different Russell-Saunders terms. Thus in Fig. 4.1, as Δ increases, a mixing of the $^3T_{1g}$ (F) and $^3T_{1g}$ (P) terms is seen. As one would expect from perturbation theory the mixing is dependent on the value of Δ, *and* the initial separation of ^3F and ^3P (15B in Racah's treatment). It is possible from analysis of the spectra to obtain values of Δ, and the Racah parameter B. It is found that the value of B is always less than that known for the equivalent free metal ion (i.e. B for $[Ni(NH_3)_6]^{2+}$ is less than B for gaseous Ni^{2+}). This corresponds to lower interelectron repulsion in the complex, and is called the *nephelauxetic* (cloud expanding) effect; it is discussed later.

There is, of course, a logical flaw in the use of the Racah parameter to discuss electron repulsion in complexes, since it was introduced to describe spherical symmetry, while the symmetry of the complex is lower. The continued use of B is thus an artificial averaging of the electron repulsion effects in the e_g and t_{2g} orbitals which we have previously shown to be differently affected by chemical bonding. The strong field treatment avoids this problem by introducing a larger number of electron repulsion parameters: nine are needed for a full description, but a reasonable description is given with three. Such experimental results that are available show that there is an appreciable difference in the electron repulsion between a pair of electrons in the e_g subshell, and between a pair in the t_{2g} subshell (i.e. the nephelauxetic effect varies). However, there are not always enough absorption bands in the spectrum of the complex to obtain all the new electron repulsion parameters, especially for strong field complexes, where the large value of Δ results in many d-d transitions having similar energies to charge transfer bands, and being obscured. The theory such as it stands at the moment is discussed by Gerloch and Slade, and Jφrgensen (1971).

Calculation of spectra

In view of the information on chemical bonding that d-d spectra may yield, a considerable amount of effort has been spent in calculation of energy levels. The first approach assumed that the splitting of the d levels arose solely from the electrostatic potential of negative ions surrounding the metal ion. Although this purely electrostatic "crystal field" theory (see Chap. 7) worked quite well for the highly ionic species to which it was initially applied, it became clear as more experimental results appeared that it had several shortcomings. However, the qualitatively more accurate molecular orbital theory was much harder to use quantitatively: quite acceptable atomic wavefunctions for the d-orbitals are available for crystal field theory, but molecular orbitals are much harder to calculate. At the moment the most acceptable approach would seem to be the *angular overlap* model of Jφrgensen and Schäffer (1965) which enables calculations to be made on a M.O. basis. For further details the reader should consult Gerloch and Slade.

The problem of calculating the patterm of energy levels is simplified by the fact that the splitting depends mainly on the symmetry, and thus the pattern is not too much affected by the physical cause we choose to give the splitting. Diagrams similar to Fig. 4.1 are given in many introductions to transition metal chemistry (e.g. Kettle (1969), Huheey), for all numbers of d electrons d^1-d^{10}. These diagrams either assume fixed values of the Racah parameters B and C (an Orgel diagram) or assume a fixed ratio B/C and plot E/B against Δ/B (a Tanabe-Sugano diagram).

Information from d–d spectra

Most of the information gained from spectra comes from the variation of the ligand field splitting Δ, and the Racah parameter B, and we now discuss their variation.

For a given metal, the order of ligands producing increasing Δ is almost invariably the same; for a given ligand the order of metal ions giving increasing Δ is also pretty well constant. These two series are referred to as the *spectro-chemical series* for ligands and central ions. Jφrgensen has pointed out that the ligand field splitting may be factorised into a ligand and a metal component: $\Delta = f{\cdot}g$ where f is a ligand and g a metal parameter; f is arbitrarily put to 1.00 for water. A summary of values is given in Table 4.1. Although the factorisation is only approximate it does enable us to discuss ligand and metal effects separately.

Table 4.1. Factorisation of the spectrochemical and nephelauxetic series (after Jørgensen)

Ligands			Metals		
	f	*h*		*g*	*k*
I^-		2.7	Mn(II)	8.0	0.07
Br^-	0.72	2.3	Ni(II)	8.7	0.12
$\underline{S}CN^-$	0.75		Co(II)	9	
Cl^-	0.78	2.0	V(II)	12.0	0.1
$NN\underline{N}^-$	0.83	2.4	Fe(III)	14.0	0.24
$(C_2H_5O)_2P\underline{S}_2^-$	0.86	2.8			
F^-	0.9	0.8	Cu(III)	15.7	
$(CH_3)_2SO$	0.91		Cr(III)	17.4	0.20
$(CH_3)_2CO$	0.92		Co(III)	18.2	0.33
C_2H_5OH	0.97		Ru(II)	20	
$C_2O_4^{--}$	0.99	1.5	Ag(III)	20	
H_2O	1.0[a]	1.0[a]	Ni(IV)	22	0.8
$SC\underline{N}^-$	1.02		Mn(IV)	24	0.5
CH_3CN	1.12		Mo(III)	24.6	0.15
C_5H_5N	1.23		Rh(III)	27.0	0.28
NH_3	1.25	1.4	Pd(IV)	29	0.7
en	1.28	1.5	Tc(IV)	31	0.3
$\underline{S}O_3^{--}$	1.3		Ir(III)	32	0.28
phen	1.34		Pt(IV)	36	0.6
$\underline{N}O_2^-$	1.4				
$\underline{C}N^-$	1.7	2.1			

[a] by definition

phen – 1,10 phenanthroline; en – ethylenediamine

$$\Delta_{oct} = f(\text{ligand}) \cdot g(\text{metal})\ \text{kK}$$
$$(1 - \beta) = h(\text{ligand}) \cdot k(\text{metal}) \quad \beta = \frac{B_{obs}}{B_{gaseous}}$$

(The ligating atom is underlined for possibly ambiguous cases).

For ligands we may note that those species such as CN^- and NO_2^- which have low-lying π antibonding orbitals which may accept electrons from the t_{2g} subshell, and which will thus increase Δ, have high f values. Those ligands which may act as π donors (halogens) have rather low f values. It was initially thought that the value of f was determined only by the atom of the ligand actually bound to the metal; however sulphur can vary from quite low (SCN^-, $(C_2H_5O)_2PS_2^-$), to quite high (CH_3–S CH_2–CH_2–S–CH_3), so next-nearest atom effects can be important. The interpretation of f values is complicated by the fact that a low value may be due to strong π donation by the ligand, or simply weak overlap, while high values may indicate strong σ donation (probably the case for NH_3) or a mixture of σ and π effects (for example NO_2^-).

The metal series is rather more interesting. As the oxidation state (and formal charge) of the metal atom increases, g rises considerably: Co(II) 9, Co(III) 18.2; Ni(II) 8.7, Ni(IV) 22. This is not easy to explain on a pure ionic model as increasing charge would be expected to shrink the d orbitals and thus decrease their interaction with the ligands. A M.O. approach would predict a greater donation of charge to the metal as the oxidation state increases as well as an increased overlap, leading to stronger bonding effects, and consequently greater Δ (see page 80). It is found experimentally that pressure increases Δ, and that metal-ligand bond lengths decrease with increasing oxidation state of the metal.

It should be noted also that g for 2nd and 3rd row transition elements is much greater than for their lst row analogues:

Co(III) 18.2, Rh(III) 27.0, Ir(III) 32; Cr(III) 17.4, Mo(III) 24.6.

Δ values are always much greater for 2nd and 3rd row complexes which are almost exclusively low spin. This can again be explained by the greater overlap of the 4d and 5d orbitals with ligands, as a result of the more diffuse character of orbitals with one and two nodes.

The *nephelauxetic effect* is generally characterised by a number $\beta = B/B_o$ where B is the Racah parameter obtained from the spectrum of the complex and B_o is the Racah parameter of the free gaseous ion. Jørgensen has shown that this also may be factorised in the form $(1-\beta) = h\,(\text{ligand}) \cdot k\,(\text{metal})$ (see Gerloch and Slade).

For ligands the series is very different from the spectrochemical series:

$$F^- < H_2O < NH_3 < en < ox^{2-} < NCS^- < Cl^- < CN^- < Br^- < I^- < (C_2H_5O)_2PS_2^-$$

This series corresponds to decreasing electronegativity (see Chap. 5) and increasing reducing power (the capacity to lose electrons). It also corresponds to an increasing overlap with the metal, implying that at least part of the effect arises from the mixing of ligand and metal orbitals.

For metals, the series is less well defined, and is quite closely related to the spectrochemical series; however, there is no sharp increase in k values on moving from 3d to 4d and 5d transition metals. There is in fact a slight decrease from Co(III) to Ir(III). The nephelauxetic effect increases with the oxidation state of the central ion.

There are two effects commonly used to explain the nephelauxetic effect, central field effects, and symmetry restricted covalency. Central field effects are based on the fact that the transfer of negative charge from the ligands to the metal, and the negative potential of the ligands acts to reduce the effective charge on the metal atom. The lower the effective charge, the less strongly the d electrons are drawn together, and thus the weaker the repulsion between them. Symmetry restricted covalency reflects the the formation of M.O.s by the e_g and t_{2g} d-orbitals – this involves a spreading out of the one electron orbital over the ligands with a consequent decrease in the inter-electron repulsion. Since the e_g and t_{2g} will form different M.O.s, one would expect them to show different nephelauxetic effects as a result of symmetry restricted covalency – this is shown by the three different electron repulsion parameters discussed before (page 135).

The Jahn-Teller effect

Hitherto we have ignored any distortion from a high symmetry such as octahedral or tetrahedral; however Jahn and Teller proved a theorem in 1937 which requires that "any non-linear system where there is orbital degeneracy has a vibrational mode which will lower the symmetry of the system and also lower the energy of the system". Thus any octahedral complex which has orbital degeneracy – that is, one in which the ground state is not of A or B type symmetry – will be expected to distort to lower its symmetry and give a more stable system. As an example consider a complex with a full t_{2g} subshell and one electron in the e_g shell. This has the groundstate ^2E, which is degenerate, and may be regarded as involving either occupation of the d_{z^2} or the $d_{x^2-y^2}$ orbital. There exists a vibration which involves the simultaneous closer approach of the two ligands along the z axis, and the moving outwards of the 4 ligands along the x + y axes. This will increase the overlap of the z axis ligands, and decrease that of the x and y axis ligands. The d_{z^2} orbitals thus becomes more antibonding and the $d_{x^2-y^2}$ less so. This corresponds to a lifting of the degeneracy of the ^2E groundstate (Fig. 4.2). Clearly, if the single electron of the species is in the $d_{x^2-y^2}$ orbital, the energy of the complex will be lowered and the distortion favoured in accordance with the Jahn-Teller theorem.

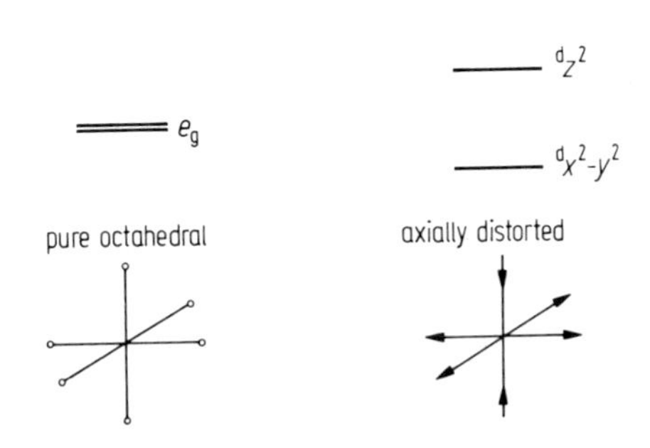

Fig. 4.2. The splitting of the e_g orbitals under distortion

The Jahn-Teller theorem is general, and also qualitative, not predicting the magnitude of the distortion. For a large distortion the orbital energy should be sensitive to changes in the metal ligand bond length – i.e. it should be strongly bonding or antibonding. Thus distortions due to the Jahn-Teller effect are normally greater when the degeneracy is in the occupation of the e_g* levels than the t_{2g} levels of an octahedral complex. If the t_{2g} levels are involved in bonding, for example in $V(CO)_6(3d^5)$, groundstate $^2T_{2g}$, the structure is distorted, the t_{2g} levels being bonding as a result of π donation to the carbonyl groups. In complexes of the lanthanides, the 4f orbitals are almost completely non-bonding, and show orbital degeneracy, but no Jahn-Teller effect.

If the Jahn-Teller distortion energy is sufficiently great, the distortion shows up in the metal-ligand bond lengths. This is particularly common in the case of Cr^{2+} (high spin $3d^4$) and Cu^{2+} ($3d^9$) compounds where there is a degeneracy arising from one 'hole' in the e_g* levels. CrF_2 and CuF_2 both show approximate octahedral symmetry but the bond lengths are not equal:

CuF_2	4F at 1.93 Å	2 at 2.27 Å
CrF_2	4F at 2.00 Å	2 at 2.43 Å

The 4 short and 2 long bonds distortion is the most common, but there are other modes (K_2CuF_4 shows 2 short and 4 long bonds). The effect is also seen in d-d spectra.

$[Ti(H_2O)_6]^{3+}$, $3d^1$, has only one d-d absorption in pure octahedral symmetry, corresponding to $t_{2g} \rightarrow e_g$. However, in the excited state, the Jahn-Teller effect splits the two e_g levels, causing one conformation (e.g. 4 long + 2 short, or 4 short + 2 long bonds) to be favoured. Depending on the conformation state during the zero point vibrations[1], one sees a transition to both the e_g levels. Since they have different energies, the transition appears as two peaks rather than one. In $[Ti(H_2O)_6]^{3+}$ the bands are at 17.4 and 20.3 kK showing that the magnitude of the Jahn-Teller effect is much less than the value of Δ.

Cases where the distortion may be measured (e.g. by X-ray crystallography), are said to show the static Jahn-Teller effect. Where the stabilisation distortion is similar to the vibrational zero-point energy, no permanent distortion is seen, but there is a distortion of the complex which changes direction rapidly and gives a time average of zero. This is the dynamic Jahn-Teller effect; its magnitude is small, and it is difficult to detect.

The Jahn-Teller theorem is not specific to metal complexes, and there is evidence for a Jahn-Teller distortion of the CH_4^+ ion (groundstate $^2T_2$) from photoelectron spectroscopy; however, orbital degeneracy is comparatively rare in compounds not containing partially filled d or f shells. The *second-order Jahn-Teller effect* is the mixing of orbitals resulting from the lower symmetry of a species after distortion. We showed earlier (page 66) that for a linear triatomic molecule the central atom s and p orbitals transform respectively as σ_g and π_u, but on bending both have a_1 symmetry, and consequently can mix together – this mixing by distortion is an example of the second-order Jahn-Teller effect – it is usually invoked to explain large distortions and the mixing of levels which were initially of different symmetry.

b) Charge Transfer Spectra

Apart from d-d transitions, most complexes show intense absorption bands resulting from excitation of electrons from orbitals localised mainly on the metal to an orbital localised mainly on the ligands, *or* vice versa. In the case of WCl_6 where the metal t_{2g} orbitals are unoccupied (Fig. 3.8 a), transitions from ligand orbitals such as the t_{1u} π bonding orbitals are allowed by the dipole moment mechanism, and give rise to intense absorptions (for WCl_6 at 22.4, 26.4 and 29.9 k). In view of the localisation of the orbitals concerned, these bands are referred to as *charge transfer* bands. In general, at least one of the possible charge transfer bands is allowed by the dipole moment selection rules, and consequently the absorptions are very strong, obscuring any d-d transitions nearby.

Their position naturally depends on the relative energies of the metal and ligand orbitals. If we consider ligand to metal transfer (as for WCl_6), the higher the oxidation state of the metal, the lower the energies of the d orbitals. Thus for $[IrCl_6]^{3-}$ which has the M.O. diagram of Fig. 3.8 a with 6 electrons in the metal t_{2g} shell, there is a transition t_{1u} $(\pi) \rightarrow e_g$ * at an energy of 48.5 kK; the complex $[IrCl_6]^{2-}$ with the con-

[1] By the Franck-Condon principle the instantaneous position of the ligands does not change during the electronic transition. Thus the complex may be distorted at one particular point during the vibrations, causing a splitting of the e_g* levels.

figuration t_{2g}^5 has an equivalent t_{1u} $(\pi) \rightarrow e_g^*$ transition at 43.1kK; the fall in energy resulting from the lower energy of the e_g levels in a system with fewer d-electrons, and thus less electron repulsion. If we consider the bromide complex $[IrBr_6]^{3-}$, the equivalent absorption is seen at 36.8 kK indicating the t_{1u} levels of bromine to be less strongly bound than those in chlorine. These transitions have molar extinction coefficients of about 2 x 10^4 compared with about 50 for the d-d transitions in $[IrCl_6]^{3-}$; the d-d transitions in $IrCl_6^{2-}$ are masked by the t_{1u} $(\pi) \rightarrow t_{2g}$ transfer band.

Jørgensen (1971) has shown that there is a fair amount of additivity in the positions of these charge transfer bands, and the frequency can be expressed for X→M (ligand to metal) charge transfer as:

$$\nu_{corr} = [\ \chi_{opt}(X) - \chi_{opt}(M)\ ] \cdot 30\,kK \qquad (4.1)$$

where ν_{corr} is the frequency corrected for electron repulsion effects, and χ_{opt} is the *optical electronegativity.* The correction of ν is necessary to avoid anomalies arising from electron repulsion effects. For a high spin d^5 ion, for example, electron transfer

Table 4.2. Optical electronegativities

(a) Ligands

F^-	3.9	Cl^-	3.0
H_2O	3.5	Br^-	2.8
NH_3	3.3	N_3^-	2.8
SO_4^{--}	3.2	$\underline{S}CN^-$	2.8
CH_3OH	3.1	$acac^-$ [a]	2.7
NCO^-	3.0	I^-	2.5

[a] $acac^-$= acetylacetonate.

(b) Metals	K	γ	δ
		$\chi_{opt}(M(K)) = \gamma + q\delta$	
$3d^q$	IV	2.05	0.22
$4d^q$	III	1.1	0.2
	IV	1.6	0.2
$5d^q$	III	1.0	0.2
	IV	1.4	0.2
	V	1.8	0.24
	VI	2.0	0.3
$4f^q$	III	0.4	0.12
	IV	2.05	0.2
$5f^q$	IV	1.0	0.25
	V	1.7	0.2
	VI	2.2	0.15

will involve spin pairing, while this is not necessary for d^4 ions – the d^5 species will thus absorb at a higher frequency even if the $t_{1u} - e_g{}^*$ gap is the same. If the frequency can be corrected, then $\chi_{opt}(M)$ shows a steady rise with the number of d electrons in the ion, and has the form for the d^q ion $\chi_{opt}(M(K)) = \gamma + q\delta$ where γ and δ are the same for a given oxidation state K in a given period, and increase with K. Examples of this and other χ_{opt} values are given in Table 4.2.

Ligand to metal electron transfer corresponds to simultaneous oxidation of the ligand and reduction of the metal, and the energies of the absorption bands and the values of χ_{opt} agree with this: oxidising metal ions such as Mn(VII) show electron transfer in the visible part of the spectrum (e.g. in $KMnO_4$); similarly, the readily oxidised iodide ion has a lower χ_{opt} than the fluoride ion.

The treatment of metal to ligand transfer (e.g. from the t_{2g} to empty t_{1u} ligand orbitals in Fig. 3.8 b) is essentially similar, but in this case the metal should be in a low oxidation state to lower the energy of the transfer – an example is the acetylacetonate $(CH_3{-}CO{-}CH{-}CO{-}CH_3)^-$ ion ($acac^-$) which shows metal-ligand (or inverted) transfer for $[Ti(acac)_3]$, but ligand metal transfer for the more oxidising $[Fe(acac)_3]$.

Finally we should note that electron transfer is *not* specifically associated with transition metal compounds: absorption spectra due to a transition from an orbital localised on one atom to one localised on an adjacent atom are found in all branches of chemistry. For example, some of the far ultra-violet absorptions of alkali halides have been ascribed to transfer of an electron from the halide ion to the metal ion. Similarly the yellow colour of $[PbCl_6]^{2-}$ comes from a charge transfer band in the ultra-violet.

c) Electronic Spectra of Lanthanides and Actinides

As before we may classify the transitions involved as electron transfer, or internal transitions, in this case $f \rightarrow f$ and $f \rightarrow d$ transitions. The electron transfer spectra are little different from those of the transition metals, and optical electronegativities may be derived exactly as for the transition metals (Jørgensen 1971). Our main interest is in the transitions involving electrons in the f orbitals.

Certain lanthanide and actinide ions, notably Ce(III), Sm(II), Eu(II), Tb(III), Yb(II), U(III) and Np(III) shown strong transitions which are quite sensitive to chemical environment. These are Laporte allowed $4f \rightarrow 5d$ (lanthanide) or $5f \rightarrow 6d$ (actinide) transitions. They are only of sufficiently low energy to be visible if the $nf - (n+1)d$ gap is quite small. This gap increases with increasing oxidation state and it will be noted that many of the ions quoted above are strongly reducing, such as Sm(II).

Most spectroscopic studies have been concerned with $f \rightarrow f$ transitions. These are very insensitive to chemical environment, changing little with change of ligand, very weak ($\epsilon_0 \sim 0.01 - 10$), and very narrow (there is no vibrational coupling with the ligands). Their weakness arises from the fact that they are Laporte forbidden and so little involved in bonding that the means of gaining intensity invoked for d-d transitions (vibronic coupling; overlap with ligands) are no longer important. The f-f transitions of the lanthanides thus represent the extreme 'weak field' case and are very little different from the spectra expected of gaseous ions with a spherical environment.

Electron repulsion gives rise to a series of Russell-Saunders terms with separations of the order of $10^4 cm^{-1}$. These terms are then split by spin-orbit coupling according to their J values. These splittings are smaller, of the order of 2000 cm^{-1}. Finally, any ligand field effects will split the separate J levels up according to their M_J values – however, as this effect is typically only 100 cm^{-1} or so, it is very difficult to detect in the spectra. Figure 4.3 shows the series of Russell-Saunders terms, and gives a rough idea of the magnitude of electron repulsion and spin orbit effects for an f^2 species, Pr^{3+}. Ligand field splittings are difficult to detect, but a nephelauxetic effect can be detected, although its magnitude is much smaller than for 3d species, typically 4 percent compared with 10–40 percent for the 3d ions. It presumably arises from the 'central field' effect.

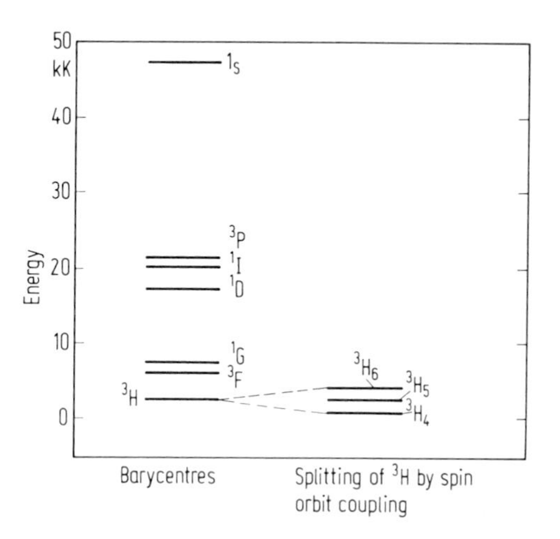

Fig. 4.3. The energy levels of a $4f^2$ ion (Pr^{3+})

These remarks apply equally to the actinide elements, but the 5f orbitals show rather more evidence of involvement in chemical bonding: the nephelauxetic effect is more pronounced, and the bands are broader and about 30 times more intense. A complication is that the spin-orbit coupling constant ζ_{5f} is about double that of the equivalent 4f ion, and the electron repulsion parameters decrease. The spin-orbit splittings are thus greater than or equal to the electron repulsion splittings and the Russell-Saunders scheme may no longer be used, even as an approximation.

The strict adherence to the selection rules of these electronic spectra enables one to observe unusual transitions; a magnetic dipole transition has been observed in lanthanide complexes, while an electric quadrupole transition was recently reported for $Cs_2UO_2Cl_4$. Luminescence is also common for lanthanide complexes – referring to Fig. 4.3, an excitation to the ^3P level of Pr^{3+} could decay directly to the ^3H ground-state, or via a long lived intermediate state such as ^1D). Because of the spin multiplicity difference, some of these excited states can be quite long-lived. This is finding use in laser construction, since glasses containing rare earth ions (particulary neodymium) can be 'optically pumped' easily. Rare earth fluorescence is also for some of the cathodoluminescent phosphors used for colour television.

d) Electronic Spectra of Compounds of Main Group Elements

The electronic spectra of compounds of main group elements have been little studied by comparison with those of transition and lanthanide and actinide elements. It is unusual to find electrons in weakly bonding, antibonding, or non-bonding orbitals, and absorptions in the visible part of the spectrum are much less common than for the compounds discussed before. We may discuss the spectra observed under following titles:

1) Molecular spectra
2) Charge transfer spectra
3) Rydberg spectra

Molecular spectra involve transitions between molecular orbitals in covalently bonded molecules such as the triatomics discussed in the first section of Chap. 3. The interpretation of each spectrum depends on the structure and composition of each particular molecule, and it is not possible to generalise as, for example, for d-d transitions or to concentrate on a single central atom. The spectra of covalently bonded solids may often be best discussed in terms of a band structure. Examples of compounds showing such spectra are the halogen vapours, and NO_2 (brown) (all gaseous), yellow $Se_4{}^{2+}$, red $Te_4{}^{2+}$, and orange S_4N_4.

Charge transfer spectra are quite common, and the case of $[PbCl_6]^{2-}$ and its salts has already been mentioned. The fact that the orbitals involved in the transitions are localised implies that charge transfer spectra are observed in compounds with a fair degree of ionic character.

It has been stated that non-bonding orbitals are rarer in compounds of the main group elements; however, in many compounds of the heavy post transition-metal elements (e.g. the series Tl–Pb–Bi) the outermost s-orbital is only weakly involved in bonding (in the series above, Tl^+, Pb^{2+} and Bi^{3+} all have the configuration $[Xe]4f^{14}5d^{10}6s^2$). If the s-orbital is occupied, then transitions to the empty p-orbital above it are allowed, and directly analogous to the allowed $s \rightarrow p$ Rydberg transitions of atomic spectra. Such transitions are indeed seen in compounds of Tl(I), Pb(II), Bi(III), and also In(I), Sn(II), Sb(III). They are, however, very different from the transitions observed in the corresponding gaseous ions, occuring at much lower energies, and being quite sensitive to the local environment of the atom (see Jørgensen (1962) p.185). It seems that the absorption bands of ionic crystals may be due either to charge transfer between oppositely charged ions ($M^+X^- \rightarrow MX$) or to Rydberg excitation of one of the constituent ions (e.g. $Cl^-(3p^6) \rightarrow Cl^-(3p^5 4s^1)$). These Rydberg like transitions are analogous to the $4f \rightarrow 5d$ transitions shown by some of the lanthanides.

e) Optical Activity

Many molecules have a symmetry so low that they are not superimposable on their mirror image; the formal symmetry requirement for this is the absence of an improper axis of rotation – a frequently encountered (but rather less general) alternative requirement is the absence of a centre or plane of symmetry. An example of a mirror image pair which may not be superimposed is shown in Fig. 4.4. The two different forms of $[Co(en)_3]^{3+}$ are said to be *enantiomers*. A molecule which has two enantiomers is said to be *chiral* or dissymmetric.

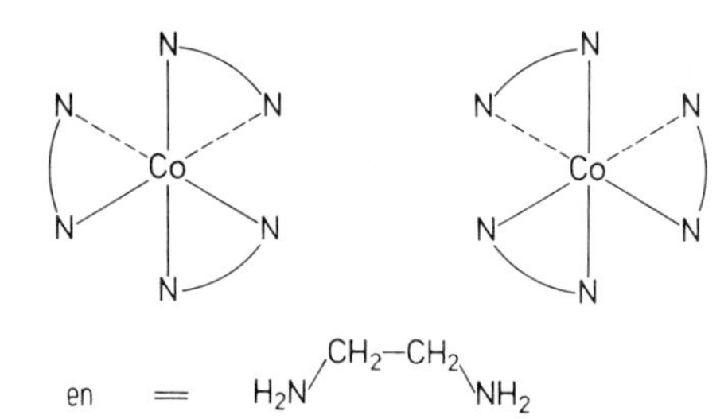

Fig. 4.4.
The two enantiomers of $[(Co(en)_3]^{3+}$

The interest of chiral molecules derives from the fact that separate samples of the two enantiomers will rotate the plane of polarisation of a beam of plane polarised light passing through them in opposite directions. This behaviour constitutes optical activity, and systems where there are unequal numbers of the two enantiomers of a chiral molecule will show optical activity. The fact that species such as $[Co(en)_3]^{3+}$ could show optical activity was put forward as a proof of their octahedral coordination before X-ray crystallography was developed. Similarly, tetrahedral and square coordination could be distinguished if one or the other would be expected to show optical activity.

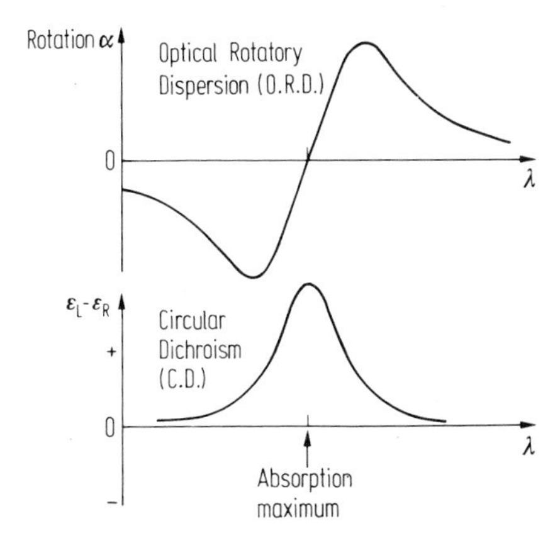

Fig. 4.5.
O.R.D. and C.D. curves for an absorption band

The main interest of optical activity is in the spectra of transition metals as measured by the *Cotton* and *Faraday* effects. The rotation produced by a given enantiomer is not constant with wavelength, and, in particular changes rapidly with wavelength near an absorption, passing through zero at the absorption maximum. The plot of the rotation α against wavelength λ shown in the upper part of Fig. 4.5 is known as an *optical rotatory dispersion* (O.R.D.) curve. As an absorption is approached, the absorption of left hand circularly polarised light is not the same as that of right hand circularly polarised light. If we denote the two extinction coefficients by ϵ_L and ϵ_R then the plot of $\epsilon_L - \epsilon_R$ shows a maximum at the absorption maximum (Fig. 4.5, lower half) and is known as a *circular dichroism* (C.D.) curve. The phenomena of change in rotation with wavelength (O.R.D.) and differential absorption (C.D.) are related and are generally described as the Cotton effect. In the two figures we have drawn a positive Cotton effect – the equivalent diagrams for a negative effect can be produced by a rotation of 180° about the wavelength axis.

The Cotton effect is becoming increasingly studied as interest grows in rates of racemisation (the interconversion of two enantiomers, and the equalisation of their abundance) and the difference in reactivities. The frequent occurence of chiral molecules in biological systems has led to study of the Cotton effect by bio-inorganic chemists. C.D. data may be used to determine the configuration of ligands about the central atom – for further details see the review by Gillard.

The Faraday effect, discovered in 1845, has only recently found wide application. When any compound (not necessarily a chiral species) is placed in a magnetic field with a component parallel to a beam of light passing through the sample, circular dichroism is observed. A simple physical explanation may be given. Any electric dipole allowed transition involves a change in angular momentum ΔJ of 1 unit. The magnetic field splits the levels involved in an electronic transition according to their M_J values (the Zeeman effect). The magnetic field gives rise to a slight energy difference between the transitions corresponding to $\Delta M_J = +1$ or -1. However, the transition $\Delta M_J = +1$ corresponds to an absorption of a quantum of right-hand circularly polarised (r.c.p.) light, and $\Delta M_J = -1$ to absorption of a quantum of left hand circularly polarised (l.c.p.) light. Because these two transitions are now separated (by the magnetic field) ϵ_L and ϵ_R have different maxima, and their difference gives a *magnetic circular dichroism* (M.C.D.) curve.

M.C.D. curves are proving extremely useful in assigning electronic transitions, often a difficult task. McCaffery and Schatz give an introductory review.

f) Polarisation of Absorption Spectra

In octahedral symmetry, the electric dipole moment operator which produces the transition between two electronic states has t_{1u} symmetry. If the symmetry of the environment is lowered to D_{4h} (for example, by a Jahn-Teller distortion) the electric dipole moment operator now transforms as the I.R.s $b_{2u} + e_u$; the b_{2u} component corresponds to the component of the incident radiation polarised parallel to the z axis, and the e_u component to that polarised in the x-y plane. The transition moments for the two different polarisations will clearly be different (since the dipole moment operators are different) and consequently one might expect a dependence of the absorption on the polarisation of the incident light. This is indeed found to be the case: in species of low symmetry, the absorption bands are polarised. As the polarisation is measured relative to the molecular axis, it is necessary for the molecular axis to be held in a fixed orientation as in a crystal. Thus the study of the polarisation of absorption bands is confined to the spectra of crystals. Sutton discusses the polarisation of the absorption bands of $[Ni(NH_3)_4(NO_2)_2]$ with D_{4h} symmetry, and Jaffe and Orchin give a more general treatment.

Strong polarisation effects are particularly common in minerals where transition metals are frequently found in low symmetry environments. Crystals of tourmaline containing iron absorb light polarised perpendicular to the z axis so strongly that the transmitted light is almost plane polarised parallel to the z axis. The variation in absorption of polarised light can be very useful in the assignment of spectra.

g) Mixed Valence Spectra

There exist many chemical compounds containing atoms of the same element in different oxidation states; a good example is Prussian Blue $KFe[Fe(CN)_6]$ containing Fe^{3+} ions, and $[Fe(CN)_6]^{4-}$ complex ions containing iron in the +2 oxidation state. Many of these *mixed valence* complexes show very strong optical absorptions – for example $[Fe(CN)_6]^{4-}$ is pale yellow in aqueous solution, and $[Fe(OH_2)_6]^{3+}$ is almost colourless, yet $KFe[Fe(CN)_6]$ has an intense blue colour.

A detailed discussion of mixed valence compounds and their spectra is given by Robin and Day, and the subject is discussed only briefly here. In a similar way to the charge transfer complexes discussed in Chap. 3, . a small amount of a charge transfer state is mixed into the ground state wavefunction of the ligand. Thus, if the two different environments of the same element are A and B, containing the element in oxidation state 2 and 3 respectively, then the wavefunction of the ground state is not $|\psi(A(II))\ \psi(B(III))\ \psi(\text{other atoms})|$ as might be expected,[2] but

$$\left\{\cos\alpha\ [\psi(A(II))\ \psi(B(III))] + \sin\alpha\ [\psi(A(III))\ \psi(B(II))]\right\}\ \psi\ (\text{other atoms})$$

where α is a measure of the admixture of the charge transfer state into the ground state function.

We may now ascribe the absorption shown by such compounds to a transtion to the corresponding excited state:

$$\left\{\cos\alpha\ [\psi(A(III))\ \psi(B(II))] + \sin\alpha\ [\psi(A(II))\ \psi(B(III))]\right\}\ \psi\ (\text{other atoms}).$$

Thus Prussian Blue has a wavefunction corresponding to $KFe(III)\ [Fe(II)(CN)_6]$ with a small amount of $KFe(II)\ [Fe(III)(CN)_6]$ mixed in – a consequence of this is that neither iron atom has an *exactly* integral oxidation state – this is not very serious as we are more interested in the mixed valence compound as a whole than in the individual atoms in it.

Robin and Day classified mixed valence compounds by the value of α, the mixing coefficient. α depends on (i) the energy difference between the two states (A(II)B(III) and A(III)B(II)) – clearly if the state A(III)B(II) has a much higher energy, there will be little advantage gained in mixing with the ground state, and (ii) the extent of the interaction between A and B sites: if the A and B sites are adjacent and overlap strongly, or if they are linked by one intermediate ligand forming a fairly covalent bond with each site, the delocalisation and mixing will be much more favourable than if A and B are ionic sites separated by some distance in a crystal lattice. Thus α is larger for sulphides than oxides.

Class I mixed valence compounds contain two very different environments for A and B. $\alpha \approx 0$ and there is no interaction. These solids are insulators and show no charge transfer band below 27 kK (although they may show spectra typical of their con-

---

[2] We have assumed for simplicity that site A and site B do not overlap or mix with other atoms in the molecule – as for example in a pure ionic compound.

stituent species). An example is Co_3O_4, a spinel (see p.117) with Co^{2+} high spin ions in a tetrahedral environment and Co^{3+} low spin ions in the octahedral sites.

Class III mixed valence compounds have large α values and extensive delocalisation. They are subdivided into classes IIIA and IIIB. In class IIIA delocalisation is found within a discrete molecular unit (e.g. I_3^-, Bi_5^{3+}). They show mixed valence absorption and it is not possible to identify constituent ions (i.e. I_3^- cannot be described as $I^+ + 2I^-$). In Class IIIB the delocalisation is spread throughout the whole crystal lattice, and a band structure description is the most suitable. Class IIIB compounds show metallic conductivity and may show strong magnetic exchange (see later in this chapter). A good example is the non-stoichiometric nickel oxide discussed in Chap. 3 – here the Ni^{2+} and Ni^{3+} sites are identical.

Class II compounds fall in between Classes I and III. α is small but non-zero. They show mixed valence spectra in the visible part of the spectrum (14–27 kK), but, if they are not obscured, it is possible to see the spectra of the constituent species, or to detect them by other methods. The two sites are not too dissimilar, and there is often a bridging ligand (as in $[Fe^{II}Fe^{III}(CN)_{11}]^{6-}$ with a single cyanide bridging ligand). They may show magnetic exchange at low temperatures, and are semiconductors. Prussian Blue is a Class II mixed valence compound.

B. Magnetic Properties

Most molecules and ions have no resultant electronic angular momentum in their groundstate. The most important exceptions to this statement are the compounds of d and f block elements where a d or f shell is partially occupied. In Chap. 1, it was mentioned that electronic angular momentum produces a magnetic moment, and the study of the magnetic properties may thus provide information about the electronic structure.

The magnetic moment will arise from the contributions of the orbital angular momentum and the spin angular momentum. If the groundstate has orbital angular momentum l_z and spin angular momentum s_z, both along the z axis, then the magnetic moment in that direction is

$$\mu_z = l_z + g \cdot s_z \text{ Bohr magnetons.} \quad (4.2)$$

The factor g arises from a relativistic effect, and is very close to 2.

If we wish, for example, to predict the magnetic properties of a transition metal complex such as $[Fe(NH_3)_6]^{2+}$, Fig. 3.8 a, we must first determine the groundstate, taking into account electron repulsion, ligand field, and spin-orbit coupling effects. If there is no resultant angular momentum, the complex will be diamagnetic, and the only effect of an applied magnetic field will be to induce a small opposing magnetic field in the sample, giving a small, negative magnetic susceptibility; this is *diamagnetism*. If there is a resultant angular momentum, the groundstate will be degenerate, the number of states corresponding to the number of possible orientations of the angular momentum with respect to an applied magnetic field. On applying a magnetic field, those orientations with rather than against the field will have lower energy, and the corresponding states will have a greater occupation, giving a net magnetisation of the sample, and a positive susceptibility. This is *paramagnetism*.

Two further points should be borne in mind when considering paramagnetism:

(i) The difference in energy between states aligned with or against the magnetic field is small (for most laboratory magnetic fields, less than 1 cm^{-1}), and there is thus a thermal occupation of several of the states of higher energy. This may be treated by the Maxwell Boltzmann distribution.

(ii) We have assumed that the magnetic moment associated with each metal atom does not interact with the others. If this is not the case the substance is said to show *magnetic exchange* (see later).

A full treatment of paramagnetism therefore involves a good deal of calculation, and goes beyond the scope of this book, the treatment given here being mainly qualitative. As is frequently the case in inorganic chemistry, we may however simplify the treatment in certain cases to give simple relationships which have proved useful sources of chemical information.

Magnetic moments and magnetic susceptibility. The molar magnetic susceptibility χ_m is related to the magnetic moment μ by:

$$\chi_m = \frac{N\mu^2}{3kT} \tag{4.3}$$

where N = Avogadro's number, k = Boltzmann's constant, T = the absolute temperature, and μ is measured in Bohr magnetons (0.927 10^{-20} $erg.gauss^{-1}$). As we are normally interested in the paramagnetic susceptibility, it is necessary to correct χ_m for the diamagnetism of closed electronic shells of the system (e.g. in diamagnetic ligands). The measurement of magnetic susceptibilities is discussed in the references at the end of the chapter.

a) Orbital Angular Momentum

In Chap. 1 it was stated that the orbital magnetic moment in a given direction **z** was proportional to the orbital angular momentum in that direction, given by the quantum number *m*. In non-spherical symmetry however, the quantum number *m* is not always well-defined. We may see this by writing out the d orbitals as in Eq. (1.16):

$$e_g \begin{cases} |2,0\rangle & & z^2 \\ 1/\sqrt{2}\ (|2,+2\rangle & + |2,-2\rangle) & x^2-y^2 \end{cases}$$

$$t_{2g} \begin{cases} -i/\sqrt{2}\ (|2,+2\rangle & - |2,-2\rangle) & xy \\ 1/\sqrt{2}\ (|2,+1\rangle & + |2,-1\rangle) & xz \\ -i/\sqrt{2}\ (|2,+1\rangle & - |2,-1\rangle) & yz \end{cases} \tag{4.4}$$

For each of these orbitals the value of *m* given by the operator $\mathcal{L}_z$ is zero. However, if xz and yz are degenerate, they may mix in a magnetic field to give the eigenfunctions $|2,+1\rangle$ and $|2,-1\rangle$ which will have magnetic moments. In a octahedral complex the d_{xz} and d_{yz} orbitals will be degenerate, so we may expect an orbital magnetic moment if the $|2,+1\rangle$ eigenfunction is populated to a greater extent than the $|2,-1\rangle$ eigenfunction (or vice-versa); this will occur in a magnetic field if the t_{2g} orbitals are only partially occupied.

In octahedral complexes the e_g orbitals lie at higher energy than the t_{2g} and the magnetic field is not usually strong enough to mix the d_{xy} and $d_{x^2-y^2}$ orbitals, so the magnetic moment due to the $m = \pm 2$ levels is zero, and is said to be *quenched* by the ligand field. If the complex has symmetry lower than octahedral, the d_{yz} and d_{xz} orbitals may no longer be degenerate, and will no longer be mixed by the magnetic field. The orbital contribution to the magnetic moment is then completely quenched by the low symmetry ligand field.

In principle, the Jahn-Teller theorem predicts a distortion of any system with an orbital contribution to its magnetic moment, but the magnitude of the distortion is generally too small to make the orbital contribution vanish. Low symmetry in a complex might be expected to result in anisotropy in the magnetic properties, and this is indeed the case. If we imagine an octahedral complex tetragonally distorted along the z axis, with the t_{2g} subshell incompletely filled (e.g. $t_{2g}{}^5$), the d_{yz} and d_{xz} orbitals will still be degenerate, and there will still be an orbital contribution to the magnetic moment, but this is fixed relative to the z axis defined by the distortion. The contribution to the susceptibility of this orbital magnetic moment will thus depend on the angle of the field to the z axis, and the susceptibility will be anisotropic. In general, magnetic anisotropy is found whenever the symmetry is lower than cubic; however, the effect is much greater when there is a considerable orbital contribution to the magnetic moment, and this contribution is fixed relative to the molecular axes.

It has been mentioned that the magnitude of the magnetic field is usually too small to mix levels which are separated by the ligand field splitting energy Δ, or by very large distortions. There is however a very small amount of mixing between levels as the field increases (for example mixing of d_{xy} and $d_{x^2-y^2}$) giving rise to a slight paramagnetism. The energy of the upper level is so high that it is not occupied thermally, and the magnitude of the effect is thus dependent only on the amount of mixing into the lower level. The effect is dependent only on the field strength, and is referred to as *Temperature Independent Paramagnetism* (T.I.P.) – it is generally a fairly small effect.

b) Spin Angular Momentum

In the previous section we saw that orbital angular momentum may be quenched by a ligand field, especially if the symmetry is low. A theorem due to Kramers shows that, in a system with an odd number of unpaired electrons, it is impossible to quench the spin angular momentum completely. It is therefore interesting to consider the effects of the spin angular momentum alone, and this gives rise to the 'spin-only' treatment.

We assume initially that the orbital contribution is either zero or is completely quenched by the ligand field, and that spin-orbit coupling effects may be ignored (see later). We may now use Hund's rule which tells us that the groundstate will be that with the maximum value of S. The effective magnetic moment $\mu_{eff.}$ is now given by

$$\mu_{eff.} = 2\sqrt{S(S+1)} = \sqrt{n(n+2)} \text{ Bohr magnetons} \quad (4.5)$$

where n is the number of unpaired electrons.

In the presence of a strong ligand field, spin-pairing to give a low-spin configuration may be favoured; this will produce a reduction in n, and a consequent decrease in $\mu_{eff.}$ and χ_m. The value of μ can thus be used as an indication of high or low spin configuration.

Let us consider an example: for Na_3FeFe_6 $\mu = 5.85$ B.M., and for $K_3Fe(CN)_6$ $\mu = 2.25$ B.M. Both contain the $3d^5$ Fe^{3+} ion, and so we may deduce that Na_3FeFe_6

is high spin and $K_3Fe(CN)_6$ is low spin. This is quite in accord with the relative ligand strengths of cyanide and fluoride ions as shown by the spectrochemical series. The high spin ion has one electron in each t_{2g} orbital, and therefore we would expect no orbital contribution. The low spin ion has the configuration $t_{2g}{}^5$ and one of the t_{2g} orbitals is only singly occupied: an orbital contribution is thus possible, and the value of μ is indeed higher than the 1.73 B.M. predicted by a spin-only calculation.

The spin-only formula is most useful for first row transition metals, and for these compounds the magnetic susceptibility is a useful means of distinguishing high and low spin compounds, as the change in μ for a high spin – low spin changeover is generally great enough not to be confused by the presence of a small orbital contribution.

c) Spin Orbit Coupling

In Chap. 1, the interaction of spin and orbital angular momenta, or spin-orbit coupling, was discussed. If this effect is at all important, then the separation of spin and orbital contributions is unlikely to work. For the first (3d) series of transition metals, spin-orbit coupling is much smaller than the ligand field and electron repulsion effects, and we may use, to a first approximation, the 'uncoupled' treatment above. This simplification does *not* however hold for the other transition series, and the lanthanide (4f) and actinide (5f) series, as the spin-orbit coupling constant ζ increases steadily with Z, the atomic number.

The magnetic properties are determined by the electronic ground state. In spherical symmetry this corresponds to the lowest Russell-Saunders term. As the spin-orbit coupling increases, the Russell-Saunders term is split into levels with different J values, where J the total angular momentum is the vector sum of L and S. In such a case the magnetic moment is given by:

$$\mu_{eff} = g\sqrt{J(J+1)}; \quad g = 1 + \frac{J(J+1) - L(L+1) + S(S+1)}{2J(J+1)} \tag{4.6}$$

The reader will see that if $L = 0$, then $J = S$, and the spin only term results.

For non-spherical symmetry (arising from a ligand field for example), it is necessary to use a different treatment as J values are no longer useful. There are two choices: 1) a group theoretical treatment using the double groups (spinors) mentioned in Chap. 1, where each spin-orbit coupled electronic state transforms as an I.R. of the double symmetry group. 2) either apply the spin-orbit coupling as a perturbation to our previously determined states *or* apply the ligand field as a perturbation to previously determined spin-orbit states (in spherical symmetry). Both methods are complicated, and we give here only the qualitative effects of approach 2 a.

(i) The spin-orbit coupling splits the degeneracy of the ground state if there are spin and orbital angular momenta present. Instead of one state there are now two or more states and spectroscopic transitions to this state will be split, or, since the separate lines are not usually resolved, will be broadened by the spin-orbit coupling.

(ii) The spin and orbital angular momenta can either be coupled parallel to each other, increasing the total μ, or anti-parallel, decreasing μ. If they are antiparallel in the lowest spin orbit coupled state, then μ will be reduced below the spin only value, in some cases to zero. If the temperature is raised, then the thermal population of the higher spin-orbit coupled states increases, until, in the limit, the total effect of the spin-orbit coupling is zero, since the excited states with spins and orbital angular momenta coupled parallel will be occupied to the same extent as the lower energy states with antiparallel coupling. We would therefore expect an increase in μ on raising the temperature as shown in Fig. 4.6. If the spin-orbit coupling is large then the value of $\frac{kT}{|\lambda|}$[3] is small and the value of μ changes with the temperature, while if $|\lambda|$ is small the 'plateau' region B is reached at quite low temperatures and the magnetic susceptibility changes little with temperature.

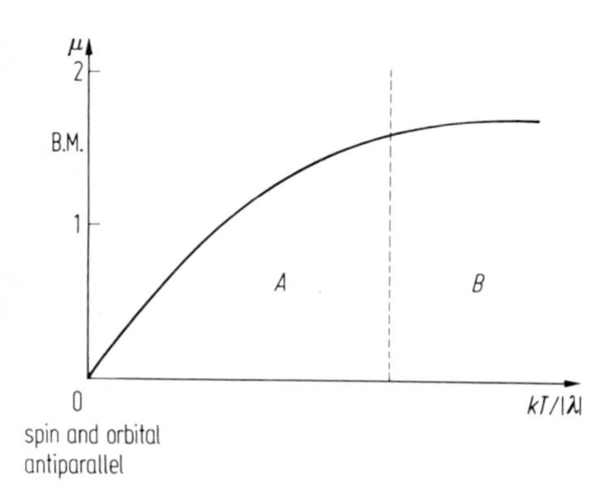

Fig. 4.6. The variation of μ with $\frac{kT}{|\lambda|}$ for a d^1 species. Such figures are usually referred to as Kotani diagrams

For the $3d^n$ transition metals, the spin-orbit coupling is small, and at all but the lowest temperatures the magnetic behaviour corresponds to region B. For second and third row transition metals, λ is much larger and the magnetic behaviour is better described by region A. Thus simple 'spin only' treatments of complexes of the heavy metals are very unreliable.

(iii) Spin-orbit coupling, by mixing the spin and orbital parts of a wavefunction increases the intensity of formally 'spin forbidden' electronic transitions. The two states involved in a transition do not have well-defined values of S, and consequently ΔS is not well defined.

(iv) Just as electron repulsion parameters are affected by the environment of the ion (the nephelauxetic effect) the spin-orbit coupling parameter is sensitive to chemical environment, and is reduced below the free ion value.

d) Calculation of Paramagnetic Properties

We have discussed the factors affecting the magnetic properties and may now summarise the method of calculation: first it is necessary to calculate the magnetic moments associated with each of the possible states of the system, second to calculate the thermal occupation of these states (using the Boltzmann distribution) and sum the product

---

[3] λ is the many electron spin-orbit coupling constant. For electrons in the same shell $\lambda = \zeta/2S$.

of the population and the magnetic moment of a given state to give the bulk property. The second step is simple classical physics, but the first step involves using ligand field theory. One proceeds by a series of perturbations, applied in order of decreasing magnitude. Thus for a weak field example we could start with pure atomic d functions, apply the electronic repulsion interaction to obtain the wavefunction of the lowest Russell-Saunders state. The next step is to apply the ligand field – there are two possibilities here: 1) we may mix in some ligand orbital character – this is normally only done for a strong field case; 2) we may simply multiply the wavefunction by an orbital reduction factor k, which is less than unity, to allow for any covalence – this is a simpler procedure, although its physical meaning is clearly rather dubious.

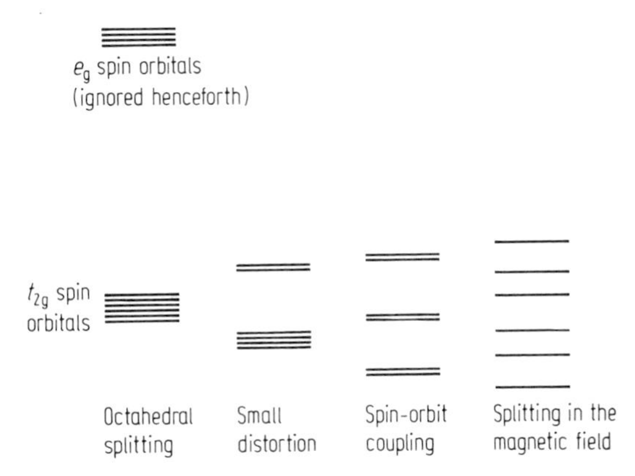

Fig. 4.7. The gradual lifting of the degeneracy of spin orbitals (not to scale). There will be a large number of states of the system arising from different occupations of these orbitals

We now have a wavefunction, or set of degenerate wavefunctions for the groundstate of the metal atom in the complex; we still have to apply the spin orbit coupling and magnetic field perturbations. It is necessary here to decide whether any excited states will be mixed into the groundstate by these two remaining perturbations. If the complex is pure octahedral, then the excited states involving the e_g orbitals will be at least $10^4 cm^{-1}$ above the groundstate for a first row transition metal; the spin orbit coupling will typically be 300 cm^{-1}, and the effect of the magnetic fields normally used even smaller ($\sim 1 cm^{-1}$); it is thus unlikely that they will mix these excited states into to groundstate, and they may be ignored. If there is a slight distortion (perhaps as a result of the Jahn-Teller effect) there may be low lying excited levels (say at a few hundred cm^{-1}) which may be mixed in, or even partially occupied thermally, and these may not be neglected (kT at 300 °K $\sim$200 cm^{-1}).

The spin orbit coupling perturbation must then be applied to first and second order, considering first the lifting of degeneracy of the groundstate and then any mixing in of excited states. Finally, we should consider the effects of the magnetic field – this nomally need only be done to first order, splitting the groundstate into terms with different alignments to the magnetic field. If the second order term is considered, the mixing in of states which gives rise to temperature independent paramagnetism appears.

The way in which the individual spin orbitals are split by the various effects described above is shown in Fig. 4.7. Note that for species with more than one d electron we must consider electron repulsion, and thus discuss terms or states rather than the energies of individual orbitals.

Relative magnitudes of effects

The relative magnitudes of the effect discussed are not the same for all d^n and f^n configurations. In Table 4.3 we give examples of how the magnitudes of these effects vary. Thus for lanthanides we see that electron repulsion > spin orbit coupling > ligand field effects, the ligand field effects being so small as to be almost swamped at room temperature (kT = 200 cm^{-1}) – we may ignore ligand field effects, and use Eq. (4.6) relating $\mu_{eff.}$ to J.

For 4d and 5d elements, the ligand field effects are stronger than electron repulsion effects, suggesting a strong field approach to be suitable.

Table 4.3. Relative magnitudes of effects

Metal	Electron repulsion	Ligand field	Spin-orbit coupling	Spin-pairing energy parameter D
$3d^n$	12000	15000	400	6000
$4d^n$	8000	22000	1000	4000
$5d^n$	5000	25000	2500	3000
$4f^n$	15000	50	1500	6000
$5f^n$	10000	100[a]	2500	3000

(all energies in cm^{-1})

[a] this value is for M(III) only – ligand field effects are *much* larger for higher oxidation states.

N.B. This table gives only typical values. The variation within a given series, or with oxidation state, is considerable.

e) Magnetic Exchange

Thus far we have considered the magnetic moments of individual complexes, and for bulk susceptibilities and mean magnetic moments we took an average of all states which might be occupied. Magnetic exchange phenomena arise when the magnetic moments of individual complex molecules interact and couple – it is then no longer possible to consider each molecule as being separate and unaffected by the others. Magnetic exchange phenomena thus affect the bulk properties of the solid. The magnetic moments of the individual complex species are all coupled together within a certain region (or domain); if the solid is very finely ground, the size of the individual particles can become so small that there are not enough spins present to show the normal magnetic exchange interaction, and the particles show normal paramagnetic behaviour – this phenomenon is known as *superparamagnetism*.

We discuss exchange phenomena under three headings: ferromagnetism, antiferromagnetism and ferrimagnetism.

Ferromagnetic species have the magnetic moments coupled parallel to each other, giving a large total magnetic moment, and a large magnetic field inside the solid. This leads to a large positive susceptibility. Ferromagnetism is generally found in metals and alloys, and is best known in iron metal. Ferromagnetic materials materials become paramagnetic above a certain temperature (the Curie point) above which the thermal energy is too great to maintain the alignment of spins. Antiferromagnetic species have their spins coupled antiparallel to each other so that each magnetic moment is balanced by an equal and opposite moment. This leads to a zero magnetic susceptibility. Again the coupling breaks down above a certain temperature (the Néel temperature) and normal paramagnetic behaviour is shown. A decrease in μ with decreasing temperature is sometimes taken as an indication of antiferromagnetism, but can often be ascribed to spin-orbit coupling effects reducing μ. It should be noted that antiferromagnetic 'pairing' involves no actual mixing of orbitals (as for formation of a molecular orbital) but may be regarded as a 'through space' interaction in which the state with moments opposed is favoured.

Antiferromagnetism is common in compounds (such as FeF_2, Néel temperature 77 °K, K_2IrCl_6 3.08 °K). The mechanism of exchange is not well understood, but it may be remarked that antiferromagnetism is common in closely packed structures with electronegative atoms (such as F, Cl, O) present.

Ferrimagnetism is the simultaneous manifestation of ferro- and anti-ferromagnetism; in a solid with two different sites occupied by paramagnetic species, the paramagnets in one site may couple ferromagnetically with themselves and antiferromagnetically with the other paramagnetic site. If there is an imbalance in the population of the two sites, this leads to a net alignment of the spins and a large magnetic field in the solid and a large susceptibility. Solids showing such ferrimagnetism are referred to as ferrites. An example is the spinel structure which has two possible metal cation sites within a cubic close packed array of oxygen atoms. AB_2O_4 has, for each 16 oxygen atoms, 8 octahedral sites and 16 tetrahedral sites; in Fe_3O_4 the tetrahedral and half the octahedral sites are occupied by Fe^{3+} ions, and Fe^{2+} ions occupy the other eight sites. The magnetic moments of the two Fe^{3+} sites cancel out, leaving the coupled moments of the Fe^{2+} ions – giving the large magnetic susceptibility of magnetite Fe_3O_4. Many artificial ferrites have now been prepared, some of which have extremely useful magnetic properties.

f) Spin Crossover Complexes

We have seen that a complex is high spin or low spin according to the relative magnitude of electron repulsion and ligand field effects, and we may imagine circumstances in which these two effects might be roughly equal. In such cases the complex may be found in both high and low spin states, depending on the experimental conditions, and is said to exhibit *spin crossover* behaviour. Thus the octahedral $3d^6$ complex $[Fe(phen)_2(NCS)_2]$ has a magnetic moment corresponding to high spin Fe^{2+} at room temperature, and is diamagnetic (low spin Fe^{2+}) at low temperatures. The change from high spin to low spin behaviour (or vice-versa) can be produced by changing temperature or pressure, and presumably results from a change in the metal-ligand bond lengths, and the possible occupation of electronically excited states.

Spin crossover occurs in various ways. For the high spin – low spin crossover of Fe^{2+} mentioned above, there appear to be two phases (one high spin, the other low spin) co-existing near the crossover temperature, and distinct Mössbauer spectra are seen for the different sits. By contrast, the Fe^{3+} spin crossover systems such as $[((CH_2)_4NCS_2)_3Fe]$ show only one site with magnetic moment and Mössbauer spectrum corresponding to a time average of high and low spin forms, the relative proportions of which may be fitted to a Boltzmann distribution. Here the high spin – low spin spin transition appears to take place in a time shorter than 10^{-7} sec.

Most spin crossover compounds contain iron and cobalt, although some 5 coordinate nickel compounds show similar behaviour. Spin crossover complexes of cobalt and iron appear to be important biologically, where the chemical behaviour of one spin state of the compound, and the rate of interconversion may govern the behaviour of the complex as a whole. Some interesting examples of spin crossover under applied pressure are discussed by Drickamer and Frank.

Bibliography

Day, P. (senior reporter): Electronic structure and magnetism of inorganic compounds. Specialist periodical reports. London: the Chemical Society.

Drickamer, H.G., Frank, C.W.: Electronic transitions and the high pressure chemistry and physics of solids. London: Chapman and Hall 1973

Figgis, B.N., Lewis, J.: A review of magnetochemistry. Prog. Inorg. Chem. *6*, 37 (1964)

Gerloch, M., Slade, R.C.: Ligand field parameters. Cambridge: Cambridge University Press 1973

Gillard, R.D.: The Cotton effect. Prog. Inorg. Chem. *7*, 215 (1966)

Gillard, R.D.: Optical rotatory dispersion and circular dichroism. In: Physical methods in advanced inorganic chemistry, Hill, H.A.O. and Day, P. (eds.) London, New York, Sydney: J. Wiley 1968

Goodenough, J.B.: Magnetism and the chemical bond. London, New York, Sydney: J. Wiley 1963

Goodwin, H.A.: Spin crossover of six-coordinate Fe(II) complexes. Coord. Chem. Revs. *18*, 293 1976)

Griffith, J.S.: The theory of transition metal ions. A very mathematical treatment. Cambridge: Cambridge University Press 1971

Hawkins, C.J.: Absolute configuration of metal complexes. London, New York, Sydney: J. Wiley 1971

Jaffe, H.H., Orchin, M.: Symmetry, orbitals and spectra. London, New York, Sydney: J. Wiley 1971

Jørgensen, C.K.: Absorption spectra and chemical bonding in complexes. Oxford: Pergamon Press 1962

Jørgensen, C.K.: Modern aspects of ligand field theory. Amsterdam: North Holland, and New York: Elsevier 1971

Kettle, S.F.A.: Coordination compounds. London: Nelson 1969

Lever, A.B.P.: Inorganic electronic spectroscopy. Amsterdam, London, New York: Elsevier 1968

Mabbs, F.E., Machin, D.J.: Magnetism and transition metal complexes. London: Chapman and Hall 1973

Mitra, S.: Magnetic anisotropy. Prog. Inorg. Chem. *22*, 309 (1977)

Murrell, J.N., Kettle, S.F.A., Tedder, J.M.: Valence theory 2nd Ed. London, New York, Sydney: J. Wiley 1970

Reisfeld, R., Jørgensen, C.K.: Lasers and excited states of rare earths. The spectra of lanthanides and their applications. Berlin, Heidelberg, New York: Springer 1977

Robin, M.B., Day, P.: Mixed valence chemistry. Adv. Inorg. Chem. Radiochem. *10*, 247 (1967)

Schatz, P.N., McCaffery, A.J.: An introduction to M.C.D. Quart. Rev. *23*, 554 (1969)

Selwood, P.W.: Magnetochemistry 2nd Ed. London, New York, Sydney: J. Wiley 1956

Sutton, D.: Electronic spectra of transition metal complexes. London: McGraw-Hill 1968

Problems

1. (i) Using the formula of Eq. (1.31), and a value for the spin pairing parameter D of 7kK, calculate the difference in spin pairing energy between high and low spin $[Co(OH_2)_6]^{3+}$.
(ii) Calculate the value of Δ for this complex (using Table 4.1), and predict whether the complex will be high or low spin.
(ii) Predict whether $[Co(NH_3)_6]^{3+}$ will be high or low spin.

2. Cobalt(II) salts dissolved in water give pale pink solutions, but alcoholic solutions are dark blue. Suggest an explanation.

3. What effect, if any, will the nephelauxetic effect have on spin pairing energy?

4. Comment on the changes in absorption maxima or colour in the following series:

(i)	VO_4^{3-}	CrO_4^{2-}	MnO_4^{-}
	colourless	yellow	purple
$\nu_{max.}$ (kK)	36.9	26.8	18.6
(ii)	CrO_4^{2-}	MoO_4^{2-}	WO_4^{2-}
	yellow	colourless	colourless
$\nu_{max.}$ (kK)	26.8	43.2	50.3
(iii)	WO_4^{2-}	WS_4^{2-}	WSe_4^{2-}
	colourless	yellow	orange
$\nu_{max.}$ (kK)	50.3	25.5	21.6
(iv)	WCl_6 black		
	$W(CO)_6$ colourless		

5. Propose origins for the spectra or colours of the following species:
(i) $[Fe(OH_2)_6]^{3+}$ – very feeble absorption in the visible, strong absorption in the ultra-violet.
(ii) $[FeF_6]^{3-}$ colourless; $[FeCl_4]^-$ yellow; $[FeBr_4]^-$ red.

6. The difference in energy of two M.O. as deduced from the difference in their ionisation potentials is often much smaller than that deduced from charge transfer spectra. Why?

7. The first two absorptions corresponding to spin allowed d-d transitions are given below for some d^3 ions

	Energy (kK)	
$[V(OH_2)_6]^{2+}$	12.35	18.5
$((C_2H_5O)_2PS_2)_3Cr$	14.4	18.9
$[CrF_6]^{3-}$	14.9	22.7
$[Cr(OH_2)_6]^{3+}$	17.4	24.6
$[Cr(NH_3)_6]^{3+}$	21.55	28.5
$[MnF_6]^{2-}$	21.75	28.2
$[MoCl_6]^{2-}$	19.75	23.9

Comment on the variation of these absorption maxima.

8. NiO is green if prepared carefully, but becomes black upon heating in air. Why?

9. Comment on the colours of the following compounds:
NaCl colourless; $PrCl_3$ pale blue green; PbS black, with a metallic lustre; Fe_3O_4 black; $((C_2H_5)_4N)_2ZnCl_4$ colourless; $((C_2H_5)_4N)_2MnCl_4$ pale green; $Cs_4Sb_2Cl_{12}$ dark blue.

10. Even before X-ray crystallography was well established, it was possible to identify the *cis* and *trans* isomers of $[Co(en)_2Cl_2]^+$. How?

11. The compound shown below afforded the first definite proof that Pt(II) complexes were square planar. What is the basis of this proof?

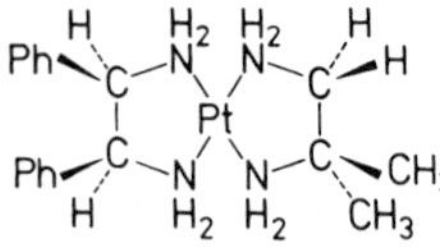

12. Calculate μ for a f^2 configuration using (i) the spin-only formula and (ii) using Eq. (4.6), taking the J value of the groundstate from Fig. 4.3.

13. Would you expect an orbital contribution to the magnetic moment of octahedral Cr^{3+} complexes?

14. Would you expect the magnetic moment to be anisotropic in $[V(C_5H_5)_2]$ or $[Mn(C_5H_5)_2]$?

15. Why do many high-spin Co^{2+} complexes show pronounced magnetic anisotropy?

16. Many biological systems contain iron-sulphur proteins in which two or more iron atoms are coordinated tetrahedrally by bridging sulphur ligands. Although the iron atoms appear to be in the high-spin form, the proteins often have very low or zero magnetic moments. Propose an explanation.

5. Alternative Methods and Concepts

We have previously discussed chemical bonding exclusively in terms of a L.C.A.O. molecular orbital model. This is by no means the only approach, although it has gradually established itself as the most popular for inorganic chemists. In this chapter we discuss other models of chemical bonding, both theoretical and phenomenological, consider the thermodynamics of inorganic chemistry, and finally discuss some of the concepts frequently encountered in inorganic chemistry.

A. Alternative Models of Chemical Bonding

a) *X–α Methods[1]

It is of course possible to calculate molecular orbitals without using atomic orbitals as a starting point – this was in fact one of the first approaches to calculation of a molecular wavefunction for H_2. The most recent development in this field is the X–α method originally proposed by Slater. In Chap. 2 it was stated that J (coulomb) and K (exchange) integrals may be defined to allow for electron repulsion, but that their calculation was a formidable task; this is particulary true of exchange integrals, since the 'classical' coulomb term may be related to the potential arising from the total electronic charge density. Slater proposed the elimination of the exchange integrals and the substitution of a potential proportional to the cube root of the electron density to allow for the exchange interaction. This is clearly an approximation, and its merits are discussed by Slater in his review (1972). The potential proposed is:

$$V_{X\alpha\uparrow}(\mathbf{r}) = -6\alpha\left[\frac{3}{4\pi}\varrho\uparrow(\mathbf{r})\right]^{1/3}$$

where $\varrho\uparrow(\mathbf{r})$ is the electronic density having spin $\uparrow$ at $\mathbf{r}$. The coefficient α is a variable parameter, chosen to give the best fit with Hartee-Fock calculation calculations for atoms.

In summary, the multicentre integrals J and K may be replaced by potential terms related to charge densities. This simplifies the Schrödinger equation considerably, and it is now a relatively straightforward task to solve it numerically. When applied to atoms

[1] This is the logical place for this section, but it may easily be skipped by anyone not interested in calculations.

it is found that the best agreement with Hartree-Fock total energy is found with an α of the order of 0.7. If the α value is chosen to give this good agreement, it is then found that the atomic wavefunctions calculated by the two methods are similar.

The elimination of complicated integrals made it attractive to extend this method to molecules. We describe here the most popular approach, the multi-scattering X–α method. A given molecule is divided into three potential regions for the purpose of the calculation (see Fig. 5.1):

1) The extramolecular region, outside the molecular cluster
2) The atomic region – spheres surrounding the constituent nuclei of the molecule
3) The interatomic region, between the atomic spheres and the extramolecular region.

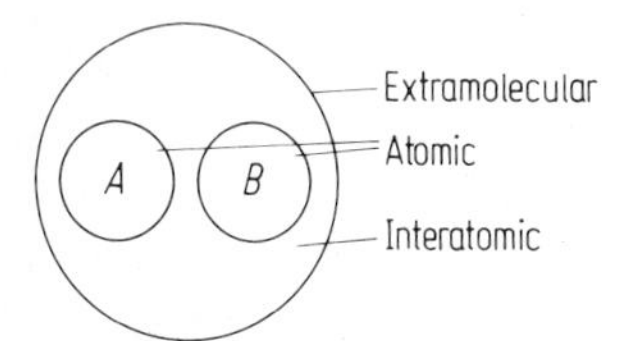

Fig. 5.1. The potential regions for a diatomic molecule A–B

Assumptions are made about the potentials in each region (e.g. spherical in the atomic region, constant in the interatomic region) and the Schrödinger equation solved numerically in each region with the restriction that the wavefunction and its first derivative are continuous at regional boundaries. Because of the use of 'scattered wave' formalism as a mathematical tool, the method is referred to as multiple scattering X–α (MS–X–α) or X–α –SW (scattered wave). The process is repeated using the new wavefunction to calculate the potentials until no further change in wavefunction is found (i.e. a self consistent field (S.C.F.) method – Chap. 1). The whole subject is discussed in detail by Johnson.

The advantages of MS Xα calculations are:

1) Their speed – for complicated molecules they often take only a few minutes of computer time compared with several hours for Hartree-Fock methods.

2) Predictions of optical and electronic transitions. The MS Xα method calculates these transition energies in a different way to Hartree-Fock methods, and to some extent allows for the effects of reorganisation of all the electrons in the molecule. The predictions are consequently more accurate.

Against this there are certain disadvantages:

1) The arbitrary division of the molecule into separate potential regions is not very attractive, the more so as the results of the calculation are sometimes very sensitive to this partition. Against this, there is no attempt to describe the interatomic region, fairly distant from the nuclei, in terms of atomic orbitals.

2) The sensitivity to molecular symmetry. X–α calculations only converge quickly and efficiently for molecules with fairly high symmetry. This is one reason why X–α calculations are so common for crystalline solids.

Finally, since the L.C.A.O. approach is abandonned we lose the capacity to predict some of the molecular properties from a knowledge of the atomic orbitals involved.

This cuts both ways, since it is possible to compare the numerically calculated wave function with L.C.A.O. results. Thus, the Xα results for the SO_4^{2-} ion and for SF_6, show no sulphur 3d like character in the M.O. This is an interesting result since the 3d orbitals are of a suitable symmetry for bonding and their involvement has previously been proposed (page 265), and suggested by a L.C.A.O. calculation – this apparent mixing of orbitals was mentioned in chapter 2 (page 63). By contrast, the Xα calculation suggests fairly strong Mn-O π bonding involving the manganese 3d orbitals in MnO_4^-.

b) Valence Bond Theory

The valence bond (V.B.) theory of chemical bonding dominated theoretical inorganic chemistry for a long time, largely as a result of Pauling's classic book 'The Nature of the Chemical Bond'. However, in recent years, inorganic chemists have been more interested in fields where the valence bond model is less simple to use than the molecular orbital approach, and the V.B. model is now rather out of fashion. In this section we discuss only the fundamentals and the difficulties of the V.B. approach. Full details of the V.B. model and comparisons with the molecular orbital approach are given in most theoretical chemistry texts (Atkins; Murell, Kettle and Tedder; Coulson) and, in particular, Coulson gives a discussion of most aspects of molecular structure from both points of view.

As for the M.O. model, we may usefully begin our discussion of the valence bond theory by discussion of molecular hydrogen. If two hydrogen atoms (a and b) are too far apart to interact with each other, then the total wavefunction of the system will simply be:

$$\psi = \phi_a \cdot \phi_b \tag{5.1}$$

The valence bond model takes this wavefunction as the basis for the molecular wavefunction as the two atoms are brought together. The interaction of the two atoms is treated as a first-order perturbation (Chap 1), changing the energy, but not the wavefunction which for H_2 still has the form of Eq. (5.1).

However, if we label the electrons 1 and 2, it is clear that the functions $\phi_a(1)\,\phi_b(2)$ and $\phi_a(2)\,\phi_b(1)$ have equal probability so a better wavefunction would be:

$$\psi = \frac{1}{N}\left[\,\phi_a(1)\,\phi_b(2) \;\pm\; \phi_a(2)\,\phi_b(1)\,\right] \tag{5.2}$$

where N is a normalisation constant (Chap. 1)
– this is a natural result of the indistinguishibility of electrons in the molecule. We may now consider the effects of the Pauli exclusion principle and the effects of the electron spins. If the spin part of the wavefunction is antisymmetric e.g.

$$\frac{1}{\sqrt{2}}(\alpha(1)\,\beta(2) - \alpha(2)\,\beta(1))$$

then the orbital part must be symmetric for the total wavefunction to be antisymmetric with respect to the exchange of electron coordinates. Thus the total wavefunction would have the form:

$$\psi = \frac{1}{N}\left[\,\phi_a(1)\,\phi_b(2) + \phi_a(2)\,\phi_b(1)\,\right]\left[\,\alpha(1)\,\beta(2) - \alpha(2)\,\beta(1)\,\right] \tag{5.3}$$

Alternatively, if the spin part is symmetric, the orbital part must be antisymmetric:

$$\psi = \frac{1}{N}\ [\ \phi_a(1)\,\phi_b(2) - \phi_a(2)\,\phi_b(1)\ ] \begin{cases} \alpha(1)\,\alpha(2) \\ \frac{1}{2}\,(\alpha(1)\,\beta(2) + \alpha(2)\,\beta(1)) \\ \beta(1)\,\beta(2) \end{cases} \tag{5.4}$$

The wavefunction (5.3) corresponds to a singlet molecular wavefunction, and the wavefunction (5.4) to a triplet wavefunction. If it is borne in mind that the atomic orbitals overlap on approach, the function (5.3) corresponds to a build-up of electronic charge between the two nuclei, and (5.4) to a diminution of the internuclear charge density. It follows that (5.3) is a bonded state while (5.4) is a repulsive state. If is therefore concluded that the ground state of H_2 is a singlet (as was found for H_2 by M.O. methods). Eq. (5.3) is actually quite a good wavefunction for H_2.

If we consider the wavefunction (5.3) closely, it will be seen that the possibility of both electrons being on the same atom at the same time is excluded – the correlation effects preventing the close approach of two electrons have thus been allowed for, if not overallowed for; the wavefunction (5.3) is a 'pure covalent' representation of the bonding with the electrons being equally shared between the two atoms. If we wish to consider the possiblity of having both electrons on the same atom we should add in a function of the form $\phi_a(1)\,\phi_a(2)$. There is however an equal probability of both electrons being on atom a or atom b, so we should add

$$\phi_a(1)\,\phi_a(2) + \phi_b(1)\,\phi_b(2) \tag{5.5}$$

instead. These two wavefunctions correspond to ionic forms of the molecule ($H_a^-H_b^+$ and $H_a^+\,H_b^-$ respectively). If we refer to the function (5.5) as ψ_{ion} and (5.3) as ψ_{cov}, the best total wavefunction is thus

$$\Psi = \frac{1}{N}\,(\psi_{cov} + \lambda\,\psi_{ion})\,\psi_{spin} \tag{5.6}$$

The function (5.6) is the final form of the molecular wavefunction for H_2 in the groundstate. It is in fact identical to the M.O. wavefunction for H_2 *if* correlation effects have been allowed for by configuration interaction.

More complicated systems

The V.B. model supposes that a link between two atoms arises from the interaction of an atomic orbital on each atom, leading to a pairing of spins of the electrons in the orbitals. The chemical bond is thus related to two atomic orbitals, one on each atom, and is localised between these two atoms. The one electron functions are unchanged atomic orbitals (e.g. ϕ_a and ϕ_b) and are not molecular orbitals. Let us now consider a heteronuclear diatomic molecule A–B each with one atomic orbital available for chemical bonding. We may proceed as for H_2, and form a pure covalent function similar to function (5.3). We now need to consider the addition of the ionic function (5.5). If the atomic orbital ϕ_B is much more strongly bound than ϕ_A (e.g. if A is H, and B is Cl) then we would expect the electrons to spend more time on B than A. We may write the wavefunction (neglecting the spin part) as

$$\Psi = C_1\ \psi_{cov} + C_2\{\phi_A(1)\,\phi_A(2)\} + C_3\{\phi_B(1)\,\phi_B(2)\} \tag{5.7}$$

The second part of the function corresponds to a state A^-B^+ and the third to A^+B^-; we would expect C_2 to be less than C_3 and may even approximate it to zero. It should be noted, however, that we are describing the A–B bond in terms of a mixture of three different forms – pure covalent and the two possible ionic forms. This is in contrast to M.O. theory where the link was described in terms of one, possibly polarised, M.O. – the only terms to be added being the configuration interaction terms to allow for electron correlation.

The construction of a molecular wavefunction by the mixing of possible forms is fundamental to valence bond theory – the phenomenon of mixing is referred to as *'resonance'*. It is a device to overcome the problem that the valence bond theory allows only 'extreme' forms of bonding (pure covalent, electrons having equal probability density on each atom, or pure ionic, with both electrons on the same atom). This problem is avoided by L.C.A.O. M.O. theory which allows any number of atomic orbitals to mix in any proportion. The various forms of the chemical bond A–B which are considered in (5.7) (A–B covalent, A^+B^-, A^-B^+) are called resonance forms; the resulting molecule is described as a *resonance hybrid.*

It is vital to remember that resonance is a mathematical artefact, and that the resonance forms proposed are simply basis function (with no physical existence) from which the molecular wavefunction is constructed. This fact is all too easily forgotten: two particularly common errors made by students are (i) the belief that the greater the number of resonance forms (however wildly improbable on energetic grounds) the greater the stability of the molecule, and (ii) the discussion of molecular behaviour in terms of only one resonance form which may have minimal importance in the final wavefunction.

Multiple bonds (as in N_2 and O_2) present no grave problem for the V.B. model – thus for N_2 one merely forms separate bonds between the three groups of p orbitals of nitrogen. It is however less straightforward to explain the triplet ground state of O_2, although it is possible.

As a typical triatomic we studied MgF_2 (Chap. 2). If for the moment we assume that the magnesium atom uses only the 3s orbital, it is clear that there must be at least partial ionisation of this orbital before it can interact with one of the fluorine 2p orbitals. Three resonance forms suggest themselves immediately:

$$F^-Mg^+\!-\!F \qquad F\!-\!Mg^+F^- \qquad F^-Mg^{2+}F^-$$

To preserve the symmetry of the molecule, the first two forms must have equal weight. It is only possible to produce a 'pure covalent' form if the magnesium atom is hybridised to produce two sp hybrids, each occupied by one electron. Each hybrid atomic orbital may now interact with a fluorine orbital to give a pure covalent bonded

$$\underset{(sp)}{F\!-\!Mg\!-\!F}$$

The final wavefunction is thus a mixture of four resonance forms, and is rather less elegant than the M.O. functions.

The necessity to resort to hybridisation of atomic orbitals to obtain sufficient 'atomic orbitals' to establish pure covalent bonds in valence bond theory is one of the disadvantages of the method. Thus for SF_6, a molecular orbital approach (ignoring any π bonding) forms delocalised M.O.s with the fluorine $2p\sigma$, and the sulphur 3s and

3p orbitals (see Fig. 5.2). The valence bond theory requires (among the many possible resocance forms) a sp^3d^2 hybrid to be formed, using the 3d orbitals. The situation is even worse with the transition metals and their complexes (see below).

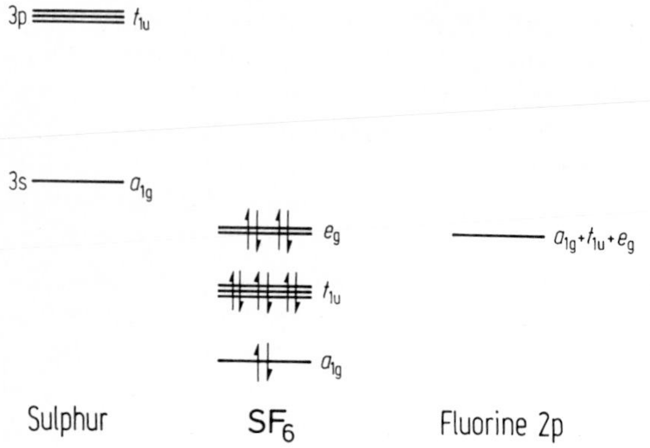

Fig. 5.2.
The bonding M.O. of SF_6

Disadvantages of the valence bond theory

In this section some of the disadvantages for the inorganic chemist of the V.B. model are given. This is not out of vindictiveness, but more to justify the short coverage given here; some textbooks of inorganic chemistry still maintain the V.B. model for descriptive inorganic chemistry.

1) Localisation. The V.B. method deals essentially with one link (between two atoms) at a time. This is a localised bond approach which has advantages in explaining the relative constancy of certain bond lengths or energies, especially in organic chemistry. However, to explain delocalised bonding the V.B. method involves a large number of resonance forms; this is particularly common in inorganic chemistry.

2) Polarity and atomic hybrids. M.O. theory expresses polarity fairly simply by the variation of atomic orbital participation in a M.O., which may be related to overlap integrals and the energies of the atomic orbitals involved. By contrast, the calculation of the amount of a given resonance form in the final resonance hybrid is by no means intuitively obvious. Similarly, hybridisation of atomic orbitals is a convenience in M.O. theory, but a necessity for V.B. theory – the involvement of some rather suspect atomic orbitals in bonding is the result.

3) Transition metal compounds. For a low spin octahedral complex such as $[Fe(CN)_6]^{4-}$ ($3d^6$) the V.B. model assumes the t_{2g} d-orbitals to be nonbonding, with a valance bond formed between each ligand and one of six d^2sp^3 hybrids formed from the e_g d orbitals (in $[Fe(CN)_6]^{4-}$ $3d^24s4p^3$ hybrids). However, this approach clearly fails for a high spin $3d^6$ complex ($[Fe(H_2O)_6]^{2+}$) since the 3d e_g orbitals are occupied by two unpaired electrons – in this case a 4s $4p^3$ $4d^2$ hybrid is proposed – this involvement of the 4d orbitals is not very happy, and the spin crossover complexes discussed in Chap. 4 must feel particularly worried about this. The involvement of 4p orbitals is expected (from a M.O. approach) to be fairly small anyway.

4) Electronic spectra. Electronic spectra are readily explained in M.O. theory by one electron jumps between molecular orbitals, each molecular orbital comprising 2 one electron spin orbitals. Each molecular orbital has associated with it a well defined one electron energy. The V.B. approach is less simple, since the one electron functions are unchanged atomic orbitals, and there are no antibonding orbitals to which electrons can be excited. One can only discuss spectra in terms of the energy of different *states* of the whole molecule. While this is a more realistic approach, given the importance of electron repulsion in electronic spectra, it is a very great deal more complicated, and necessitates a knowledge of the wavefunctions of both states involved. The difficulty in explaining spectra is one of the main reasons for the neglect of V.B. theory. In his book on the absorption spectra of complexes Jϕrgensen (1962) discusses this in greater detail.

There are two fields in which V.B. theory has some following: accurate calculations on small molecules, and weak chemical bonds. It was stated that for H_2 the V.B. and M.O. wavefunctions are identical when all effects are allowed for, and in some cases the V.B. calculation is more suitable than a full M.O. calculation with configuration interaction. Moffit's method of atoms in molecules allows the V.B. method to be corrected with experimental data and gives good values for dissociation energies. This field has recently been reviewed by Gerratt. For weak chemical bonds it is reasonable that an approach which keeps the atomic orbitals of the interacting species essentially unchanged, (rather than forming mixed molecular orbitals) should be more realistic, and the V.B. method is still used in this area.

c) The Ionic Model

The ionic model is the most important phenomenological model of chemical bonding. It regards a given chemical substance as composed of two species, one carrying a negative charge, and the other a positive charge; the substance is held together by the electrostatic attractions between the unlike charges. One immediately predicts structures where ions are surrounded as closely as possible by ions of an opposite charge and as far away as possible from ions of the same charge – that is, fairly close packed solid structures. The ionic model is in fact only applicable to solids, although it may be applied in a qualitative way to solutions and gaseous ion pairs. The ionic species may be polyatomic (e.g. NH_4^+, $[Co(NH_3)_6]^{3+}$, PF_6^-, SO_4^{2-}) but we will consider only monatomic ions – the extension to polyatomic species is trivial.

If we consider the alkali metal halides (the classic examples for the ionic model), the chemical bonding involves the ionisation of the alkali metal to give M^+, the addition of an electron to a halogen atom to give X^- followed by the gain of electrostatic energy as a result of the approach of the oppositely charged ions to give a continuous crystal lattice. Even for the alkali metal with the weakest bound valence electron (caesium), and the halogen with the highest electron affinity (chlorine), the first *two* steps are endothermic; they are however offset by the very large amount of electrostatic energy liberated on formation of the crystal. If we allow for the initial atomisation of the metal and halogen, the whole process may be represented by a thermodynamic cycle, the *Born Haber* cycle (Fig. 5.3).

The energy liberated by the approach of oppositely charged ions to form the crystalline solid is called the *lattice energy.* According to the ionic model, this is the stabilisation effect favouring chemical bonding. If the constituent ions of an ionic solid are regarded as hard spheres within which is contained the entire ionic charge, then, once the structure is known, the lattice energy may be calculated using classical electro-

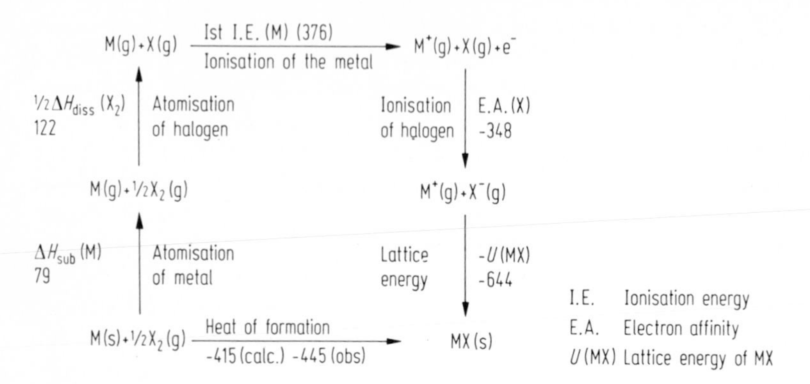

Fig. 5.3.
Born Haber cycle for the formation of MX, an ionic salt. The figures (in kJ/mole) are for CsCl

statics to sum the attraction and repulsion energies between all the ions. Furthermore, if we accept the hard sphere hypothesis, we may use the radius-ratio rule to predict the structures and coordination adopted (see Chap. 3). It should be noted that the hard sphere hypothesis is not a quantum mechnically valid approximation as we could expect the wavefunction (and hence the electronic density) to tail away asymptotically to zero; this problem is discussed further below.

The lattice energy

Since the ionic model regards lattice energy as the determining stabilisation factor in chemical bonding is worth while to discuss it in some detail. If a 'hard sphere' ionic model is used the calculation is relatively straightforward. A review of lattice energy calculations is given by Waddington, and precise details may be found in many books, so only the fundamentals are discussed here. The fundamental equation for the lattice energy has the form

$$U = \frac{N_o M z_+ z_- e^2}{r} \tag{5.8}$$

where N_o = Avogadro's number
M = the Madelung constant
z_+ = the charge on the cation
z_- = the charge on the anion
e = the electronic charge
r = the closest distance between two ions of opposite charge.

The Madelung constant M allows for the sum of attractive and repulsive interactions between the ions, and clearly depends on the way in which the ions are disposed (i.e. the crystal structure). Thus for the NaCl structure (page 117) M = 1.748, for CsCl M = 1.763, for the fluorite structure (c.c.p. with all the tetrahedral holes occupied, as in fluorite, CaF_2) M = 2.519, for the idealised rutile structure (f.c.c. with half the octahedral holes occupied as in rutile, TiO_2) M = 2.408.

Equation (5.8) has two disadvantages: it is necessary to know the interionic distance r, and the crystal structure to know M. We discuss the first problem later (see ionic radii), but the second problem may be partially overcome by an approximation due to Kapustinskii who noticed that the Madelung constant M is approximately equal to 0.84 times the number of ions in the smallest neutral unit of the compound, ν (i.e. $\nu = 2$ for NaCl, ZnS, 3 for TiO_2, CaF_2, 4 for LaF_3 etc.).

Equation (5.8) is purely electrostatic, and assumes the ions to be perfect hard spheres. As was mentioned above this is an unjustifiable assumption; and one of the first corrections to Eq. (5.8) is to allow for the repulsive interactions between ions of *opposite* charge – that is the repulsions between the electron densities of the two ions which stop the two ions coalescing. This may be corrected for in several ways (see Waddington and Johnson) – one of the most common is the Born-Lande equation

$$U = \frac{N_0 M z_+ z_- e^2}{r} \left(1 - \frac{1}{n}\right) \qquad (5.9)$$

corresponding to the introduction of a repulsive energy of the form $\frac{B}{r^n}$ The value of n may be determined from measurements of the compressibility of the solid in question. n varies between 5 and 12 – a mean value of 9 is frequently used, corresponding to a diminution of the lattice energy by 10 percent or so. If we substitute this approximation, together with Kapustinskii's approximation for the Madelung constant into (5.9), we obtain

$$U = \frac{1.08 \times 10^3 \, \nu z_+ z_-}{r} \ \text{kJ/mole} \qquad (5.10)$$

(r is measured in Ångstrøm)

This formula has the attraction of being extremely general, especially if r can be substituted by the sum of two ionic radii (see next section). It is not the most refined equation for the calculation of lattice energies, but even so is capable of predicting heats of formation by a Born Haber cycle (Fig. 5.3) within 10 percent, without the need for extensive data on crystal structure and compressibility. This, by the standards of chemical calculations, is extremely high accurary, although the refined models are even more accurate. Another feature of (5.10) which is even more useful for the inorganic chemist is the capacity to predict trends in the lattice energy very easily. For example, small anions such as O^{2-} and F^- will give small interionic distances, and consequently high lattice energies – this enables them to stabilise in crystalline solids ions which require a great deal of energy for their formation (i.e. high oxidation states).

Ionic radii

If a hard sphere model is adopted, it is natural to assign an 'ionic radius' to a given ion. These values may be used to calculate interionic distances in solids, and also, using the radius-ratio rules, to predict the most favoured coordination. Several methods have been used to extract such series of ionic radii:

a) Crystallographic. In compounds with a small cation surrounded by much larger anions (e.g. LiI, LiBr, MgSe) the large ions are presumed to be just touching with the small cation 'rattling' in between (see Fig. 3.46). The size of the unit cell is then dependent only on the size of the large anion r_-, and r_- may be calculated, and then used as a constant in other compounds of that anion.

b) Quasi-theoretical. The interionic distance between two oppositely charged univalent ions (assumed to be in contact) is split up according to the ratio of the effective atomic numbers (page 35) experienced by the outer valence electrons of the ions. Using simple arguments, Pauling derived ionic radii (or crystal radii) for other ions (see Huheey). The completeness of Pauling's set of ionic radii has given them considerable popularity.

c) Thermochemical. By calculating the lattice energy from experimental data, and using Eq. (5.10) or some other variant, value of r may be obtained. If one of the ionic radii is known, the other may be obtained by simple subtraction. These radii (often referred to as thermochemical radii) are useful for complex species such as SO_4^{2-}, IO_4^-, etc.

At the moment, method a) appears to be the most popular. All methods are somewhat dependent on the assumptions made, and the same method can give rise to several different scales. In view of the approximation of the hard sphere model, and the

Table 5.1. Some ionic and thermochemical radii

Ion	Radius pm	Ion	Radius pm	Ion	Radius pm
Li^+	74	Co^{2+}	74	OH^-	140 t
Na^+	102	Ni^{2+}	69	SH^-	195 t
K^+	138	Cu^{2+}	73	CN^-	182 t
Rb^+	149	Zn^{2+}	75	$CH_3CO_2^-$	159 t
Cs^+	170	Ag^+	115	CO_3^{2-}	185 t
Mg^{2+}	72	Sn^{2+} (8 coord)	122	NO_3^-	189 t
Ca^{2+}	100	Pb^{2+}	118	ClO_3^-	182 t
Sr^{2+}	113	La^{3+}	105	ClO_4^-	236 t
Ba^{2+}	136	O^{2-}	140 P	SO_4^{2-}	230 t
Al^{3+}	53	F^-	133 P	PO_4^{3-}	238 t
Tl^+	150	Cl^-	181 P	$[Mg(H_2O)_6]^{2+}$	235 t
Tl^{3+}	88	Br^-	195 P	$[Mn(H_2O)_6]^{2+}$	235 t
V^{2+}	79	I^-	216 P	$[Ni(H_2O)_6]^{2+}$	229 t
Cr^{3+}	62	S^{2-}	184 P	$[Ni(NH_3)_6]^{2+}$	258 t
Mn^{2+}	83	Se^{2-}	198 P	NH_4^+	143 t
Fe^{2+}	78			$N(CH_3)_4^+$	300 t
Fe^{3+}	65				

All radii in picometers (pm) : 100 pm = 1 Å.
Values marked P are Pauling crystal radii; values marked t are thermochemical radii. All other values are crystallographic (Shannon and Prewitt) for octahedral coordination unless otherwise marked. Radii for transition metal ions are for high spin complexes.

inevitable presence of a small degree of covalency which may be sufficient to influence the choice of structure (see Sect. 3.E), the radius-ratio rule is not very reliable. Crystallographic data show that the ionic radius is in fact coordination dependent, increasing with increasing C.N. Much more useful are the predictions of interionic distances which may be used to calculate lattice energies, and estimate the stabilities of compounds. If interpreted with a little common sense, these predictions can be very useful, but care must be taken to avoid mixing up radii derived by different methods, as this can lead to an accumulation of the systematic errors of each method.

Experimental evidence in favour of the ionic model

Apart from the reasonable success of calculations of heats of formation based on the Born Haber cycle (Fig. 5.3) there is other experimental support for an ionic or partially ionic model.

1) Structural evidence The packings adopted are the arrangements expected to minimise repulsion between similarly charged ions, and maximise attractions between oppositely charged ions. The internuclear distances are at least approximately additive (e.g. the cell parameters of bromides of the alkali metals are 0.16 ± 0.02 Å greater than the chlorides and so on), and ionic radii predict internuclear distances fairly accurately. Finally, the most accurate X-ray work shows that the electron distribution around each constituent atom is close to spherical, and that the density falls to a minimum between adjacent unlike nuclei (see Morris).

2) Spectroscopic and magnetic evidence. In many cases solids show spectroscopic and magnetic properties similar to those known or predicted for ions of the elements present in the solids. Thus lanthanide salts LnX_3 or Ln_2O_3 show the magnetic moments and spectra expected for the Ln^{3+} ion.

3) Physical properties When pure and stoichiometric, ionic salts are electrical insulators, even though they may have partially filled atomic orbitals (e.g. transition metal salts); on fusion, they become good conductors as a result of ionic transport. Any conduction in the solid is due to to ionic 'hopping' or crystal defects.

4) Behaviour in solution. Ionic salts are frequently soluble in polar solvents where they behave as dissociated ions, conducting electricity strongly. Solvation effects observed agree with an ionic model.

A critique of the ionic model

Having sketched the outlines of the model, we may now discuss its general utility in inorganic chemistry. First, it should be emphasised that however simplified the treatment of lattice energy, and whatever the assumptions involved in defining ionic radii, the ionic model gives extremely useful predictions of heats of formation of a large number of compounds (often within 10 percent), using a limited set of data. No other model of chemical bonding is capable of predicting heats of formation so simply or accurately. Furthermore, the fact that the simple expressions can be improved, without too great an effort, to give even more accurate energies (see Phillips and Williams) suggests that the model is not too far from being true. The ionic model gives useful predictions for almost all compounds of the alkali and alkaline earth metals, the lanthanides, and the simple salts of the first row transition metals.

The basis of the model is that the energy needed to form the ions is recouped by the lattice energy of the resulting solid. Thus KCl_2 is predicted to be unstable since the large second ionisation energy of potassium is not offset by a greatly increased lattice energy in KCl_2 (by comparison with KCl). The simple salt CaCl is predicted to be unstable, not with respect to the constituent atoms but with respect to the reaction $2CaCl \rightarrow Ca + CaCl_2$. Here the second ionisation energy is offset by a greatly increased lattice energy. It should be noted that instabilities with respect to other ionic salts and not to the constituent atoms are very common. It should further be noted that there is no requirement for a 'closed shell' ionic configuration although it is of course the case for alkali and alkaline earth compounds; indeed the ionic model is very successful in predicting stabilities of oxidation states of the first row transition metals which do *not* have closed shell ions (see Johnson).

The use by the ionic model of experimentally determined data (such as ionisation energies, electron affinities and X-ray crystallographic data) makes the model semi-empirical. Thus the use of experimental ionisation energies overcomes the problems that a molecular orbital approach would have in calculating electron repulsion and correlation effects. It has been claimed that even in the most favourable cases (such as CsCl) covalency contributes at least 10 percent of the binding energy of the solid, yet by neglecting any shared electronic density (i.e. covalency) and assuming a complete separation of charge, the model automatically overestimates the ionic binding energy, and the neglect is at least partially compensated. This self-compensating effect is shown graphically by the example first given by Phillips and Williams for lithium metal. It the metal is assumed to have the structure Li^+Li^-, then the predicted heat of atomisation is 92 kJ/mole, against the observed 162 kJ/mole. In percentage terms, the error is large, but 70 kJ/mole is not an excessively large energy, and the model is so obviously false that so small an error is surprising. We may deduce that reasonably accurate thermodynamic predictions do not necessarily support an ionic formulation.

Apart from the obvious cases such as Li metal, can we predict when the model is appropriate? The classic example is silver chloride and the corresponding iodides and bromides, all of which have experimental heats of.formation significantly more negative than the calculated values. The additivity of bond lengths is also upset by the interionic distances in these silver halides, the silver ion appearing to shrink along the series AgF, AgCl, AgBr, AgI. These two properties may be taken as indications of considerable non-ionic character, not altogether surprisingly in view of the known sensitivity of silver halides to visible light. Fajans has proposed three rules for the predictions of substantial covalent character:

1) When the cation is small and highly charged. Under these conditions the cation will be able to move closer to, and will attract charge more strongly from, the anion – corresponding to the introduction of covalent character. Thus simple aluminium salts such as $AlCl_3$ which, according to the ionic model contain Al^{3+}, ionic radius 0.5 Å. show several features typical of covalent solids, notably in their ease of sublimation. Physical data (such as boiling points) show a steady progression from the almost pure ionic NaCl through $MgCl_2$ and $AlCl_3$ to the highly covalent $SiCl_4$. By contrast the trivalent lanthanide ions with a much greater ionic radius (around 1.0 Å) show strong ionic character in their chemistry.

2) When the anion is large and carries a high charge. The larger the anion, the more diffuse the electron clouds will become (as a result of electron repulsion), and conse-

quently overlap with the surrounding cation will increase. Thus AgF is described well by the ionic model, but the other halides show increasingly strong deviations. Similarly sulphides are usually appreciably more covalent than chlorides although the ions are isoelectronic.

3) When some of the electrons left on the cation are weakly shielding. If we consider magnesium and zinc, the ionic radii are fairly close (~0.6 Å) but the electron affinity of the Zn^{2+} ion is much greater than that of Mg^{2+} as a result of poor shielding by the 3d electrons in Zn^{2+}. In M.O. terms we would expect a greater delocalisation of charge towards the Zn^{2+} ion in such conditions. The chemistry of zinc is appreciably more covalent than that of magnesium, a fact of some significance in the biochemistry of these elements.

It should be fairly clear that Fajans' predictions are in good agreement with those from a simple M.O. theory based on overlap and orbital energy considerations.

From a molecular orbital point of view, the ionic model excludes all mixing between atomic orbitals, and assumes all overlaps to be zero. The two 'molecular obitals' formed by the interaction of two ions are the anion atomic orbital ϕ_a (doubly occupied) and the cation atomic orbitals ϕ_c (unoccupied); the energy of ϕ_a and ϕ_c are shifted from the free atomic orbital energies by the electrostatic energy of interaction of the ions, and by electron repulsion effects arising from the redistribution of electrons. The effect of the field of the cation on ϕ_a is to lower its energy. Referring to the secular determinant Eq. (2.6), the 'off-diagonal' elements $\mathcal{H}_{ij}$ are zero, and only the diagonal $\mathcal{H}_{ii}$ elements are shifted by the interaction of the two elements. We may thus expect large deviations from the ionic model when the off-diagonal ($\mathcal{H}_{ij}$) elements are expected to be large – when the orbital energies are close, and when high overlap is expected.

Solvation energies show some interesting parallels with lattice energies: the smaller and more highly charged the ion, the higher the solvation enthalpy. Thus both lithium and fluoride ions have high solvation energies. It is possible to treat the solubility of ionic salts using a thermodymanic cycle involving lattice and hydration energies, but as the enthalpy terms are often small, entropy effects (see below) are often important. The subject is discussed in detail by Johnson, and Phillips and Williams.

In summary, the ionic model provides a useful approach to a great deal of simple inorganic chemistry. It has the advantage that it is very easy to predict when it is going to fail or give a poor result (for example, for binary compounds where orbital energies are close, or in calculating the stabilities of metal complexes with neutral ligands). The ionic model is extremely useful even if it is not always a valid approach.

d) Valence Shell Electron Pair Repulsion

This is not, strictly speaking, a model of chemical bonding, but the basis of a method of predicting molecular geometry. Its origins lie in valence bond theory and empirical observation, and it seems appropriate to discuss it at this point. It was initially noted by Sidgwick and Powell that the structures of many simple molecules in which the bonding is essentially covalent may be explained if the electron pairs of the central atom are assumed to repel each other. Thus for methane, with four electron pairs forming four single bonds, a tetrahedral arrangement is predicted. In ammonia, which also has eight electrons (or 4 pairs) in the valence shell of the nitrogen atom, a tetra-

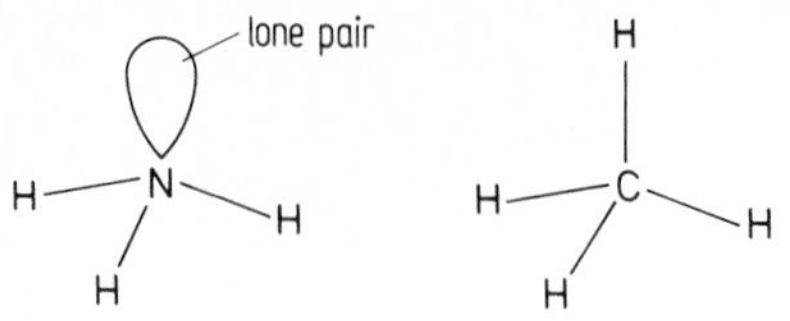

Fig. 5.4. The structures of methane and ammonia

hedral arrangement is again predicted but now one of the apices of the tetrahedron will be a lone pair, giving a pyramidal shape for the molecule (Fig. 5.4).

The general rule for predicting the stereochemistry around a given atom (assuming, in the first instance, only single electron pair bonds to be involved) is to calculate the number of electrons (and hence the number of electron pairs) in the valence shell of the atom, and then dispose them as far apart from each other as possible. This approach predicts that atoms with the same number of electrons in their valence shell (in the molecule) will have the same ionic stereochemistry – this has already been mentioned for the specific example of EH_4 compounds in Chap. 2. Table 5.2 gives the arrangements predicted for various numbers of electron pairs. It will be noted that there are two possible arrangements for 5 electron pairs – the trigonal bipyramidal is the more common.

Table 5.2. Structures predicted by VSEPR theory for a given number of electron pairs

No. of electron pairs		Structure		Example
2		Linear	—E—	MgF_2
3		Triangular (planar)	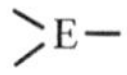	BF_3
4		Tetrahedral	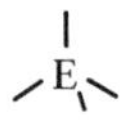	CF_4
5		Trigonal bipyramidal (tbp)		PF_5
	or	Square-based pyramid (sp)	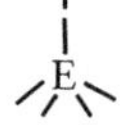	$InCl_5^{2-}$
6		Octahedral		SF_6
7		Pentagonal bipyramidal		IF_7

The predictions of the basic stereochemistry are simple and, for s and p block elements, are found to agree well with experiment – this is in fact the major attraction of the theory. A theoretical justification is rather less easy. The explanation usually advanced is based on the Pauli principle (see page 24): it follows from the Pauli principle that two electrons having the same spin ($+\frac{1}{2}$ or $-\frac{1}{2}$) tend to avoid each other.

Thus if we have 2n electrons in an atomic valence shell, then n electrons with $m_s = +\frac{1}{2}$ will keep as far away from each other as possible, and will tend to be found in regions close to the apices of a polyhedron with n vertices. Any ligands approaching the central atom will be able to share more electron density (i.e. from a covalent bond) if they approach these n vertices. At the same time the n electrons with $m_s = -\frac{1}{2}$ will approach the n electrons with $m_s = +\frac{1}{2}$ so that they too will be directed towards the ligands, giving a resultant electron distribution of electron pairs directed towards the apices of a polyhedron with n apices. We thus arrive at the arrangements shown in Table 5.2.

The disadvantage of this approach is that we have regarded spin correlation (the mutual avoidance of electrons with parallel spins) as the dominant feature in determining the structure; in some cases it is possible to ascribe the failure of the model to the failure of this assumption.

The above is the basis of the *valence shell electron pair repulsion* (VSEPR) model for the prediction of molecular structure as originally proposed by Sidgwick and Powell. The model was refined by Nyholm and Gillespie to take account of the fact that lone pairs are different from bonding pairs. They proposed that since lone pairs are localised on the central atom and not shared with another atom as a result of covalent bonding, they will repel each other more strongly than for example two bonding pairs. This leads to the order of interactions lone pair: lone pair > lone pair: bonding pair > bonding pair: bonding pair.

This simple extension is remarkably successful and very useful. If the molecule SF_4 is considered, we would expect that, with 10 electrons in the valence shell (six from sulphur and one each from the fluorines), the structure would be based on a trigonal bipyramid (or less probably, a square pyramid). There are five pairs of electrons but only four fluorine atoms bonded to the sulphur, so we would expect one lone pair. If this lone pair occupies one of the axial positions of the trigonal bipyramid structure (Fig. 5.5 a) there will be three lone pair: bond pair repulsions at 90°, but if the equatorial position is chosen (Fig. 5.5 b), there will only be two at 90°. The repulsions at low angles (e.g. 90° rather than 120°) are more important and structure of Fig. 5.5 b is adopted.

The molecule ClF_3 is iso-electronic with SF_4 and has a structure based on the trigonal bipyramid, but with two lone pairs in the equatorial positions (i.e. at 120° to each other). Nyholm and Gillespie's approach also allows a prediction of possible distortions of molecular structure: thus, if lone pair: bond pair interactions are stronger

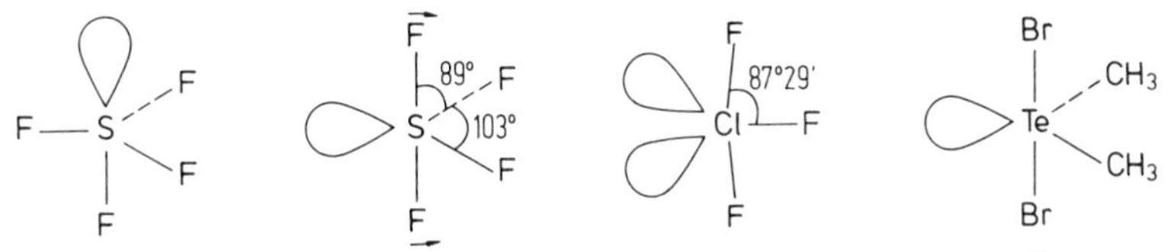

Fig. 5.5 a – d

than bond pair: bond pair interactions, we would expect the axial fluorines in SF_4 to be slightly pushed away from the lone pair; this is in fact observed and the axial F–S–F angle is 178° and not 180°. An even stronger effect would be expected for ClF_3, and is observed – the axial F–Cl–F angle is 175° (Fig. 5.5 c).

The treatment of double bonds is quite simple – the double bond can only take up one bond site, but has 4 electrons – thus its repulsion of other electron pair bonds will be greater. Thus in SOF_4 (12 valence electrons, 1 double bond (S=O)) the structure is basically trigonal bipyramidal with the double bond in an equatorial position.

It is interesting to study the variation of the structure with change of ligand. Thus the F–N–F bond angle in NF_3 is 102.1°, appreciably smaller than the equivalent angle in ammonia (106.6). This may be explained by assuming the fluorine ligand to draw more charge towards itself, and consequently reduce interaction between the bond pairs close to the nitrogen atom. A similar argument may be used to explain the variation of H–E–H angles along the series

NH_3	PH_3	AsH_3	SbH_3
106.6°	93.8°	91.83°	91.3°

where the increasing donation of electronic charge density from the heavier elements to hydrogen is presumed to reduce the bond pair – bond pair repulsions. There is, however an alternative explanation (see critique). Similar arguments explain the fact that in trigonal bipyramidal compounds such as TeX_2R_2 where X is an electron withdrawing ligand the X groups tend to occupy the axial positions (Fig. 5.5 d), minimising the bond pair: bond pair interactions at 90°.

A critique of the VSEPR model

The VSEPR model is extremely successful for elements of the p block of the periodic table, and has found extensive application in non-metal chemistry. An excellent survey is given in the book by Gillespie. Its success in predicting structures is all the more remarkable when it is realised that the energy differences between the various possible structures are often quite small. Ammonia 'inverts' (corresponding to the nitrogen atom flipping above and below the plane of the hydrogen atoms) with a frequency of 10^{11} sec^{-1}. The barrier to this inversion is only 25kJ implying the planar intermediate form of NH_3 to be only slightly less stable (see also page 62). In fact, failures of the VSEPR model often arise from effects generally regarded as weak. Thus crystal packing effects may cause the $TeCl_6^{2-}$ ion to be pure octahedral in $(NH_4)_2$ $TeCl_6$ although the 14 valence electrons would imply a distorted structure. A mixed valence compound $(NH_4)_4(SbBr_6)_2$ contains $SbBr_6^{3-}$ (14 electrons) with octahedral symmetry although the $SbBr_6^{3-}$ ion appears to be distorted in solution. Steric effects may also be important: if sufficiently large groups are substtuted for CH_3 in $(CH_3)_2$ $TeBr_2$ (Fig. 5.5 d), it is possible to bend the bromine atoms back towards the lone pair.

The difficulty with $TeCl_6^{2-}$ introduces another problem – the definition of the valence shell. If the 5s orbital of tellurium is assumed to be chemically inert then the valence shell of $TeCl_6^{2-}$ contains only 12 electrons and octahedral geometry is predicted. The problem of definition of the valence shell is even more acute for transition metal and f block elements, and here the VSEPR model is best left well alone, since in most cases it is *not* possible to regard compounds as involving localised covalent electron pair bonds. There is no experimental evidence for directed lone pairs as there is for main group elements. For the elements with partially filled s and p shells, it appears to be generally possible to include s and p electrons in the valence shell. However the s electrons appear to become much shyer towards chemical bonding as one moves to the bottom right hand corner of the periodic table. It appears from Mössbauer spectra for iodine compounds that the s orbitals only play an important part in the bonding when the iodine is in a high oxidation state (+VII), and is bonded to strongly electron withdrawing ligands such as fluoride or oxygen.

Similarly, while $TeCl_6{}^{2-}$ is an octahedral complex with apparently inert 5s orbitals, $TeF_6{}^{2-}$ is unknown, and $TeF_5{}^{-}$ is square pyramidal with strong evidence from the Mössbauer spectrum that a lone pair occupies the sixth octahedral position (in accord with VSEPR predictions). The strong electron withdrawing properties of fluorine are presumably responsible for the apparent involvement of s electrons in the bonding in XeF_6 which is non-octahedral (as predicted by VSEPR). (While we are discussing the probability of participation of atomic orbitals it should be noted that VSEPR theory does *not* presuppose any hybridisation of the central atom).

The lessening involvement of s orbitals in bonding can also explain the variation in bond angles of EH_3 compounds discussed above. If the p orbitals alone are used by antimony in SbH_3, one would expect the bond angles to be 90° – very close to the true value of 91.3^o. The gradually increasing bond angle as one ascends the series to NH_3 may be taken as an indication of increasing s orbital participation; this would equally predict the lone pair to have more directional character (more like an sp^3 hybrid than a pure s orbital) and consequently be more basic; NH_3 is in fact the only strong base of the series.

Pearson has proposed that the involvement of the s orbtials may be described on the basis of the *second order Jahn-Teller effect* (Chap. 4) – a molecular vibration may distort a complex whose structure involves only p orbitals. The distorted structure may allow mixing of p and s orbitals of the central atom as a result of the lower symmetry, and this may stabilise the molecule, 'locking' it in its new arrangement. This mixing in is plausible since it will be recalled that excitation of electrons from 'non bonding' s orbitals to antibonding M.O. may be observed for s^2 ions (e.g. Pb (II) at relatively low energies (page 144).

When the bonding is highly delocalised, VSEPR theory, based as it is on localised electron pairs, is unreliable. Thus $C(CN)_3{}^{-}$, $C(NO_2)_3{}^{-}$, $(SiH_3)_3N$ all have 8 electrons in the valence shell of the central atom, yet have planar trigonal coordination, presumably as a result of a delocalised π bonding M.O. Similarly, bridging oxygen and fluorine atoms show a wide range of bond angles, and may even show linear coordination. There is little evidence of any tetrahedral character even though there are 4 electron pairs in the valence shell. This is presumably due to delocalised (i.e. at least 3 centre) bonds.

One of the useful features of VSEPR theory is that, by its prediction of electron distribution, it also allows certain predictions of reactivity. Clearly a structure with a directed lone pair can act as a Lewis base; conversely a molecule with no lone pairs and a fairly open coordination (e.g. trigonal) can accept a pair of electrons to adopt a tighter coordination (in our example, tetrahedral) i.e. act as a Lewis acid. Typical examples of Lewis bases are $SnCl_3{}^{-}$, $P(CH_3)_3$, $SO_3{}^{2-}$ (all tetrahedral); of Lewis acids BF_3 (trigonal → tetrahedral), SbF_5 (trigonal bipyramidal → octahedral). An interesting example of change of structure to give a 'tighter' coordination is the solid state form of PCl_5 : PCl_4^+ PCl_6^- – respectively tetrahedral and octahedral.

In conclusion, the VSEPR model provides a very simple means of predicting structures for a large number of compounds of p block elements, and it has in fact stimulated a great deal of research in this area. Its shortcomings are relatively easy to see and one may anticipate where the model will fail. The use of M.O. theory to predict structures is often a great deal more complicated but by virtue of the greater amount of information given, is sometimes more useful. Thus the simple molecule methylene CH_2 has six valence electrons, and is predicted by VSEPR to be trigonal (angle H–C–H 120° or slightly less). Consultation of the Walsh diagram in Fig. 3.2 predicts the spin paired form to be bent, but that the triplet (with configuration $a_1^2b_2^2a_1^1b_1^1$) will be more nearly linear as the b_1 orbital is not stabilised by bending. The spin paired form has bond angle 102.4°, while the triplet has bond angle 136° in agreement with this. In a recent review Burdett has applied simple M.O. theory to a series of structures normally treated by VSEPR methods, obtaining essentially similar results.

B. The Use of Thermodynamics in Inorganic Chemistry

Thermodynamic data is essential for any quantitative treatment of chemical stability, and consequently, for the prediction of chemical reactions. The use of thermodynamic cycles enables one to break down a chemical reaction into a series of steps, which although artificial, are more readily discussed from a theoretical viewpoint. We have already seen an example in the Born-Haber cycle for the heat of formation of ionic salts. In some respects thermodynamic cycles are a way of uniting empirical data, a new way of presenting data; even at this empirical level, it is very useful since variations from one element to another may more easily be discussed in terms of the individual steps of the reaction than by consideration of overall energies of reaction. Many examples of this are given in the book by Johnson. In this section we are concerned only with the use of thermodynamics to present data in a suitable form for theoretical discussion.

The equilibrium constant for any chemical reaction is related to the overall free energy change during the reaction. The most commonly used free energy is the Gibbs' Free Energy G, for systems at constant pressure; for systems at constant volume the Helmholtz Free Energy A is used. At constant pressure

$$\Delta G^o = -RT \ln K_p \tag{5.11}$$

where K_p is the equilibrium constant and ΔG^o is the standard free energy change of the reaction. ΔG^o may be expanded to give

$$\Delta G^o = \Delta H^o - T\Delta S^o \tag{5.12}$$

where ΔH^o is the standard enthalpy change and ΔS^o is the standard entropy change. It should be noted that ΔH^o and ΔS^o are not invariant with temperature although most frequently they are assumed to be constant – this is not a particularly grave approximation in view of the accuracy of thermochemical data and the approximations used in calculations. It is often assumed that ΔH^o is of dominating importance in determining ΔG^o – this approximation is very frequently applicable, but is sometimes misleading, especially when ΔH^o is close to zero. Some examples when the ΔS^o term is of fundamental importance are discussed at the end of the section, and for the moment we will discuss only ΔH^o terms. Generally speaking, enthalpy data are much more readily available than entropy data.

Heats of formation and bond energies

The standard heat of formation of a given compound is the enthalpy change of the system when 1 mole of compound is formed from its constituent elements in their standard state at the temperature of the measurement. A positive heat of formation (ΔH_f^o) implies that the compound is unstable with respect to decomposition into its constituent elements. The fact that many compounds with positive heats of formation may be prepared demonstrates one of the pitfalls of thermodynamics – a compound may be unstable from a thermodynamic point of view, yet react so slowly as to be effectively stable. In the presence of a suitable catalyst, such compounds may decompose explosively; good examples we have already mentioned are WMe_6 (see below) B_2H_6, SiH_4, PH_3 and NO. Compounds which are thermodynamically unstable but react infinitely slowly and hence do not reach equilibrium are described as being kinetically stable.[2]

It should also be noticed that a negative heat of formation does not imply absolute stability. Although $MnCl_3$ has $\Delta H_f^o = -460 \pm 20$ kJ/mole, that of $MnCl_2$ is -482 ± 4 kJ/mole, and $MnCl_3$ is only stable with respect to the decomposition $MnCl_3 \rightarrow MnCl_2 + \frac{1}{2} Cl_2$ in the presence of a partial pressure of chlorine. Similarly, CaCl (mentioned earlier in the chapter) would almost certainly have a negative ΔH_f^o, but is unstable with respect to $2CaCl \rightarrow Ca + CaCl_2$. In discussing thermodynamic stability, it is vital to identify the most probable decomposition product or reaction.

The dominance of localised bonding theories in the early years of quantum chemistry led naturally to the idea of *bond energy,* the energy associated with one localised chemical bond, and consequently the energy required to break the bond. Thus for a diatomic molecule, the bond energy is simply the dissociation energy. For a molecule such as methane CH_4, the bond energy is one quarter of the enthalpy change on going from one g-atom of dissociated gaseous carbon atoms and 4 g-atoms of dissociated gaseous hydrogen atoms to 1 mole of methane.

For organic compounds very good values for heats of formation may be obtained by summing the appropriate bond energies, and the heats of combustion and formation of simple alkanes and other homologues series increase linearly with the number of CH_2 groups present. The additivity of bond energies in inorganic chemistry is less satisfactory, especially where oxidation states or coordination numbers change, or where partial multiple bond character is suspected. Thus, the P–Cl bond energy calculated from PCl_3 is 325 kJ/mole; calculated from the heat of information of PCl_5 it is 268 kJ/mole.

---

[2] Thus the fact that almost all organic compounds are unstable to combustion implies that they should decompose in contact with air. The very high kinetic stability shown by organic compounds is rare in inorganic chemistry.

Inorganic chemists should thus be very wary of assuming strict additivity of bond energies. Nevertheless, for compounds where some sort of molecular unit can definitely be recognised it is useful to have a general idea of bond energies involved. The fact that the boron-oxygen and boron-fluorine bonds are so strong (523 and 645 kJ/mole respectively) explains the instabilitiy of boron compounds with respect to B–O or B–F bonded systems. By contrast, tungsten in the +6 oxidation state has a very high affinity for oxygen which dominates the chemistry of this oxidation state; however most tungsten-oxygen compounds are solids and are closely packed, with variable coordination number. It is thus difficult to define a specific tungsten-oxygen bond energy.

Thermodynamic cycles

A simple example is given in Fig. 5.6. The heat of formation of WCl_6 and the heats of atomisation of tungsten and chlorine permit the calculation of the WCl bond energy in this compound to be 330 kJ, roughly comparable with the P–Cl bond energy in PCl_3.

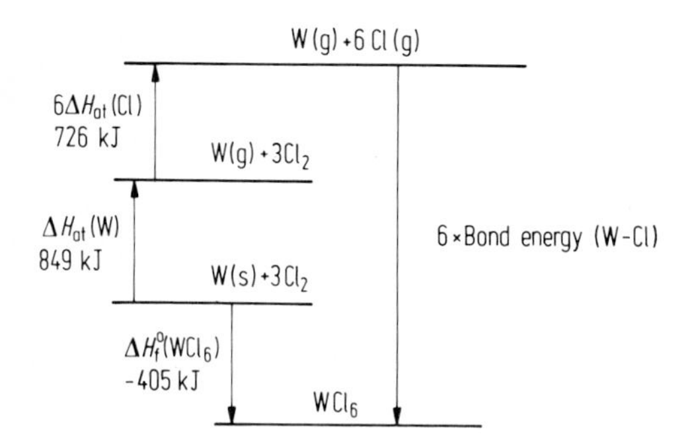

Fig. 5.6

We may see the use of thermodynamic data by considering the formation of $W(CH_3)_6$ (discussed in Chap. 3) from WCl_6 by the reaction

$$WCl_6 + 6CH_3Li \rightarrow 6LiCl + W(CH_3)_6$$

The enthalpy of formation of CH_3Li is approximately +60 kJ/mole, of LiCl –409 kJ/mole. If the enthalpy of formation of $W(CH_3)_6$ is x, the enthalpy of the reaction will be

$$x - (6 \cdot 409) + (405 - (6 \cdot 60) = x - 2409 \text{ kJ/mole}$$

Thus unless $W(CH_3)_6$ has an almost impossibly high enthalpy of formation the reaction will be strongly exothermic. In fact ΔH_f^0 for $W(CH_3)_6$ is +738,7 kJ/mole, and the bond energy W–C is 159 kJ/mole. We may draw two conclusions from this:

(i) the existence of a stable M.O. configuration (Fig. 3.5) does not immediately confer stability on a compound. In this case the bonds are not especially weak, but the heats of atomisation of W, C and H are particularly high, leading to a positive heat of formation from the elements.

(ii) If one of the products of the reaction is particularly stable (in this case LiCl) the formation of another, less favourable product may be achieved. Thus several cyclic polyphosphines, arsines and mixed ring compounds may be prepared by reactions of this type

$$\frac{n}{2}\ RPH_2 + \frac{n}{2}\ RPCl_2 \rightarrow (RP)_n + nHCl$$

The P–P bonds are rather weak, but the reaction is 'pushed over' by the formation of the strong HCl bond (bond energy 431 kJ/mole).

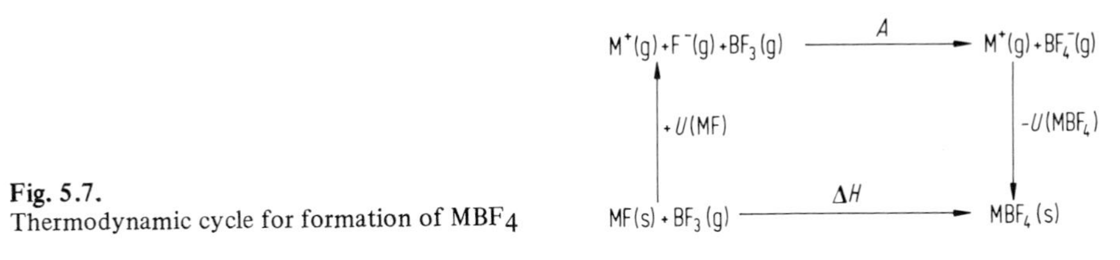

Fig. 5.7. Thermodynamic cycle for formation of MBF_4

One may also blend the ionic model in with our calculations. Thus for the reaction

$$MF\ (s) + BF_3\ (g) \rightarrow MBF_4\ (s)$$

since both MF and MBF_4 may reasonably be regarded as ionic solids, one may draw the cycle shown in Fig. 5.7 involving two lattice energy terms and the heat of association A between F^- and BF_3. ΔH is now given by

$$\Delta H = A - U(MBF_4) + U(MF).$$

Now U(MF) will always be greater than $U(MBF_4)$ since the fluoride ion is much smaller than BF_4^-. The reaction will be most strongly favoured if M is large, since then the term $(U(MF) - U(MBF_4))$ will be smallest – one would thus predict $CsBF_4$ to be more stable than $LiBF_4$. This example may readily be generalised to the statement that large anions will be stabilised in the solid state by large cations and vice versa – a useful preparative rule.

As a final example of the use of thermodynamic cycles, we may consider the nitrogen trihalides NX_3 (Fig. 5.8). The heats of formation of the fluoride and the chloride are –106 and + 257 kJ/mole respectively. Nitrogen trichloride is an explosive liquid at

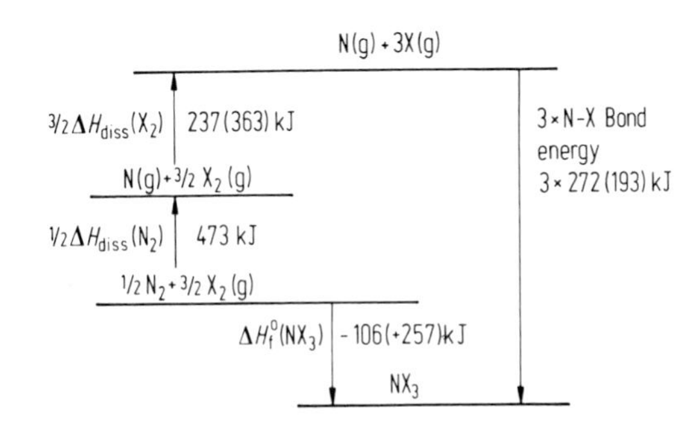

Fig. 5.8. Thermodynamic cycle for formation of NX_3. The figures in parentheses refer to the chloride, the others to the fluoride

room temperature (this fact caused the chemist Dulong to lose three fingers and an eye) while NF_3 is stable. Examination of Fig. 5.8 shows that while the N–F bond energy is greater than the N–Cl bond energy this fact alone does account for the thermodynamic stability of NF_3 – the energy required to produce the free halogen atoms is 126 kJ/mole less for fluorine than chlorine. In our discussion of the relative stabilities of NF_3 and NX_3, the relatively weak F–F bond should also be considered. In general it appears probable that many fluorine compounds are 'stabilised' by the weak F–F bond of free F_2.

Entropy effects

We may now consider some of the cases where entropy changes in a chemical reaction are important – we have already mentioned the presence of defects in crystals as a

Table 5.3

(a) Enthalpy and entropy changes for the complexation of one ligand

Metal ion	Ligand	Temp.°C	Ionic strength	ΔH° kJ/mole	ΔS° J/mole/°C
Ce^{3+}	SO_4^{2-}	25	0	20	131
Mn^{2+}	SO_4^{2-}	25	0	14	99
Fe^{3+}	SO_4^{2-}	25	0	26	163
Co^{2+}	SO_4^{2-}	25	0	7.5	70
Ni^{2+}	SO_4^{2-}	25	0	14	91
Zn^{2+}	SO_4^{2-}	25	0	17	102
Ba^{2+}	EDTA	20	0.1M KNO_3	−21	80
Ca^{2+}	EDTA	20	0.1M KNO_3	−27	113
Cd^{2+}	EDTA	20	0.1M KNO_3	−38	184
Co^{2+}	EDTA	20	0.1M KNO_3	−18	251

(b) Total enthalpy and entropy changes for complex formation

Complex		Temp.°C	Ionic strength	ΔH° kJ/mole	ΔS° J/mole/°C
$Ni(NH_3)_6^{2+}$		25	1M NH_4NO_3	− 88	−130
$Ni(en)_3^{2+}$		25	1M KNO_3	−117	− 42
$Ni(dien)_2^{2+}$		25	0.1M KCl	−106	0

EDTA see Fig. 6.21

dien $H_2N-CH_2-CH_2-NH-CH_2-CH_2-NH_2$

Data from Sillen and Martell (eds.): 'Stability Constants' and Supplement (see Biography) result of the favourable entropy gain. The expression for ΔG^o would lead us to predict that the ΔS^o term will be important when ΔH^o is near zero, or when T is large. One or other of these conditions is found in each of the three cases we discuss.

1) Solubilities of ionic solids. Although the heats of hydration of gaseous ions are always negative, they are almost balanced by the lattice energy (i.e. energy necessary to produce the gaseous ions from an ionic solid). This is not all that surprising since one is merely removing oppositely charged ions from the environment of a given ion, to be replaced by a polar solvent molecule: for a cation one exchanges the electrostatic energy due to a neighbouring anion for the electrostatic energy due to the negative end of a dipole. Ionic solids are rarely soluble in non-polar solvents.

Clearly the entropy gain on breaking up an ionic lattice is large, but the entropy of hydration of an ion is negative if the ordering of solvent molecules around the ion is great. This will arise if the ion is small or carries a high charge; thus anhydrous $Al_2(SO_4)_3$ has a very negative entropy of solution. The balance between entropy and enthalpy terms is quite delicate, but the entropy term is often dominant. For the solution of NaCl (at 25°) $\Delta H^o = +3.8$ kJ/mole, $T\Delta S^o = 13$ kJ/mole. The subject is discussed in greater detail by Phillips and Williams, and Johnson. We will note only here that detailed argument gives quite satisfactory predictions for general solubilities: solids composed of small oppositely charged ions (e.g. LiF) or ions carrying high charges ($BaSO_4$) are often poorly soluble, while compounds with relatively large ions (Na^+, K^+, Cs^+, ClO_4^-, NO_3^-) are generally soluble. Note however that very large monovalent ions will often precipitate other large ions (e.g. $KClO_4$, $[(CH_3)_4N]_2PtCl_6$).

It should be noted however, that hydrated salts must be treated separately from anhydrous salts. The hydrated salts have more favourable entropies of solution, but less favourable enthalpy terms.

The solubility of ionic salts depends of the balance of small enthalpy and entropy terms. It is thus scarcely surprising that a change of solvent can appreciably affect the solubility – the behaviour of ionic salts in non-aqueous solvents (e.g. the less polar liquid ammonia) is very different from that in water.

2) Stability constants of complexes. These may be measured as the stepwise stability constants K_n, the equilibrium constant for the reaction

$$ML_{n-1} + L \rightleftharpoons ML_n$$

or as overall formation constants β_n, the equilibrium constant for the reaction

$$M + nL \rightleftharpoons MLn$$

They have been much studied, for their theoretical interest as much as for their practical use in selectively complexing certain metals. ΔH values for the complexation reaction are often low, for reasons similar to those for enthalpies of solution, and entropy terms are again important. This is particularly true when the complexing ions carry high charges – free in solution they will produce considerable ordering of the solvent and thus have high negative entropies. Complexed together, the charge carried will be the difference between the two charges, and the size of the complex will clearly be greater than the component ions. Solvent molecules will thus be ordered less strongly and the entropy of the solution will increase (see Table 5.3).

Entropy effects are particularly marked in the formation of chelates or multidentate complexes where the ligand is bound to the metal at several points, and a large number of complexed solvent molecules are released by the incoming ligand. This is well shown by the data in Table 5.3 for complexes of Ni^{2+}. The entropy change becomes increasingly more positive in the series

$$[Ni(H_2O)_6]^{2+} + 6NH_3 \rightleftharpoons [Ni(NH_3)_6]^{2+} + 6H_2O$$
$$[Ni(N_2O)_6]^{2+} + 3en \rightleftharpoons [Ni(en)_3]^{2+} + 6H_2O$$
$$[Ni(H_2O)_6]^{2+} + 2dien \rightleftharpoons [Ni(dien)_2]^{2+} + 6H_2O$$

with each step the translational entropy lost by the incoming ligands is smaller (since there are fewer), while the translational entropy of the products is high as a result of the six water molecules produced. This so-called 'chelate effect' is particularly strong for the tetrabasic acid EDTA (ethylenediamino-tetra-acitic acid) where thé entropy gain is also augmented by the high negative charge this ion carries – complexes with highly charged cations thus have very high positive entropies of formation. The very high stability of EDTA complexes is used in complexometric analysis. An introduction to stability constants is given by Rossotti.

3) Effects of high temperatures. If the value of ΔS is large, it will increasingly dominate ΔH as the temperature increases. This is particularly important where the reactants are solids and the products are gaseous, or when the reaction produces an increase in the number of free molecules (with a consequent entropy gain). A classic example of this is the oxidation of carbon. There are two possible products:

$$C(s) + O_2(g) \rightarrow CO_2(g) \quad \Delta H^o_{298} = -394\ KJ/mole$$
$$\Delta S^o_{298} = +2.85\ JK^{-1}mole^{-1}$$
$$2C(s) + O_2(g) \rightarrow 2CO(g) \quad \Delta H^o_{298} = -222\ KJ/mole$$
$$\Delta S^o_{298} = +179\ JK^{-1}mole^{-1}$$

The very high entropy increase of the second reaction makes it increasingly favourable as the temperature rises. Solid carbon becomes increasingly more reducing as the temperature rises, and the extraction of metals from their oxides by the reaction

$$MO(s) + C(s) \rightarrow M + CO(g)$$

is of great industrial importance. If the temperature is high enough for the metal to be formed as a liquid or gas the entropy change for the reaction is even more favourable.

Thermodynamics and spectra

It should be noted that thermodynamic data may not be compared with spectroscopic data. Thus in Chap. 3 it was mentioned that the ionisation potential of WF_6^- is 5.2 eV. However the charge transfer absorption of this complex is probably above 50 kK (by optical electronegativities or extrapolation from other compounds), over 1 eV higher – it thus appears to be easier to remove an electron completely, than move an electron from a bonding (halogen) to an antibonding (metal) orbital. We may explain this by reference to Fig. 5.9 which shows the potential wells for the three electronic states we are interested in. WF_6 has no antibonding orbitals occupied and will have the shortest W–F bond length; WF_6^- has one electron in the t_{2g}^* (π antibonding – see Fig. 3.8 a) and has a slightly longer bond length, and the excited state obtained after the charge transfer absorption $(WF_6^-)^*$ has two electrons in the t_{2g}^* levels and a hole in one of the bonding or non-bonding halogen orbitals; it will thus have an even longer bond length.

We may now apply the Franck-Condon principle which requires the relative positions of the nuclei to stay fixed during the transition; the 'vertical' excitation thus populates one of the excited vibrational states of the $(WF_6^-)^*$ state; the additional vibrational energy causes the excitation energy to be greater than the 'ionisation energy' or difference between the WF_6 and WF_6^- vibrational ground state *after* suitable change of internuclear distances.

The vibrationally excited $(WF_6^-)^*$ will lose energy either by falling back directly to the electronic ground state, or by losing some vibrational energy and falling back in a stepwise process. The additional vibrational energy needed for the excitation is sometimes referred to as a *Franck-Condon* barrier. Generally, there is a spread of excited vibrational levels to which the molecule may be excited and so the absorption band has the form of several overlapping bands separated by the vibrational level spacing.

One final point may be made from Fig. 5.9: the energy of excitation is distributed among the electronic and vibrational levels of the whole molecule – it is thus possible to absorb energy greater than an individual metal-ligand bond without actually breaking any bonds. Indeed the excitation as shown in Fig. 5.9 is greater than the total dissociation energy of WF_6^-, but rather than dissociate the molecule as a whole, the system will decay to the vibrational ground state of $(WF_6^-)^*$ which is a bound state – i.e. the molecule is stable.

While the value of the ionisation energy of WF_6^- is well defined thermodynamically, it should still be borne in mind that chemically the WF_6^- unit does not exist alone – there will always be a cation, or solvent species present, and that this will radically affect the stability of WF_6^- relative to WF_6.

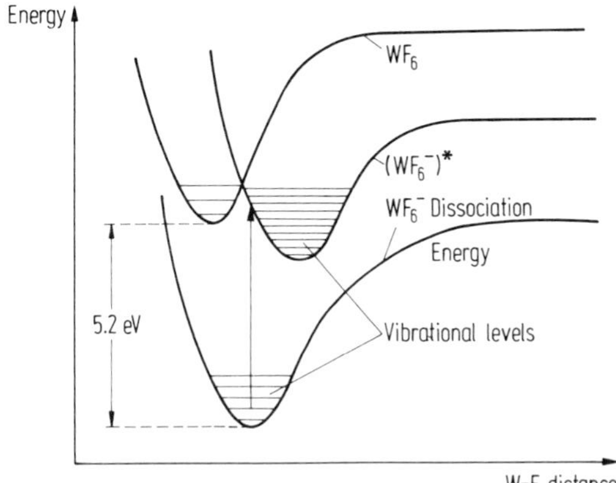

Fig. 5.9

C. Useful Concepts in Inorganic Chemistry

We endeavoured to show in the short discussion above that thermodynamics are useful for breaking down a chemical problem into a form more amenable to discussion. Generally speaking the theories of the chemical bond we have discussed are more useful for predicting ΔH than ΔS values. We shall use thermodynamic data frequently in the rest of the book. In this section we discuss some useful concepts frequently encountered in inorganic chemistry.

a) Electronegativity

Electronegativity is a qualitative concept well defined by Pauling as *the power of an atom in a molecule to attract electrons to itself.* It is thus a useful concept for predicting polarity of bonds between two elements, and hence indirectly for predicting reactivities. Pauling was also the first to introduce a numerical scale of electronegativity – as he was using valence bond theory, he regarded the electronegativity as being manifested by the increase in stability due to mixing in of an ionic resonance form into the molecular wave function. He therefore produced a series of electronegativity values based on bond energy data using the relationship

$$B(\mathrm{A-B}) = 96(x_\mathrm{A} - x_\mathrm{B})^2 + \sqrt{B(\mathrm{A-A})\,B(\mathrm{B-B})} \qquad (5.13)$$

where $B(\mathrm{X-Y})$ is the X–Y bond energy in kJ/mole and x_A is the electronegativity of element A. The constant 96 was chosen to express $(x_\mathrm{A} - x_\mathrm{B})^2$ in electron volts, and the geometric mean $\sqrt{B(\mathrm{A-A})B(\mathrm{B-B})}$ was chosen to represent the 'expected' A–B bond

Table 5.4. Pauling electronegativities

H	2.20	N	3.04	Mo	2.16
Li	0.98	P	2.19	Pd	2.20
Na	0.93	As	2.18	Ag	1.93
K	0.82	O	3.44	Cd	1.69
Cs	0.79	S	2.58	W	2.36
Be	1.57	Se	2.55	Pt	2.28
Mg	1.31	F	3.98	Au	2.54
Ca	1.00	Cl	3.16	Hg	2.00
Ba	0.89	Br	2.96	Ce	1.12
B	2.04	I	2.66	U	1.38
Al	1.61	V	1.63		
In	1.78	Cr	1.66		
Tl	2.04	Mn	1.55		
C	2.55	Fe	1.83		
Si	1.90	Co	1.88		
Ge	2.01	Ni	1.91		
Sn	1.96	Cu	1.90		
Pb	2.33	Zn	1.65		

energy in the case of no ionic resonance since the geometric mean give more consistent results than the arithmetic mean.

The series of electronegativities shown in Table 5.4 allow the prediction of a large number of bond energies with fair accuracy. Thus the P–Cl bond energy is calculated as 331 kJ/mole against the experimental value of 325 kJ/mole for PCl_3. However, as we mentioned earlier the bond energy for PCl_5 is calculated to be 268 kJ/mole from its heat of formation. Qualitatively this result is scarcely surprising – the higher electronegativity of chlorine implies that the P–Cl bond will be polarised with chlorine carrying the greater negative charge. On addition of two more chlorine atoms to PCl_3 to give PCl_5 the new chlorine atoms will have to work harder to pull more charge off the already positively charged phosphorus atom; as was mentioned in Chap. 1 it rapidly becomes difficult to remove more charge from a positively charged atom. Pauling has recognised this by his introduction of the *electroneutrality principle* requiring that the fractional charge carried by any single atom in a molecule lies between +1 and –1 electronic units. This idea is more or less accepted, in the sense that covalent bonding is assumed to donate charge to atoms which would otherwise carry a high positive charge, or that species with high negative charges will tend to act as electron donors. Many chemists would argue that charges greater than +1 unit are possible, but in the author's opinion the fact that much of the valence electron density will be found in the overlap region between atoms (and will consequently be difficult to assign to one atom or another) severely limits the usefulness of such discussion.

If we return to the mathematical approach to molecular orbital theory discussed in Chap. 2, and recall that the secular determinant for a heteronuclear diatomic molecule A–B has the form

$$\begin{vmatrix} \mathscr{H}_{aa} - E & \mathscr{H}_{ab} - ES_{ab} \\ \mathscr{H}_{ab} - ES_{ab} & \mathscr{H}_{bb} - E \end{vmatrix} = 0 \qquad (5.14)$$

then the electronegativity dependence of the binding energy shows up in the difference between $\mathscr{H}_{aa}$ and $\mathscr{H}_{bb}$ which we defined rather loosely as the orbital binding energies of atoms A and B. We are thus ignoring the overlap effects which are shown in the offdiagonal elements $\mathscr{H}_{ab}-ES_{ab}$. In Eq. (5.13) this fact is partially allowed for by taking the mean bond energy of the homoatomic A–A and B–B bonds. However we can see that even in the diatomic case we are neglecting a part of the problem. The correlation between orbital energies and electronegativities is however quite good, and Mulliken has suggested a scale of electronegativity based on this idea.

We may now see some of the problems associated with the use of electronegativity. When the electronegativity difference is small ($\mathscr{H}_{aa} \approx \mathscr{H}_{bb}$) then the neglected off-diagonal elements are very important. When the electronegativity difference is very large, the compound will be very nearly ionic, and we have left the region of a polarised covalent bond between atoms, and are concerned with the Madelung attraction between ions and the electronic energies of the ions. With species containing multiple bonds, there will be more than one atomic orbital on each atom involved in the A–B bond, and they will not generally have the same Coulomb integrals ($\mathscr{H}_{aa}$ and $\mathscr{H}_{bb}$). This has led some chemists to introduce 'orbital electronegativities' for the different atomic orbitals involved.

A final stumbling block for electronegativity arguments is the localisation approximation inherent in the assumption that the polarity or strength of a bond A–B is dependent only on the elements A and B, and not on what other atoms are bonded to A or B. We have already seen that the mean P–Cl bond energy changes on addition of a Cl_2 molecule to PCl_3 to give PCl_5 – equally bond lengths change. Addition of F^- to BF_3 to give BF_4^- increases the B–F bond length from 1.31 Å to 1.41 Å. Bonding in inorganic chemistry is delocalised, and the strength or polarity of a given bond is a function of the oxidation state and coordination numbers of the atoms involved.

Experimental correlations with electronegativity

An enormous number of correlations of experimental data with electronegativity have been made, and in general, whichever electronegativity scale is used (see below) the idea of 'the power to attract electrons in a molecule' is fairly satisfactory. The bond energy Eq. (5.13) gives good results if one is careful in its use, and one does not mix too many coordination numbers or oxidation states.

Measures of polarity of chemical bonds are less easy. Dipole moments are not reliable, since it is not always clear which way they are oriented (in H_2Te for example the direction is unknown) and a simple charge times distance method of calculation

Table 5.5. N.Q.R. data and bond ionicities for simple chlorine complexes

	ν_{Cl}(MHz)[a]	ionic character (%)	Charge on central atom
$SiCl_4$	20.4	63	2.52
PCl_3	26.2	52	1.56
SCl_2	39.2	28	0.56
Cl_2	54.5	0	0
PCl_4^+	32.4	41	2.64
PCl_6^-	29.8	45	1.70
$SbCl_6^-$	23.7	57	2.42
$SbCl_6^{3-}$	12.3	77	1.62
$TeCl_6^{2-}$	15.4	72	2.32
$AlCl_4^-$	10.6	81	2.24
$NiCl_4^{2-}$	9.2	83	1.32
$((C_4H_9)_3P)_2NiCl_2$	15.99	71	1.42[b]
CCl_4	40.6	26	1.04
CH_3Cl	34.0	38	
PF_2Cl	22.4	59	
$P(Et)_2Cl$	25.5	53	

[a] Averaged where necessary.
[b] Assuming no donation from the phosphines.

fails because of the appreciable (and unequally distributed) overlap electron density.

One of the more spectacular results in this field was the discovery that the negative end of the CO dipole in carbon monoxide is on the carbon atom – explaining why metal carbonyls are M–C coordinated but violating the predictions of electronegativity. However the dipole moment is very small, and the triple C≡O bond should lead us to be suspicious about simple electronegativity calculations.

Nuclear quadrupole resonance (Chap. 8) provides a useful measure of electron distribution, and in the case of halides where the halogen atom forms only one bond, the n.q.r. data may be related to the ionic character of the bond. A plot of ionic character (deduced from quadrupole coupling constants) against electronegativity difference for gaseous diatomic halides shows a steady (if not quite linear) increase in ionic character with electronegativity difference, levelling out at an ionic character close to 1 for $|x_A - x_B| = 2$ units on the Pauling scale. For the carbon halides CX_4, the ionic character increases from the iodide (6%) to the bromide (16%) and chloride (25%) as expected.

A selection of results is given in Table 5.5. For the chlorides listed an increase in ν corresponds to increasing covalent character; Cl_2 (ν = 54.5 MH_3) is assumed to be pure covalent. All the other elements are less electronegative than chlorine and may be assumed to carry positive charges. The ionic characters and charges on the non halogen atom have been calculated using the simplest possible model. In the series $SiCl_4 \rightarrow Cl_2$ the halogen becomes steadily more covalent. For a given element increasing the oxidation state increases the covalent nature of the bond to chlorine (PCl_3 : PCl_4^+ PCl_6^-; $SbCl_6^{3-}$: $SbCl_6^-$). Along the isoelectronic series $AlCl_4^-$, $SiCl_4$, PCl_4^+ the charge on the central atom increases only slowly even though the only change made to the system is to increase the central atom nuclear charge – the higher electronegativity of phosphorus enables it to 'pull back' almost enough electron density from the chlorine atoms to balance the increased nuclear charge Z in comparison with $SiCl_4$ or $AlCl_4^-$.

The general trends discussed above are in good agreement with electronegativity predictions. However, Table 5.5 also shows that some of the criticisms made above are valid: the different charges carried by PCl_4^+ and PCl_6^- (both phosphorus (V)) show the effect of changing coordination number. Similarly the results for $NiCl_4^{2-}$ and $[(Bu^n{}_3P)_2NiCl_2]$ show that the change of ligand can substantially affect the ionic character of a bond. Even more remarkable is the change of 12 percent for the C–Cl bond between CH_3Cl and CCl_4. The values for substituted PCl_3 compounds are even more surprising – the diethyl compound shows an increased ionic character, presumably because of the weak electronegativity of the ethyl groups. This argument does not explain the very sharp increase in ionic character of the P–Cl bond in $PClF_2$ – some double bond character has been invoked to explain this.

Other scales of electronegativity

Up to now we have concentrated on the Pauling scale, but many attempts at improvement have been made and several scales have been defined using different criteria.

Mulliken defined electronegativity as the mean of the ionisation energy and the electron affinity ($\frac{1}{2}(E.A. + I.E.)$) – this scale can easily be extended to ions but gives the rather unsatisfactory result that any cation, even Cs^+, is more electronegative than atomic fluorine. However it is clear that this definition takes us close to the idea of orbital energy. A supposed refinement to this definition was to use the *E.A.* and *I.E.* for the 'valence state' (see Chap. 2) – for example, s^1p^3 for sp^3 hybridised carbon. Unfortunately this approach involves the assumption of a particular hybridisation – the disadvantages of which we have already discussed. Jaffe and colleagues have extended Mulliken's approach by expanding the ionisation energy-electron affinity curve as a quadratic expression and defining the electronegativity as its differential coefficient with respect to charge – this has the advantage of increasing the electronegativity as the charge on the atom increases.

Allred and Rochow defined a scale related to the electrostatic force on the valence electrons:

$$\chi = \left(\frac{0.359 \ Z_{eff}^*}{r^2} \right) + 0.744$$

Z^*_{eff} is the effective nuclear charge or effective atomic number discussed in chapter 1, and r is the mean covalent radius of the atom. Sanderson has defined electronegativity as the ratio of the average electron density of the atom to the most stable electronic density corresponding to the same electronic number. The density of electrons is derived from the atomic radius of the atom, and the 'most stable density' derived by interpolation between the two nearest noble gases.

Sanderson has also introduced the concept of electronegativity equalisation: in a diatomic species the atoms will exchange charge until their electronegativities are equal – this will then be the most stable charge distribution. Jørgensen introduced a similar idea in his differential ionisation energy model (see also pp 252–253), where the binding energy is expressed as the sum of the energy resulting from the transfer of a certain amount of electronic charge between atoms (using an expression similar to the Mulliken-Jaffe electronegativity scale) and the Madelung energy resulting from such charge displacement; this is discussed in Jørgensen's book 'Modern Aspects of Ligand Field Theory'. As might be expected from a model ignoring covalent bonding, the calculated heats of formation are always too positive. Jørgensen has also introduced a series of optical electronegativities (Chap. 4) which has the advantage of varying sensibly with varying oxidation states of metallic elements.

Conclusions

The qualitative idea of electronegativity defined at the beginning of this section is extremely useful. Consideration of the meaning of electronegativity in relation to LCAO M.O. theory (i) shows that electronegativity considerations alone neglect certain features of chemical bonding and (ii) gives some indication of when they may be unreliable. Nonetheless electronegativity is useful for predicting the grosser trends of bond polarisation, even if it may fall down on some of the finer points. The fact that the many different scales of electronegativity correlate fairly well with each other suggests that the concept has a reasonably sound physical basis. Of the various scales, the Pauling system, the earliest and simplest, is still the most widely used; which is actually used is really a matter of personal choice, but the use of a mixture of values from different scales should be avoided.

b) Acids and Bases

The concept of acids and bases is one of the oldest ideas in chemistry. If some definitions have been abandoned (such as the theory of acids as spiky bodies) there still remain a relatively large number of acidbase 'theories'. The various definitions permit the classification of chemical reactions as a prelude to discussion of their similarities (or differences). Thus if we regard the two reactions

$$NaNH_2 + NH_4Cl \rightarrow NaCl + 2NH_3$$

$$NaOH + HCl \rightarrow NaCl + H_2O$$

as acid base reactions (the first in liquid ammonia, the second in water) we may proceed to discussion of relative basicity, solvation effects, hardness and softness (see below) etc. We give here what we regard as the three most important definitions currently in use.

1) Brønsted Lowry acids. The definition of an acid, initially proposed independently by Bronsted and Lowry, is a species capable of donating a proton; a base is a species capable of accepting a proton. This definition is still the most popular, and is the most useful for aqueous solutions. Thus for the reaction

$$\underset{acid}{HCl} + \underset{base}{H_2O} \rightleftharpoons \underset{base}{H_3O^+} + \underset{base}{Cl^-}$$

HCl is an acid, as is H_3O^+; H_2O and Cl^- are bases. The fact that HCl is almost completely dissociated in dilute aqueous solution enables us to say that HCl is a stronger acid than H_3O^+, since it donates a proton more readily. Similarly Cl^- is a weaker base than H_2O. It should be noted that the proton is never found as a free entity – it is always bound either to a solvent molecule or an added base.

2) Lewis acids and bases. Lewis defined a base as an electron pair donor and an acid as an electron pair acceptor. In the reaction

$$BF_3 + F^- \rightarrow BF_4^-$$

the fluoride ion (base) donates an electron pair to the BF_3 acceptor (acid) group. This approach is closely related to the electron pair single bond. In molecular orbital terms, we would say there is an interaction between a filled M.O. of the base and an empty M.O. of the acceptor to give a new M.O. There will clearly be partial covalent character in the new bond, and thus a donation of charge from base to acid. This approach can fruitfully be applied with the VSEPR model for prediction of the structures of the acid base adduct. Its independence of solvent effects makes the Lewis acid approach attractive for gaseous or weakly solvated environments.

Complexes of transition metals are frequently discussed in Lewis acid/base terms. WCl_6 is then an adduct of 6 Cl^- ions (Lewis bases) and a W^{6+} ion (the Lewis acid). Similarly, ligands such as CO which are presumed to donate some charge from their σ orbitals may be regarded as Lewis bases (see Chap. 3). Since they are also assumed to accept some charge into the π^* orbitals from the metal, they may also be regarded as Lewis acids. The apparent paradox is avoided by referring to them as π acids or σ bases. The term π acid is quite frequently encountered in organometallic chemistry.

3) Solvo-acids and bases. The fact that a solvent can dissociate even to a small extent, can have a dramatic effect on the behaviour of compounds dissolved in it. The definition of a solvo acid is a species which, upon dissolution, produces the characteristic cation of the solvent. Thus, the characteristic cation of water is H_3O^+, and a species producing H_3O^+ in aqueous solution (such as HCl) is an acid. A solvo base is a species which produces the characteristic anion of a solvent – in water this is OH^-. Some typical dissociated solvents are given below:

$$N_2O_4 \rightleftharpoons NO^+ + NO_3^-$$

$$2BrF_3 \rightleftharpoons BrF_2^+ + BrF_4^-$$

$$2NH_3 \rightleftharpoons NH_4^+ + NH_2^-$$

It will thus be noted that NH_4NO_3 is an acid in liquid ammonia, but a base in N_2O_4. Gold dissolves in liquid BrF_3 to give $AuF_4^- BrF_2^+$ (an acid). Addition of AgF (giving $Ag^+ + BrF_4^-$) produces a fall in the conductivity of the solution implying the formation of $Ag^+AuF_4^-$, an acid base reaction

$$\underset{acid}{AuF_4^- BrF_2^+} + \underset{base}{Ag^+BrF_4^-} \rightarrow \underset{salt}{Ag^+AuF_4^-} + 2BrF_3$$

In such a reaction one can see close parallels to the formation of salts in water.

The apparent ambiguity of NH_4NO_3 emphasises that acidity or basicity is essentially solvent dependent. Thus in liquid ammonia, almost all Brønsted acids appear to be strong, readily losing a proton to give NH_4^+, but few bases are strong enough to remove a proton from NH_3. Liquid ammonia is thus a levelling solvent for acids (all acids appear strong) but a discriminating solvent for bases (most bases appear weak). By contrast, in a solvent such as glacial acetic acid, most acids are weak, whereas most bases appear strong, readily abstracting a proton to give $CH_3CO_2^-$. This *levelling* or discriminating capacity of solvents is useful for determining the relative strengths of two acids (for example) both of which appear to be strong in water.

As a final example of the effect of solvent upon acid bases properties we may consider $AlCl_3$. In anhydrous solvents, this acts as a Lewis acid, readily accepting Cl^-, for example, to give $AlCl_4^-$. On dissolution in water the hydrated aluminium and chloride ions are produced:

$$AlCl_3 + nH_2O \rightarrow [Al(OH_2)_6]^{3+} + 3Cl^-(aq)$$

However the hydrated aluminium ion readily loses protons

$$[Al(OH_2)_6]^{3+} + H_2O \rightleftharpoons [Al(OH)(OH_2)_5]^{2+} + H_3O^+$$
(and further steps).

and in fact is a Brønsted acid in aqueous solution, giving a strongly acid solution. This fact was used by early alchemists who frequently used alum (a hydrated potassium aluminium sulphate) as an acid in their preparations. Consideration of the solvent, and its behaviour is equally important in the discussion of hardness and softness.

c) Hardness and Softness

The concept of hardness and softness is less well defined than electronegativity, and is indeed disputed by some chemists but is an attempt to allow for some of the failings of electronegativity in thermodynamic predictions. We give here four regions of inorganic chemistry where hardness and softness are often invoked.

1) Stability constants of metal complexes. In their review in 1958, Ahrland, Chatt and Davies noted two distinct trends in the stabilities of metal complexes. In discussing the relative affinities of central atoms in a complex for ligating atoms in groups VB, VIB and VIIB of the periodic table they noted that there was invariably a large difference between the first member of a group (i.e. N, O or F) and the succeeding members. Some elements formed the most stable complex with the first member of a group, while others showed the strongest affinity for the second or later members of the group (e.g. for group VIIIB Cl, Br or I). They distinguished the two types of acceptor elements as class (a) or class (b). Ahrland gives the different affinities of metals for donor atoms as follows:

Group	Class (a)	Class (b)
VIIB	$F \gg Cl > Br > I$	$F \ll Cl < Br < I$
VIB	$O \gg S > Se > Te$	$O \ll S \simeq Se \approx Te$
VB	$N \gg P > As > Sb$	$N \ll P > As > Sb$

It was possible to classify almost all metallic or metalloid elements as (a) or (b) and the distribution shows a certain pattern; in the periodic table there are two triangular 'islands' of class (b) character: Cu/Rh, Pd, Ag/Ir, Pt, Au, Hg; Sb/Tl, Pb, Bi and elements close to the frontiers of these 'islands' show at least some class (b) characteristics. The class (a) or (b) character is dependent on the oxidation state of the central atom – a metal generally becomes more class (a) as its oxidation number increases; metals in low oxidation states are more class (b) and have strong affinities for species not normally

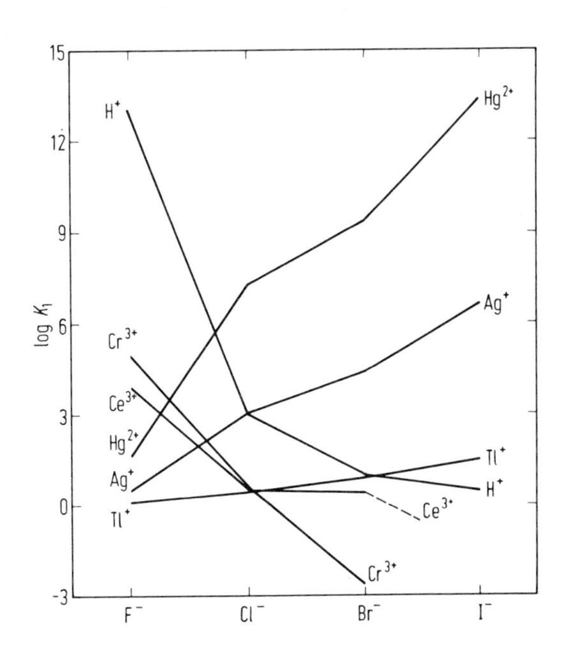

Fig. 5.10. Stability constants of halo-complexes of various cations. The values for the proton have been increased by 10 units; HF is found in solution as the strongly bound ion pair $H_3O^+F^-$

regarded as strong donors from an acid base point of view (e.g. CO). The variation in stability constant as shown in Fig. 5.10 is quite remarkable – Hg^{2+} is 10^{12} times more attracted to iodide than fluoride in aqueous solution. The class (a) / class (b) distinction has proved useful in predicting chemical affinities (and thus in selectively complexing desired metal ions).

2) Acid base equilibria. Study of a very large number of acid base equilibria of the type A + B ⇌ AB showed that the order of 'strength' of a series of acids A depends strongly on the base B they react with. It is not possible to represent the equilibrium constants K by a general relationship of the form

$$\log K = S_A S_B \tag{5.15}$$

where S_A and S_B are 'strength factors' for the acid and base A and B. Pearson (1963) introduced the concept of *hard and soft bases,* where hard bases are those where the donor atoms are hard to oxidise and have high electronegativity, and soft bases have donor atoms of low electronegativity which are easy to oxidise. One may then classify acids according to their affinity towards hard and soft bases using the rule: *hard acids bind preferentially to hard bases; soft acids bind preferentially to soft bases.*

This approach is clearly a Lewis acid/Lewis base point of view of the class (a) / class (b) classification of Ahrland, Chatt and Davies discussed above, and indeed Pearson classes class (a) metals as hard acids, and class (b) as soft acids. Similarly F^- is a hard base, and I^- a soft base. Needless to say there are borderline cases (Br^- is intermediate) as found for the class (a)/class (b) distinction. Various attempts to express softness in numerical terms (for example by rewriting Eq. (5.15) as

$$\log K = S_A S_B + C_A C_B \tag{5.16}$$

where C_A and C_B are 'softness parameters') have met with limited success – enough at least, to justify the continued use of softness as a concept. It should be noted that Eq. (5.16) emphasises that softness is *supplementary* to acid or base strength – the $S_A S_B$ parameters may well be a great deal more important than C_A or C_B. Thus, although S^{2-} is generally regarded as soft, it is a strong base, and binds the hard proton strongly to give HS^-.

3) Heat of reaction. Is is a necessary result of Pauling's electronigativity theory that the reaction

$$AB + CD \rightleftharpoons AD + CB$$

will have a negative enthalpy change if B is more electronegative than D *and* C is less electronegative than A. In other words, the reaction

$$LiF + CsI \rightarrow LiI + CsF$$

should be favoured since the most electronegative element (F) is bound to the least electronegative (Cs). In fact it is unfavourable; Pearson (1968) gives several other examples of failure of the rule. Thus, while one may excuse the above reaction on the grounds that the compounds are nearly ionic (lattice energies in fact predict the heat of reaction correctly), it is harder to explain the negative enthalpy change of

$$CH_3F + CF_3H \rightarrow CH_4 + CF_4 \qquad \Delta H = -80\ \text{kJ/mole}$$

If we accept that carbon in CF_3 is more electronegative than in the CH_3 radical, one would expect the heat of reaction to be positive (it has been estimated as +370 kJ/mole). Pearson's explanation for this result proposes that the carbon in a CF_3 group is surrounded by electronegative fluorine atoms and is consequently made harder than the carbon atom in a CH_3 group, and thus has a greater affinity for a 'hard' fluorine atom to give CF_4 and CH_4. This tendency to go to the two extremes of a possible series of compounds has been called 'inorganic symbiosis' by Jørgensen; it is associated with the preparatively useful idea that it is easier to bind soft groups to centres bonded to other soft groups than to centres bonded to hard groups.

4) Heats of formation. In a critical review of the subject matter discussed in this section Hale and Williams discussed the heats of formation of inorganic compounds and showed that for a series of binary compounds of two metals (one class (a) the other class (b)) with approximately identical ionic radii, the heats of formation for the class (b) metal compounds were consistently more negative than would have been expected by comparison with the class (a) metal compound, and were particularly favourable when the other element present could be regarded as soft (i.e. $Te^{2-} > Se^{2-} > S^{2-} > O^{2-}$ – the same affinity order as for stability constants of class (b) metals).

Theoretical aspects

We must now try to relate the class (a)/class (b) or hard/soft behaviour discussed above to the models of chemical bonding we have previously discussed. One of the simplest ways of looking at softness or class (b) behaviour is that it represents the off-diagonal element ($\mathscr{H}_{ij} - ES_{ij}$) in the secular determinant which (as we commented earlier) is overlooked by electronegativity arguments. This seems all the more probable as many soft-soft (or soft-class (b)) interactions involve cations with high ionisation energies and anions with low electron affinities i.e. the orbital energies of donor and acceptor species may be quite close, leading, as was shown in Chap. 2, to an increased importance for off-diagonal elements. We return to this approach in discussing softness parameters later on.

Several phenomenological explanations of hardness and softness have been proposed and it is generally accepted that all can play some part in governing affinities. Hard-hard interactions are almost universally admitted to be meinly electrostatic. Most explanations therefore approach soft-soft interactions. Increased covalence is generally accepted as an important feature. Multiple bonding, and in particular synergic π back donation as discussed for metal carbonyl complexes, are also thought to be important This effect is almost certainly important for transition metals in low oxidation states where the d orbitals are energetically accessible: it is less clear how important are the unoccupied high energy d orbitals in the p-block elements (see Chap. 7).

Polarisation of soft acids and bases by each other has frequently been proposed as a source of affinity. Since many soft acids and bases are heavy ions containing a large number of electrons, one might well expect an important contribution from such polarisation and from the London energy. Unfortunately comparison of polarisabilities with softness shows little correlation; while such effects may contribute, they are clearly not determining.

Solvent effects

Most of the experimental data discussed above was obtained in solution, usually aqueous solution. The interaction of two species A and B will inevitably involve some rearrangement of the solvent molecules around A and B. If we consider the case of an interaction between a metal cation M^{n+} and an anion X^- in aqueous solution, we will need to provide energy to remove water molecules from M^{n+} and X^- so that they may bind to each other, and clearly the apparent affinity of X^- for M^{n+} will be affected by their relative affinities for the solvent since the reaction we are in fact studying is

$$X^-(H_2O)_a + M^{n+}(H_2O)_b \rightleftharpoons MX^{(n-1)+} \cdot (H_2O)_c + (a+b-c)\,H_2O$$

Water is a very hard base and strongly solvates hard cations; the replacement of water by a hard base involves little enthalpy change since there is often no great difference in affinity. We saw in the discussion of stability constants that several complexation reactions are entropy controlled with ΔH small and sometimes positive; this is typical of hard-hard interactions. In contrast, soft or class (b) species are often feebly solvated and the formation of an M–X complex may be associated with a large negative ΔH; however the positive entropy change is now much less important, if still present at all. In summary, soft-soft interactions in aqueous solution are favoured by enthalpy, hard-hard interactions by entropy.

However this produces a very strong dependence on solvent of class (a) and class (b) behaviour. Because water is a hard solvent, in aqueous solution, other bases take on more class (b), soft character by comparison with (usually hypothetical) gas phase experiments. The strong solvent dependence prohibits comparisons of behaviour in different solvents, and implies that hardness or softness is not an implicit property of an atom, but is strongly dependent on environment. Thus, in nitromethane, the complexation constants (and solubility products) of all the silver halides are roughly equal.

Softness parameters

Several attempts have been made to produce softness scales, generally for aqueous solutions of ionic species. Ahrland has proposed a parameter σ_A given by the heat of the reaction

$$M(s) \rightarrow M^{z+}(aq) + ze^-(gas)$$

divided by z, the charge on the cation. In a long review article on hardness and softness, Jørgensen (1975) points out that essentially the same parameter may be extracted from $\sigma_A = \frac{E^o + 4.5}{z}$, where E^o is the standard electrode potential referred to the standard hydrogen electrode, and the value 4.5 allows for the heat of formation and hydration of the proton. This scale has the advantage of balancing the ionisation energy of the cation against its solvation energy.

Klopman, in a more general treatment of chemical reactivity, used perturbation theory to discuss the interactions between two entities in solution. He concluded that when the orbital energies of the donor and acceptor are very different, the interaction will be essentially electrostatic (charge controlled). When the two orbitals are closely matched in energy, then the transfer of charge from donor to acceptor (i.e. covalence) becomes important (frontier controlled). This is close to what we might expect from consideration of the secular determinant (see above) and its off-diagonal terms. By considering orbital energies, and the solvation energies, he was able to calculate softness scales for anions and cations showing fairly good agreement with Ahrland's series, and with the experimentally determined classification of Table 5.6.

Table 5.6. Hard and soft classification of acids and bases

Hard	Intermediate	Soft
	Acids	
H^+, alkali and alkaline earth metal cations. Mn^{2+}, La^{3+}, Ce^{3+}, Cr^{3+}, Fe^{3+}, Al^{3+}, Sn^{4+}, Ti^{4+}, $AlCl_3$, BF_3, SO_3, CO_2, RCO^+	Fe^{2+}, Co^{2+}, Ni^{2+} Cu^{2+}, Zn^{2+}, Sn^{2+}, Bi^{3+}, SO_2	Cu^+, Ag^+, Au^+, Tl^+ $Hg_2{}^{2+}$, Pd^{2+}, Pt^{2+} Hg^{2+}, Te^{4+}, Bulk metals, I^+, I_2 Br_2, Tl^{3+}, B_2H_6
	Bases	
H_2O, OH^-, F^-, $PO_4{}^{3-}$, $SO_4{}^{2-}$, Cl^-, $CO_3{}^{2-}$, $ClO_4{}^-$, $NO_3{}^-$, NH_3, ROH, RO^-	C_5H_5N, Br^-, $NO_2{}^-$, $SO_3{}^{2-}$, N_2, $N_3{}^-$	R_2S, RSH, RS^-, I^-, SCN^-, $S_2O_3{}^{2-}$, R_3P, R_3As, CN^-, RNC, CO, C_2H_4, H^-, C_6H_6

R = alkyl group.

Conclusions

Hardness and softness (or class (a)/(b) behaviour) are useful terms for describing a certain type of chemical affinity. This affinity can be described fairly well by theoretical means, provided that solvent effects are considered. However, it should be borne in mind that many factors other than hardness or softness govern chemical affinities, and that hardness and softness are often only useful in an 'an all other things being equal' approach. Thus a transition metal in a low oxidation state may form a stronger iodide complex than a chloride. In a high oxidation state, the chloride may be more strongly bound than the iodide but both halides will be more strongly bound than in the low valence state.

These problems have led certain chemists to avoid overmuch speculation on the significance of hardness and softness and stick to empirical data such as the Ahrland, Chatt and Davies class (a)/class (b) grouping. It is probably true to say that hardness and softness are a convenient way of classifying a part of chemical intuition.

d) Oxidation Numbers and Oxidation States

The assignment of oxidation states to atoms in a molecule is a routine procedure in chemistry. The approach is generally to compare the molecule with a hypothetical ionic structure, using electronegativity considerations to decide the charge distribution. Thus in $NaIO_4$, sodium and oxygen are assumed to have oxidation numbers +1 and −2 respectively, leaving iodine, by simple book-keeping, in the +7 oxidation state. The separation into oxidation states is thus a simple way of splitting up the molecular electron count we discussed in Chap. 3. It is extremely useful in determining the stoichiometry of oxidation-reduction reactions.

However, it is clear that the oxidation number is an artificial quantity, increasingly so as the bonding in a given molecule approaches non-polar covalent. Thus the assignment of an oxidation state to the atoms in methane would be of little interest; in fact, organic chemists often think of oxidation or reduction in terms of hydrogen or oxygen transfer, rather than electron loss or gain. If we accept that in inorganic chemistry bonds are more often ionic or polar, what significance does this give to the oxidation state? Common sense (and the electroneutrality principle) tells us that the oxidation number is not anywhere near the charge on the atom, but it may be more reasonable to assume that the charge on a given atom increases monotonically with its oxidation number. Generally speaking, this is a useful approximation – it appears to be supported by the charges on phosphorus and antimony atoms in Table 5.5 – however Table 5.5 also underlines the dependence of charge on coordination number. Where coordination numbers vary, or the bonding is delocalised or relatively non-polar, the charge/oxidation state relation should be regarded with suspicion.

Spectroscopic and magnetic measurements often imply a rare earth or transition metal atom to have a definite f^n or d^n configuration. Jørgensen, in his comprehensive book on oxidation numbers and states, defines a *spectroscopic oxidation state* as being determined by the preponderant electron configuration permitting the classification of the ground and low-lying excited states. This is a useful concept, especially for transition metals, and is useful for discussing chemical as well as spectroscopic behaviour. One may identify changes in n, the number of electrons in the partially filled shell as a result of electron transfer, either chemically or spectroscopically induced. Even here, however, one may have trouble, since while central atoms may be classified according to their n value, ligands may be 'collectively' oxidised or reduced, the electron or electron hole being delocalised over several atoms.

Experimental data relating to oxidation numbers

The isomer shift parameter δ measured by Mössbauer spectroscopy has provided some useful information on how seriously to take oxidation states. The ESCA chemical shift has also been used, but the unknown relaxation effects (see Chap. 8) make this a rather ambiguous method and we will not discuss these results here. (Full details of the significance of the Mössbauer isomer shift are given in Chap. 8; here it will suffice to note that δ is proportional to the electron density close to the nucleus studied).

High spin Fe^{2+} and Fe^{3+} complexes have Fe isomer shifts separated by about 1mm/s as would be expected for ionic $3d^6$ and $3d^5$ systems. Low spin ferro- and ferricyanides have very close isomer shifts, implying the electron densities to be very nearly identical, even though they may definitely be distinguished spectroscopically as Fe(II) and Fe(III) systems. Mössbauer spectroscopy was also able to show that Turnbull's blue and Prussian blue (both $Fe_7(CN)_{18}$) contain high spin ferric and low spin ferrous ions and are identical compounds $(Fe^{3+})_4(Fe(CN)_6)_3$. In these examples, charge/oxidation state correlation would only appear to be valid for the high spin iron ions. For low spin iron, covalence (i.e. d, s, and p character in the σ bonding orbitals) causes approximate electroneutrality.

Tin forms a series of compounds containing the $SnCl_3$ moiety (such as $(OC)_4CoSnCl_3$), from $SnCl_3^-$ (presumably Sn(II)) to $SnCl_4$ (Sn(IV)). A study of a series of $LSnCl_3$ compounds shows a steady progression of the tin isomer shift from $SnCl_3^-$ to $SnCl_4$ with no drastic change due to an increase in oxidation state. Iodine shows a steady increase in charge with ocidation state form iodide (I(–I) to I(C), and then a sharp discontinuity on passing to IF_7, IO_4^- and IF_6^+, almost certainly due to a sudden sharp increase in the bonding participation of the iodine 6s orbital.

Finally, we may return to the transition metal's to mention iridium, which, when bonded to electronegative elements (oxygen or fluorine) shows a steady increase in charge from Ir(III) to Ir(IX), but in low oxidation states, when bonded to ligands of low electronegativity (hydrogen, tertiary phosphines, CO, etc.) shows almost no variation in isomer shift between Ir(–I) and Ir(III).

In the examples above, we have tried to show that an oxidation state does not *immediately* imply a particular chemical character for the atom concerned, although in certain cases (notably ionic compounds) it can do so. Iridium (–I), (I) and (III) complexes differ in stereochemistry and reactivity, and the classification according to iridium oxidation state is empirically useful, but the differences come from the total number of electrons involved in holding the molecule together. In view of the delocalised bonding of the systems, we may well be obscuring certain features of the structure by localising chemical changes as a change of oxidation state of the central atom. Oxidation numbers are useful for electron counting, and for classifying compounds and reactions fof a given element but we should be careful not to let them assume a dominating position in our thoughts. As with other concepts in inorganic chemistry we should know when they may lead us astray.

e) Conclusions

Having discussed in this chapter some alternative methods of describing chemical interactions (both theoretical and phenomenological) it may perhaps be useful to recapitulate briefly some of the features of the M.O. approach with which this book is mostly concerned.

Molecular orbitals must reflect the symmetry of the molecule, and this immediately gives a relationship between structure and bonding. Even if we have only the haziest idea of the form of the M.O.s we may still classify them correctly according to the microsymmetry of the system. Furthermore, the idea of a one electron wavefunction gives us an immediate link between electronic spectra and structure, since we can treat electronic excitation as a series of one electron jumps. If we now allow ourselves the Linear Combination of Atomic Orbitals approximation, we may use our knowledge of atomic orbitals to estimate the form of the molecular orbitals for a very large number of compounds. The LCAO M.O. approach does not invariably give the simplest explanation for a particular class of compounds, but in the author's opinion, it does give the most satisfactory simple explanation for the whole range of inorganic compounds. If one is only interested by a small set of compounds, then clearly one can take short cuts, but with the risk of losing generality. LCAO M.O. methods also give some physical basis for the qualitative concepts such as electronegativity discussed earlier.

The approach does of course have disadvantages – the prediction of structures is not always straightforward, although in certain cases useful predictions can be made. Even a relatively low level calculation (for example using Extended Hückel methods) can make quite accurate predictions. The greatest defect of the LCAO M.O. method is its treatment of electron repulsion, and in particular, correlation energy. This effectively prevents the accurate calculation of heats of reaction. The qualitative method we use here, ignoring almost all electron repulsion, is even worse, and, for example, the prediction of 'lone pairs' (as found in VSEPR theory) is not straightforward. As the problem of many body interactions is not exactly soluble by mathematics, we are bound to introduce some level of approximation in this area. If we are unable to predict accurately, the energies of electronic transitions, we may at least classify them by M.O. methods, and in certain specific cases, such as for the nephelauxetic effect (Chap. 4) or the 'third revolution in ligand field theory' deal with electron repulsion effects in an approximate manner. The limited success of such a simple model as LCAO M.O. theory suggests that chemists should also subscribe to Einstein's remark that 'God is subtle, but he is not unkind'.

Bibliography

Ahrland, S.: Stability constants and hardness and softness. Structure and Bonding *5*, 118 (1968) and *15*, 167 (1973)
Ahrland, S., Chatt, J., Davies, N.R.: Quart. Rev. *12*, 265 (1958)
Atkins, P.W.: Molecular quantum mechanics. London: Oxford University Press 1970
Ball, M.C., Norbury, A.H.: Physical data for inorganic chemists. London: Longmans 1974
Burdett, J.K.: The angular overlap model compared with VSEPR theory. Structure and Bonding *31*, 67 (1967)
Coulson, C.A.: Valence, 2nd edn. London: Oxford University Press 1961
Dasent, W.E.: Inorganic energetics. London: Penguin 1970
Gerratt, J.: Specialist Periodical Reports, Theoretical Chemistry Vol. 1, p.60 (R.N.Dixon, ed.) Current uses of valence bond theory. London: The Chemical Society 1974
Gillespie, R.J.: Molecular geometry. London: Van Nostrand-Reinhold 1972
Huheey, J.E.: Inorganic chemistry. London, New York: Harper and Row 1972
Johnson, D.A.: Some thermodynamic aspects of inorganic chemistry. Cambridge: Cambridge University Press 1968
Johnson, K.H.: The MS Xα method. Adv. Quant. Chem. *7*, 143 (1973)
Jørgensen, C.K.: Absorption spectra and chemical bonding in complexes. Oxford: Pergamon Press 1962
Jørgensen, C.K.: Oxidation numbers and oxidation states. Berlin, Heidelberg, New York: Springer 1969
Jørgensen, C.K.: Hardness and softness. Topics in current chemistry, *56*, 1 (1975)
Klopman, G.: Hardness and softness. J. Amer. Chem. Soc. *90*, 223 (1968)
Morris, D.F.C.: Ionic radii and electron distribution in ionic solids. Structure and Bonding *4*, 63 (1968) and *6*, 157 (1969)
Murrell, J.N., Kettle, S.F.A., Tedder, J.M.: Valence theory 2nd edn. London, New York, Sydney: J. Wiley 1970
Nyholm, R.S., Gillespie, R.J.: The VSEPR model. Quart. Rev. *11*, 339 (1957)
Orgel, L.E., Dunitz, J.E.: The stereochemistry of ionic solids. Adv. Chem. and Radiochem. *2*, 1 (1960)
Pauling, L.: The electroneutrality principle. J. Chem. Soc. *1948*, 1461
Pearson, R.G.: Hard and soft acids and bases. J. Amer. Chem. Soc. *85*, 3533 (1963); J. Chem. Ed. *45*, 581 and 643 (1968)
Pearson, R.G.: The failure of Pauling's bond energy equation. Chem. Comm. *1968*, 65
Phillips, C.S.G., Williams, R.J.P.: Inorganic Chemistry (2 vols.). London: Oxford University Press 1965

Purcell, K.F., Kotz, J.C.: Inorganic chemistry. Philadelphia, London, Toronto: W.B. Saunders 1977
Rossotti, F.J.C. in Modern coordination chemistry (J. Lewis and R.G. Wilkins, eds.) Stability constants. London, New York, Sydney: J. Wiley 1960
Sanderson, R.T.: Inorganic chemistry. New York: Van Nostrand-Rheinhold 1967
Sanderson, R.T.: Chemical bonds and bond energy, 2nd Ed. New York, San Francisco, London: Academic Press 1976
Shannon, R.D., Prewitt, C.T.: Ionic radii derived from crystallographic data. Acta Cryst. *B25,* 925 (1969) and *B26,* 1046 (1970)
Sidgwick, N.V., Powell, H.M.: Electron pair structural theory. Proc. Roy. Soc. A *176,* 153 (1940)
Sillen, L.G., Martell, A.E. (eds.): Stability constants, special publication No. 17 and Supplement No. 1, special publication No. 25. London: The Chemical Society 1964 and 1971
Slater, J.C.: The Xα approximation. Adv. Quant. Chem. *6,* 1 (1972); Int. J. Quantum Chem. *3,* 727 (1970)
Waddington, T.C.: Lattice energies. Adv. Inorg. Chem. and Radiochem. *1,* 157 (1959)
Williams, R.J.P., Hale, J.D.: Critical review of hardness and softness. Structure and Bonding *1,* 249 (1966)

Problems

1. (a) Calculate the lattice energies of FeI_3 and FeI_2 using Table 5.1 and Eq. (5.10).
(b) Given that for

$$1/2\, I_2(s) + e^- \rightarrow I^-(g) \qquad \Delta H^o = -189 \text{ kJ/g.ion}$$

$$Fe^{2+}(g) \rightarrow Fe^{3+}(g) + e^- \qquad \Delta H^o = +2957 \text{ kJ/g.ion}$$

draw a cycle allowing you to calculate ΔH^o for:

$$FeI_2(s) + 1/2\, I_2(s) \rightarrow FeI_3(s)$$

and comment on your result. What are the most important terms in the cycle?

2. (a) Draw a cycle for the reaction

$$Cu(s) + CuF_2(s) \rightarrow 2\, CuF(s)$$

Given that $\Delta H^o_f(Cu^+(g))$ = 1082 kJ/g.ion, the second ionisation energy of Cu = 1958 kJ/g. atom, and that r_{Cu^+} = 0.96 Å, calculate ΔH^o for the above reaction, estimating any necessary lattice energies from Eq. (5.10).
(b) $\Delta H^o_f(CuF_2) = -536$ kJ/mole. What is $\Delta H^o_f(CuF)$?

3. (a) Given that: $\Delta H^o_f(V^{4+}(g))$ = 10007 kJ/mole, $r_{V^{4+}}$ = 0.59 Å

$$1/2 Cl_2(g) + e^- \rightarrow Cl^-(g) \quad \Delta H^o = -227 \text{ kJ/g.ion}$$

calculate $\Delta H^o_f(VCl_4)$ using the ionic model.
(b) The experimental value of $\Delta H^o_f(VCl_4)$ is -598 kJ/mole. Compare this with your calculated value and comment on the fact that VCl_4 is a dark red liquid at room temperature.

4. The Madelung constants (M in Eq. (5.9)) for three of the possible structures for an ionic salt AB are:

Structure	Madelung constant	Coordination no.
Body centred cubic (CsCl)	1.76267	8
Face centred cubic (NaCl)	1.74756	6
Zincblende (ZnS)	1.63806	4

Assuming all other factors to be constant, what are the percentage changes in lattice energy as the C.N. changes form 4 to 6 to 8?

5. Use VSEPR theory to predict the structures of:
$SnCl_2$; $[SnCl_3]^-$; $[ICl_2]^-$; $[Sb(C_2O_4)_3]^{3-}$; $GaCl_3$; NSF_3.
Can you predict any Lewis acid or base properties of these compounds or ions?

6. Tellurium and selenium may form the ions $[Te_4]^{2+}$ and $[Se_4]^{2+}$ under suitable conditions in strong acids. Both ions are square; does the VSEPR theory predict this? If not, can you explain the failure?

7. Draw a thermodynamic cycle for the reaction

$$M_2CO_3(s) \rightarrow M_2O(s) + CO_2(g)$$

which will enable you to predict whether the thermal decomposition will be easier for M = Li or Cs.

8. (a) The enthalpy of hydration of the proton has been estimated as −1090 kJ/g.ion. Given that $\Delta H^o_f(H^+(g)) = +1530$ kJ/g.ion, $\Delta H^o_f(Fe^{2+}(g)) = +2739$ kJ/g.ion, and that ΔG^o (assumed equal to ΔH^o) for reaction $Fe(s) + 2H^+(aq.) \rightarrow Fe^{2+}(aq.) + H_2(g)$ is −42 kJ/mole, calculate the heat of hydration of Fe^{2+} (ΔH^o for the reaction $Fe^{2+}(g) \rightarrow Fe^{2+}(aq.)$). Compare your value with the lattice energy of FeI_2 (Problem 1).
(b) If the energy of hydration arises from the formation of six $Fe{-}OH_2$ bonds in $[Fe(OH_2)_6]^{2+}$, what is the strength of the $Fe{-}OH_2$ interaction? Do you think your result has any significance?

9. Dissolution of $NaClO_4$ in water produces a sharp fall in the temperature of the solution, even though $NaClO_4$ is deliquescent. Comment on the thermodynamics of this effect.

10. Many ligands (e.g. CN^-, NH_3, $CH_3CO_2^-$) will only complex metal cations in neutral or basic solution. Why?

11. Ce^{4+} is a strong oxidising agent in acidic solution, but its oxidising strength depends on the acid used, being weaker in sulphuric than perchloric acid, and weaker still in hydrochloric acid. Explain.

12. The Mond process for the purification of nickel involves reaction of CO with the impure metal at 50°C to give $Ni(CO)_4$ which is distilled and pyrolysed at 150°–300°C to give the pure metal. What is the thermodynamic basis of this process?

13. Why do atomic spectra (e.g. in atomic absorption analysis) consist of sharp lines, whilst molecular spectra show broad absorptions?

*14. Is there any theoretical reason for a polar bond to be stronger than a non-polar bond? To what extent does Pauling's bond energy equation (Eq. (5.13)) owe its validity to the fact that some of the most electronegative elements have very weak single bond energies (F–F 158, O–O 146, N–N 160 kJ/mole)?

15. The neutralisation of a sample of sulphuric acid dissolved in methyl ethyl ketone requires half the quantity of base necessary to neutralise the same amount of acid dissolved in water. Why?

16. Ni^{2+} forms no chlorocomplex in aqueous solution, but in absolute ethanol nickel salts react with chloride ion to form dark blue $[NiCl_4]^{2-}$. Suggest an explanation.

17. There is some evidence from photoelectron spectra to suggest that the cobalt atom

in Co^{2+} salts carries a higher charge than in some Co(III) complexes. Discuss this possibility in relation to the electronic structure of the species involved.

*18. In the calculation of molecular orbitals for benzene, the MS Xα method faced a major problem inherent in the model. Can you see what this is?

19. Geochemists frequently adopt a classification of the elements according to their origin or disposition in the earth's crust. As an example, consider:

Thalassophilic:	Cl, Br
Lithophilic:	Alkali and alkaline earth elements
	Al, Ln, Si, Zr, Ti
Chalkophilic:	Cu, Zn, As, S, Se, Ag, Cd, In, Hg, Pb, Bi
Siderophilic:	Fe, Co, Ni, Ir, Pt, Au.

Comment on the possible chemical origins of this distinction.

6. Mechanism and Reactivity

In previous chapters we have discussed the electronic structures and properties of inorganic compounds with little or no reference to the chemical reactions they undergo. In this chapter we will attempt to relate the electronic structure of a compound to the chemical reactions it undergoes.

Many reactions in inorganic chemistry take place very rapidly to give the products with the greatest thermodynamic stability. This contrasts with organic chemistry where reactions are generally quite slow, and give the kinetically stable product rahter than the most thermodynamically stable one. For this reason, mechanistic inorganic chemistry developed more slowly than mechanistic organic chemistry, but increasing interest in forming products other than the thermodynamically most stable, in speeding up or catalysing slow reactions, and the development of new methods of studying fast reactions has given a considerable boost to the subject.

This chapter may be divided into three sub-sections:

A. The connection between electronic structure and reactivity. This is the most theoretical part, in which we will attempt to show how the electronic structure of the reactants is changed into that of the products;

B. The classification of simple mechanistic steps into a series of reaction types.

C. Discussion of certain selected topics, which, as well as having particular features of interest, will illustrate points made in the first two sections.

For the sake of brevity, discussion of kinetics (the cornerstone of experimental mechanistic chemistry) has been excluded. I hope that this omission of trees will enable the reader to perceive the wood more clearly. Some reference to mechanistic organic chemistry has been made, but this is only to show the links between the two subjects, and is not essential to the understanding of this chapter.

A. A General Approach to Reactivity

All chemical reactions will have a limited rate, if only because of the necessity of approach of two molecules to produce the reaction (even in collisionally excited unimolecular reactions). This diffusion rate is extremely fast in liquid and gas phases, but much slower (though non-zero) in the solid state. In gas and liquid states reactions are normally slower than the diffusion rate, as a result of the activation energy of the reaction.

The traditional view of a chemical reaction shown in Fig. 6.1 a involves passage through an excited *transition state,* where nuclear positions in the reactants have been distorted in such a way as to favour the direct passage to the products. The lower the activation energy the more accessible the transition state, and the faster the reaction. Although the transition state does not actually exist (it is at a maximum of potential energy), it is useful to consider the structure of this hypothetical entity since we can then estimate how energetically accessible it will be. Figure 6.1 b shows a two step reaction with the formation of an intermediate, of high energy, between two transitions. Although it may have high energy, the intermediate is at the minimum of a potential well, and may be experimentally detectable during the reaction.

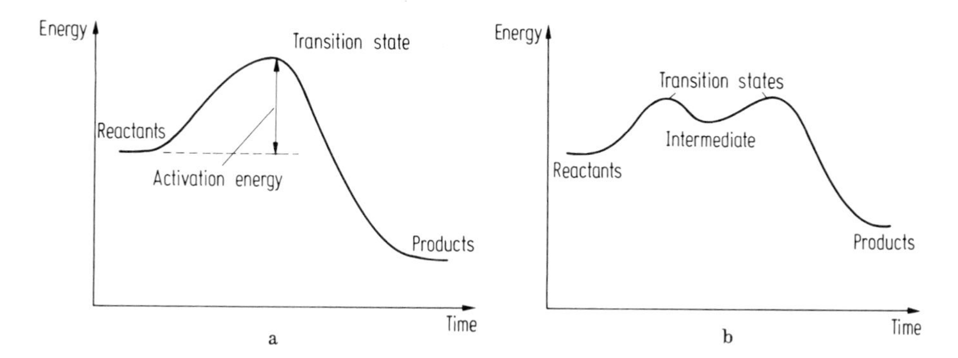

Fig. 6.1 a and b. The variation of energy of a reactant mixture viewed along the reaction coordinate. **(a)** A simple reaction. **(b)** a reaction with formation of an intermediate

The formation of the transition state involves extension or compression of various bonds in the reacting molecules, and maybe the formation or rupture of chemical bonds. There are two useful ways of looking at this – we may either concentrate on one or two localised bonds which will be broken or formed, or we may regard all the various nuclear motions (the stretching or compressing of interatomic distances) as a vibrational mode of the transition state, just as we would discuss the nuclear motions of an ordinary molecule. The first method is gerenally simpler but the second is mor general. In both cases we are seeking to estimate the energetic barrier to the formation of the transition state, by consideration of the molecular structures of the reactants.

Entropy of activation

While the activation energy is generally the dominating factor in a chemical reaction it is not invariably so. Transition state theory gives an expression for the rate constant

$$k = \kappa \frac{RTe}{Nh} e^{-E_a/RT} e^{\Delta S^{\neq}/R} \qquad (6.1)$$

where κ = transmission coefficient (generally put equal to 1 since its calculation is very complicated)

E_a = Arrhenius activation energy

$\Delta S^{\neq}$ = entropy of activation.

The enthalpy of activation in solution is given by $\Delta H^{\neq} = E_a - RT$. The entropy of activation is regarded as the difference in entropy between reactants and the transition state. One might expect that a transition state involving association between two reactant particles would have a more negative entropy of activation (as a result of loss of translational entropy) than a dissociative mechanism; there is some evidence to support this, but solvation effects can dominate $\Delta S^{\neq}$, especially when the transition state is more polarised than the reactants (as a result, perhaps, of a partial ionic dissociation). The solvent molecules are then more strongly attracted to the complex with consequent loss of entropy (see the discussion of stability constants). Although we shall concentrate on activation energies, it should be remembered that differences in activation entropies are sometimes important. Thus $[Au(dien)Cl]^{2+}$ reacts with methanol 3×10^4 times faster than the isoelectronic, isostructural $Pt(dien)Cl^+$, but the $\Delta H^{\neq}$ for the reactions are very similar, the rate differences coming from $\Delta S^{\neq}$ differences.

Bond making and breaking

We will initially discuss this topic on a localised bond model, and return to an M.O. approach later. If we accept that a localised bond involves one electron pair, then the fission of this bond can occur in two ways:

1) Each atom involved in the bond holds onto one electron, producing two radicals, each with an unpaired electron *(Homolytic fission);*

2) The bond electron pair stays localised on one of the atoms of the bond *(Heterolytic fission).* This need not necessarily involve the formation of ions – the heterolytic fission of HCl produces H^+ and Cl^-, but the fission of the B–N bond in H_3NBF_3 gives NH_3 and BF_3 with the bonding pair of the B–N bond now a lone pair on NH_3. Heterolytic fission is clearly much more probable for polar chemical bonds.

The corresponding processes of bond formation are quite straightforward – radicals may either dimerise (quite rare as their concentration is generally low), or add to a non-radical molecule to give a new radical which can either dimerise, or eliminate some other radical species, leaving behind a new molecule. The corresponding equations are

$$2A\cdot \rightarrow AA$$

$$A\cdot + B \rightarrow AB\cdot \rightarrow C + D\cdot$$

The free radical is a rather old concept in mechanistic chemistry and, historically, has always an implied reactivity. It should however be remembered that a species with one or more unpaired electrons, although a radical, is not necessarly very reactive: Mn^{2+} generally has five unpaired spins but shows no tendency to undergo radical mechanism reactions (see page 48).

Heterolytic bond formation implies the donation of an electron pair from one atom to form a chemical bond with another – that is, a Lewis acid/Lewis base reaction.

Nucleophiles and electrophiles

It is customary, when discussing heterolytic bond formation, to classify reagents as *nucleophiles,* tending to react with positive centres and acting as electron donors, or *electrophiles,* tending to act as electron acceptors and reacting with negative centres. This is an extremely useful classification for discussing reacitivity; as a general rule, species which are Lewis bases are nucleophiles, and those which are Lewis acids are electrophiles. Nucleophilicity and electrophilicity are however measured from rates of reaction, and scales of nucleophilicity are quite different from scales of basicity – this was in fact one of the starting points for hard and soft acid-base theory. Thus iodide is a weak base but quite a good nucleophile. As one might expext, nucleophilicity and electrophilicity are extremely solvent dependent.

One can, however, often make very useful predictions of reaction products simply by estimating the charge distributions in two reactant molecules and interacting the most nucleophilic atom of one reactant with the most electrophilic atom of the other. To take a trivial example, the hydrolysis of $(CH_3)_2SiCl_2$ may reasonably be expected to involve attack by nucleophilic water on the electropositive silicon atom, whose electrophilicity will be increased by the two electronegative chlorine atoms. This sort of reasoning is of course the basis of a vast amount of mechanistic organic chemistry; it is somewhat less satisfactory in inorganic chemistry where it is not always simple to identify which centres are nucleophilic or electrophilic.

Molecular orbital theories of chemical reaction

The heterolytic/homolytic approach to bond formation or breaking has several disadvantages – quite apart from the fact that it regards chemical bonds as localised, it is often difficult to relate the mechanism to the finer details of electronic structure that we have elaborated in previous chapters. It also gives us little exact understanding of how a particular bond is broken, especially for the heterolytic fission of covalent or fairly covalent bonds. Finally, it is rather hard to explain reactions where several bonds are made and broken at the same time – for example, the addition of hydrogen to a molecule may involve the simultaneous formation of two sigma bonds – such *concerted* reactions are particulary common in organic chemistry, but are by no means unknown in inorganic chemistry.

M.O. theory foresees two possible ways of breaking a chemical bond – we may either remove electrons from the bonding orbital or orbitals, or we may add electrons to the antibonding M.O. If, for the moment, we ignore electrons jumping in or out of orbitals (as found in photochemical excitations or redox reactions) we may reasonably assume that this change in population comes from the mixing of full bonding M.O. with some other empty M.O., or the mixing of an empty antibonding M.O. with some occupied M.O.

For a Lewis acid-base reaction, the approach of the base to the acid involves increasing overlap of the donor orbital with the acceptor orbital to form an occupied bonding M.O. If we consider the reaction $Cl^- + HCl \rightarrow ClH + Cl^-$ (an atom exchange reaction) we may visualise the overlap as in Fig. 6.2 a. The nucleophilic Cl^- approaches the HCl molecule and overlaps with the acceptor orbital, the empty σ^* orbital mainly localised on the hydrogen atom. This donation establishes a new HCl bond and weakens the first due to occupation of the σ^* antibonding orbital, until the original, covalently bound Cl is lost as Cl^-. The approach of the Cl^- ion thus forms a new HCl bond at the same time as the old bond is weakened.

It is possible to isolate the intermediate in this reaction; the symmetric anion $[HCl_2]^-$ (Fig. 6.2 b). The Cl–Cl distance in this anion (3.14 Å) is appreciably greater than twice the H–Cl distance in HCl (1.274 Å), as a simple M.O. treatment (cf. MgF_2 page 49) shows that there are only two electrons to assure two chemical bonds. We may also note that the reaction involves the same nuclear displacements as the asymmetric stretching vibration of $[HCl_2]^-$. This is the starting point for the second order Jahn-Teller effect (page 140) explanation of reaction mechanisms, whereby a molecular vibration lowers the symmetry of a system and allows reactant orbitals to mix and transform into the product orbitals.

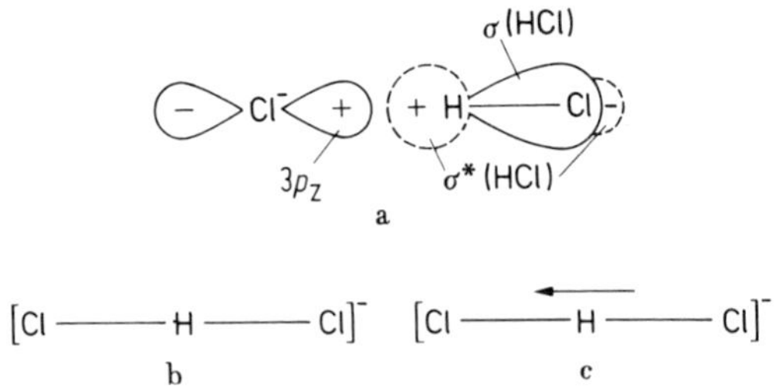

Fig. 6.2. a – c. The reaction Cl + H–Cl. **(a)** Overlap of the filled $3p_z$ orbital of Cl^- with the HCl σ orbital, **(b)** the intermediate complex, **(c)** the antisymmetric stretching mode (σ_u) leading to dissociation into Cl–H and Cl^-

It is now possible to estimate the probability of chemical reactions by considering the ease with which suitable orbitals will overlap, or, alternatively, by looking for a molecular vibration which will mix the M.O. of the reactant or reactants to give the product or products. If we can see, for example, that overlap between a donor and an acceptor orbital is impossible or very small, we may predict that the reaction will not be smooth, and will have an high energy of activation – this may be expressed by regarding the overlap criterion as a selection rule for a chemical reaction. To estimate the probability of overlap or vibrational mixing we may again use symmetry arguments.

Symmetry in chemical reactions

It will be useful to discuss exactly what we mean by symmetry as applied to reacting species. The initial symmetry of the reactants (or the final symmetry of the products) will be of little interest since it is usually lowered during a chemical reaction. It is however frequently found that some elements of symmetry are maintained during a chemical reaction. In the reaction Cl^- + HCl, cylindrical symmetry was maintained, and in the reaction $F_2 + Br_2 \rightarrow F_2BrBr$ (Fig. 6.3) we may see that a two-fold axis (along the Br–Br bond) and two perpendicular planes of symmetry parallel to this axis are maintained. It is the symmetry elements maintained throughout the reaction that are important since the orbitals involved cannot change their symmetry with respect to

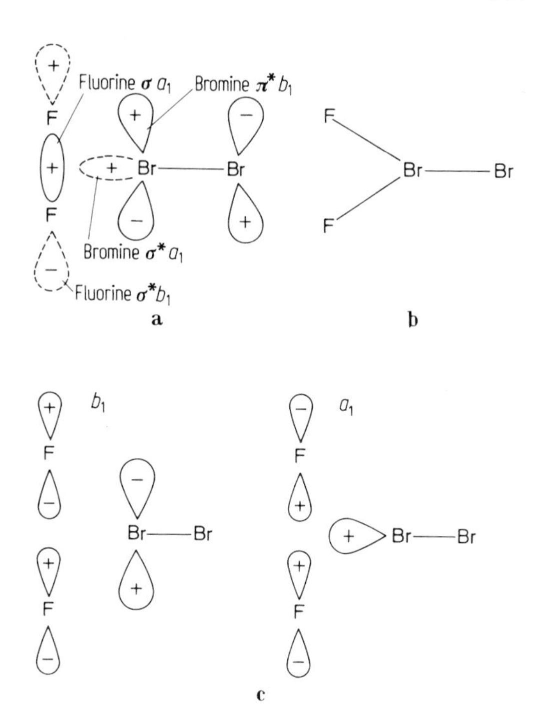

Fig. 6.3 a – c. The reaction between fluorine and bromine. **(a)** The orbitals involved, **(b)** the product, **(c)** the Br–F bonding M.O. $a_1 + b_1$

these operations. We may use this conserved symmetry to classify the orbitals and predict their mixing. It is quite easy to show (see Pearson) that any vibration responsible for the reaction or mixing must be totally symmetric (i.e. A_1) with respect to the symmetry maintained throughout the reaction. Thus, for the Cl^- + HCl reaction, the symmetry maintained is $C_{\infty v}$; the vibration corresponding to the nuclear motions in the reaction is the asymmetric stretch, which preserves cylindrical symmetry and is totally symmetric with respect to it (type σ).

Figure 6.3 shows that the occupied π^* orbitals of Br_2 may overlap with σ^* of F_2 (both b_1) and the occupied σ of F_2 may overlap with σ^* of Br_2 (both a_1) to give two Br–F bonding M.O. This reaction is indeed allowed, although the product F_2BrBr is rather reactive.

We may give a more complicated example from organic chemistry – the ring opening of cyclobutene (Fig. 6.4). Cyclobutene and butadiene both have C_{2v} symmetry. In cyclobutene we wish to break one carbon-carbon σ bond and one π bond, to produce two π bonding M.O. in butadiene. The symmetries of the relevant orbitals are shown in Table 6.1. In the C_{2v} point group we notice that the occupied orbitals of product and reactant do not correlate and that there is no means of weakening chemical bonds in cyclobutene by mixing an occupied bonding M.O. with an empty bonding M.O. (e.g. σ^* with π). We conclude that C_{2v} symmetry cannot be maintained throughout the reaction. This might also have been predicted by noticing that the hydrogen atoms of the methylene groups must be moved from out of the carbon plane (in cyclobutene) into the carbon plane (in butadiene) and that this is impossible if full symmetry is maintained. Figure 6.4 shows that it is possible to conserve either the two fold axis of symmetry (a conrotatory reaction) or the xz plane of symmetry (a dis-

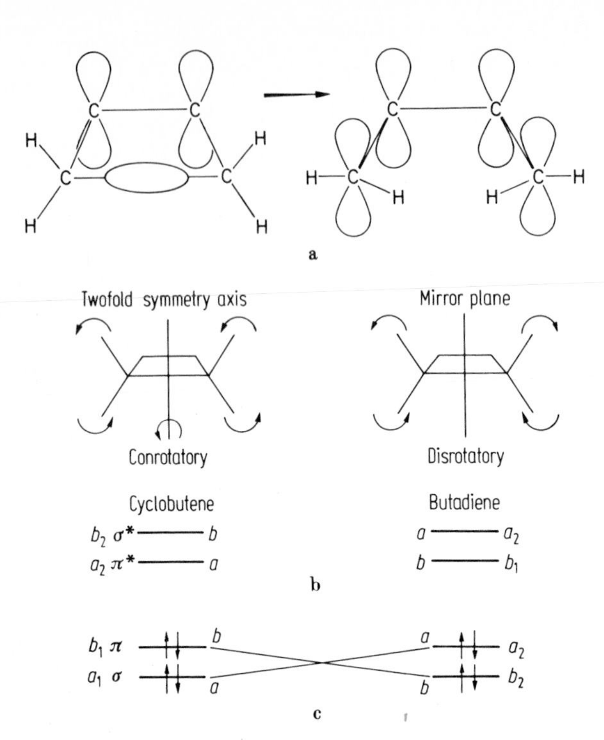

Fig. 6.4 a – c. The ring opening of cyclobutene. **(a)** The orbitals involved, **(b)** the two possible modes of ring opening, **(c)** the orbital correlation diagram for the reaction cyclobutene → butadiene following a conrotatory mode. At the edges of the diagram are shown the orbital symmetries in C_{2v} symmetry

rotatory reaction). The point groups corresponding to these paths are C_2 and C_s respectively. Inspection of the C_{2v} character table (appendix I) enables us to classify the M.O. as symmetric (type A or A') or antisymmetric (B or A") with respect to these symmetries (table 6.1). We see that if C_2 symmetry is maintained, σ and π^*, and σ^* and π can mix (and thus weaken the desired bonds) and that the occupied orbitals in

Table 6.1. Relevant M.O.s of cyclobutene and butadiene

M.O.	C_{2v}	C_2	C_s
cyclobutene			
σ^*	B_2	B	A"
π^*	A_2	A	A"
π	B_1	B	A'
σ	A_1	A	A'
butadiene			
π	$A_2 + B_1$	A + B	A' + A"

reactant and product correlate (i.e. have the same symmetry A + B). The reader should be able to see that this is not the case for the C_s point group. We thus predict a conrotatory reaction conserving C_2 symmetry as more probable. Isotopic labelling of the methylene protons confirms this prediction.

What exactly is the mechanism of lowering of symmetry that enables the reaction to take place? Since there is only one molecule involved in the reaction, it will not be the approach of a second molecule (leading to overlap) but must be a distortion of the nuclear position of cyclobutene – that is, a molecular vibration. We may predict the symmetry of this vibration since we know that it maintains the C_2 axis and must therefore be symmetric with respect to it, but that it destroys the two mirror planes and is thus antisymmetric with respect to these. It must therefore be an A_2 vibration. We might also have predicted this by second order Jahn-Teller effect arguments: if we wish to mix σ and π^* (or π and σ^*) of cyclobutene, we must find a vibration of symmetry Γ such that the direct products $\Gamma_\sigma \oplus \Gamma \oplus \Gamma_{\pi^*}$ and $\Gamma_\pi \oplus \Gamma \oplus \Gamma_{\sigma^*}$ are totally symmetric – inspection shows that the A_2 vibration will accomplish this (for the first case we have $A_1 \oplus A_2 \oplus A_2 = A_2 \oplus A_2 = A_1$). Knowing the vibrational mode which promotes the reaction, we may then find the elements of symmetry which are preserved and predict the reaction product.

This sort of argument may be used to explain the fluxionality of five coordinate species such as PF_5 and $Fe(CO)_5$ (Chap. 3). In these compounds the highest occupied orbitals have e′ symmetry and there is an empty a_1 orbital at slightly higher energy. The vibrational mode producing the scambling (Fig. 3.14) has E′ symmetry and since $E' \oplus E' \oplus A_1$ has a totally symmetric component, the vibration will allow a mixing of the e′ and a_1' orbitals which will lower the energy of the distortion and thus encourage the scrambling.

Orbital correlation

Symmetry considerations will tell us if it is possible for different orbitals to mix, but perturbation theory told us earlier that the extent of mixing will be determined by the difference in energy between the orbitals. We must therefore consider not only the symmetry but also the orbital energies of the reactants. If we wish two orbitals to be mixed during the course of a chemical reaction, their energies should be fairly close; this is the basis of Fukui's frontier orbital theory, frequently used in organic chemistry, which postulates that the majority of reactions take place at the position of, and in the direction of maximum overlap of the *highest occupied* M.O. (the HOMO) and the *lowest unoccupied* M.O. (the LUMO) of the reactant molecules. This clearly combines overlap and energetic considerations as required by perturbation theory, but its application in inorganic chemistry is limited by the fact that the frontier orbitals (i.e. HOMO and LUMO) are not always obvious, several close-lying M.O. being common. In some cases it is known that M.O. close to, but not actually the HOMO or LUMO are of dominating importance.

A second requirement is that the M.O. of reactants and products should correlate, and that the electrons displaced during the chemical reaction should move smoothly into the lowest energy orbitals of the product, so that the product is not formed in an electronically excited state.

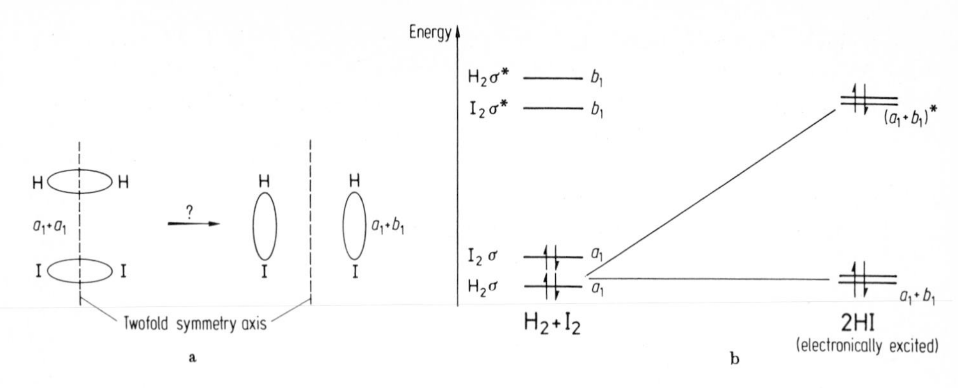

Fig. 6.5 a and b. The reaction between hydrogen and iodine. **(a)** The orbitals of reactants and products, **(b)** the correlation diagram

The reaction between H_2 and I_2 was for a long time thought to be a simple bimolecular reaction with the transition state shown in Fig. 6.5 a. The activated complex has C_{2v} symmetry which is maintained throughout the reaction. Figure 6.5 b shows that, although the two a_1 orbitals can interact and mix, the products are formed with the antibonding $a_1{}^*$ orbital doubly occupied. This will require a certain amount of energy and will produce an excited and unstable product. The H_2/I_2 reaction is now known not to take place by this mechanism. This kind of barrier is avoided if the symmetry of the bonds broken is the same as the symmetry of the bonds formed.

As a final example, let us consider the reaction $CO + Cl_2 \rightarrow COCl_2$ (Figs. 6.6 and 6.7). The most energetically accessible orbitals are the σ^* and σ of Cl_2 and the σ and π orbitals of CO. The antibonding orbitals of CO will lie at much higher energy. We assume that the p_π orbitals of the chlorine play no part in the reaction. We wish to

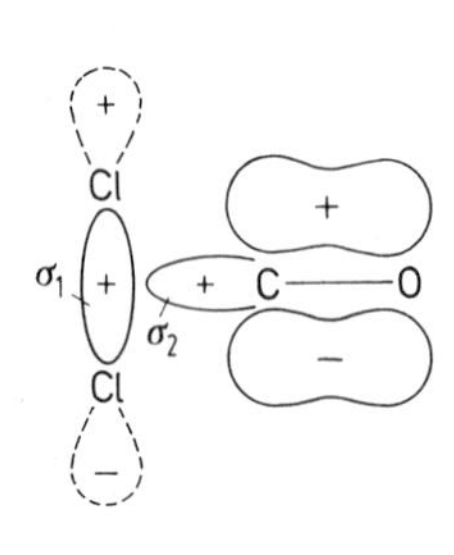

Fig. 6.6. A possible transition state for the reaction $Cl_2 + CO \rightarrow COCl_2$, showing the repulsion between two σ orbitals

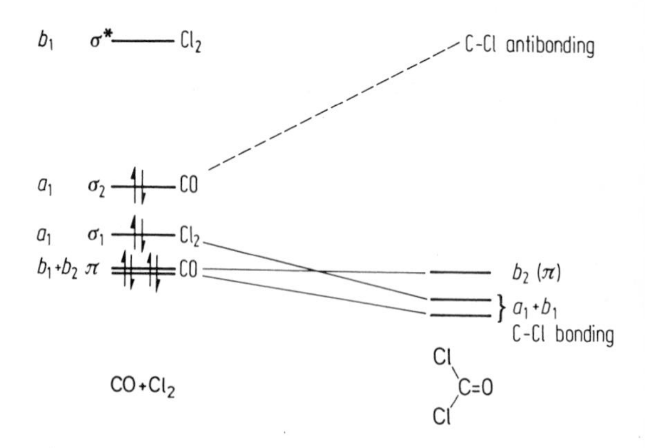

Fig. 6.7. The correlation of product and reactant orbitals for the reaction $Cl_2 + CO \rightarrow COCl_2$

break the Cl–Cl bond (a_1) and one C–O π bond (b_1) to form two C–Cl bonds ($a_1 + b_1$). Although the bond symmetries match, and transfer of electron density from π (CO) to σ^* (Cl_2) will break the bonds we wish to break, we may also see that there is a repulsion between the two σ orbitals which will give rise to the occupation of one C–Cl antibonding orbital. We may see from Fig. 6.6 that this repulsion will be stronger than the overlap between the σ^* b_1 and π (b_1) orbitals, and the reaction does not take place by this mechanism, but by a radical mechanism at high temperatures.

Conclusions

Let us summarise the points made above:

1) Bond formation or breaking arises from the mixing of atomic and/or molecular orbitals at reaction centres. This mixing may result from overlap of orbitals of approaching particles, or from mixing induced by vibrational distortion (the second order Jahn-Teller effect). We may use the symmetry conserved during the course of the reaction to predict the possibility or impossibility of mixing.

2) The mixing of orbitals will always be greatest when their energy is closest.

3) The occupied orbitals of the reactants and products should correlate. This condition requires that the electrons displaced during the reaction should move smoothly into the lowest energy orbitals of the product.

Mechanisms where mixing is difficult or where orbitals do not correlate will have higher activation energies, and are sometimes said to be forbidden.

We have emphasised the importance of mixing of orbitals in bond formation and rupture, but this needs two qualifications: firstly, the mixing criterion will clearly be most important when the bonds depend strongly on overlap, that is when they are quite covalent; secondly, the change of orbitals and energy due to mixing corresponds to a second order perturbation (Chap. 1), and we have ignored any first order term. Neutral, non-polar molecules will not interact strongly unless some mixing takes place, but for reactions between ionic or polar species, the electrostatic potentials of charged fragments may appreciably change the energy of the system without orbital mixing; this corresponds to the first order perturbation term. Thus the approach of Cl^- to $H^{\delta+} - Cl^{\delta-}$ will involve Cl^- . . . . . $H^{\delta+}$ attraction and Cl^- . . . . . $Cl^{\delta-}$ repulsion without any overlap. Discussion of reactions of the many highly ionic species in inorganic chemistry in terms of overlap is somewhat pointless. However, most reactions of ionic species take place in solution where the ions are dissociated, and strongly solvated. Overlap and mixing approaches are particularly useful for non-ionic species such as organo-metallic compounds.

Low symmetry in reactants does not always prevent the application of symmetry/overlap arguments. In Fig. 6.8 the overlap between σ and π^* in case (i) is identically zero. In case (ii) the absence of a plane or axis of symmetry removes the symmetry

Fig. 6.8

restraint, but the approach of orbital σ is still on a nodal plane of π^* and we may expect the overlap to be fairly small and maybe negligible. We may often use the nodal symmetry of the M.O. in which we are interested rather than the absolute symmetry – from this point of view case (i) and case (ii) are identical. Indeed, reactions forbidden for ethylene are not normally much more favourable for propylene which has lower symmetry.

It should always be borne in mind that symmetry conserved throughout a reaction should be useful i.e. allow the separation of M.O. according to their symmetry properties. Similarly, the vibrational modes proposed as reaction coordinates should have an appreciable amplitude at the reactive atoms and not involve only those atoms unaffected by the reaction.

Generally speaking, the predictions of allowed or forbidden reactions are followed in the sense that allowed reactions will have lower energies of activation than similar reactions which are forbidden. It is not possible to compare very different types of reaction since allowed reactions do not necessarily have low absolute heats of activation, nor do all forbidden reactions have high or experimentally inaccessible activation energies. Predictions are more qualitative than quantitative, and they frequently permit the explanation of the stereochemistry of the product. Small molecules containing light elements are often the most successful field of application of these theories, as is the study of the possible catalysis of reactions of such species.

Other selection rules for chemical reactions may be added to those discussed above. One of the more obvious is that the total spin should not change during the reaction ($\Delta S = 0$). The necessity of changing spin will slow down a chemical reaction, but as with absorption spectra, a little bit of spin-orbit coupling can get round this problem. Other rules are more empirical: the useful principle of least motion states than the lowest activation energy for a reaction is the one that requires the least motion of the nuclei and the least disruption of the original electronic distribution. This is intuitively reasonable but suffers from the difficulty of measuring the motions involved – nonetheless it is useful as a rule of thumb. Pearson suggests rewriting it to demand the retention of the maximum number of symmetry elements. The principle of least motion gives a useful starting point for the suggestion of a mechanism which may then be tested for orbital symmetry matching etc. Another useful rule which is borne out by confirmed mechanisms is that multiple bonds are broken one at a time and thus that two distinct steps will be needed to break a double bond.

It is clear that the picture of chemical reactions outlined in this section is very convenient for calculation of most favoured reaction paths and many such calculations have been made sometimes giving good agreement with experimental results. On the qualitative level, it has proved possible to give rationalisations of mechanisms which are otherwise difficult to explain, starting from fairly simple assumptions of the molecular structure.

B. Analysis of Reaction Mechanisms

We may break up almost any complex chemical reaction into a series of simple steps which may frequently be distinguished experimentally (for example by the trapping of intermediates). We may then relate the simple steps to the molecular structure as outlined in the previous section. It is extremely useful to have a system of classification of elementary steps so that we can systematically investigate the susceptibility of a given compound to various different reactions. The system given here is close to that given by Tobe in his book on reaction mechanisms.

An atom in a molecule will be susceptible to two forms of chemical change: a change in the coordination shell or a change in the oxidation state. Tobe also considers a third possibility, of reactions between the ligands bonded to the atom – this is very useful when discussing the coordination chemistry of metals, but it is clear that this third group may involve reactions where the central atom undergoes no chemical change, the reactive centres being elsewhere. Many reactions can be classed under two or more headings – this is a slight inconvenience, resulting from the fact that our classification is a human artefact, not a neutral division. We will adopt the following classifications:

A. Change in coordination shell
1. Addition – increase in coordinate number
2. Elimination – reduction in coordination number
3. Substitution – exchange of a ligand with a free ligating species
4. Change in geometry – a change in ligand-atom-ligand bond angles without actual breaking of a ligand-atom bond – for example a square planar-tetrahedral conversion
5. Change in ligand arrangement – for example a cis-trans isomerisation
6. Reaction between ligands (e.g. addition) involving a change in coordination shell.

B. Electron-transfer reactions
1. A simple loss or gain of electrons involving no change in coordination shell
2. A redox reaction involving a change in coordination shell

We may now discuss these elementary reactions separately. We shall label M the atom being oxidised or reduced, or whose coordination shell is being altered, the ligands involved being denoted L.

A. 1. *Addition* $ML_n + L \rightarrow ML_{n+1}$

Here we will consider only addition of one ligand to the coordination shell – we will discuss later the important oxidative addition reactions where two ligands are

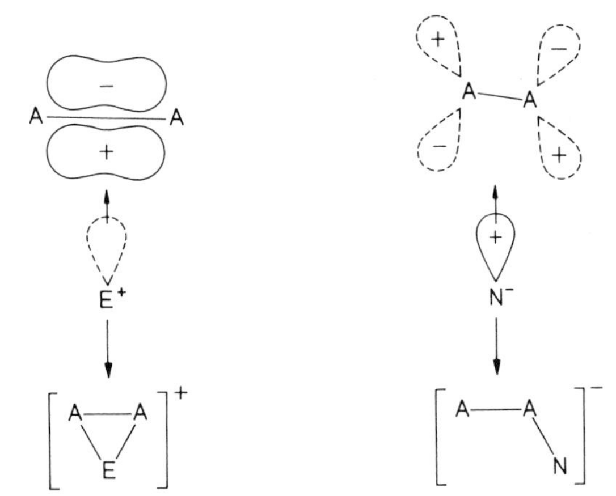

Fig. 6.9. Attack on a double bond

added in a concerted manner. For the formation or rupture of one bond we may conveniently use the homolytic/heterolytic description. Electrophilic and nucleophilic reactions are essentially the same, the distinction being which atom is regarded as substrate. Clearly it is necessary to have a doubly occupied orbital on one atom and an empty orbital on the other. In general the expansion of the coordination shell will involve a change in the positions of ligands already present – in the reaction $BF_3 + F^- \rightarrow BF_4^-$ the F–B–F angle changes from 120° to 109° and the B–F bond length increases. This overall change in structure can affect the reactivity: BF_3, which is harder to distort than BCl_3, is a weaker acceptor than BCl_3 (as measured by equilibrium constants). Expansion of the coordination shell can also be excluded by steric effects.

The acceptor or donor orbital may be a molecular orbital. In Fig. 6.9 the interactions of an electrophile with a π orbitals and of a nucleophile with a π^* orbital are shown – the structures of the products are different in the two cases as a result of the different spatial distributions of the M.O. involved. Radicals may act similarly to nucleophiles *or* electrophiles – if the partially filled orbital has high energy the radical will tend towards nucleophilic behaviour, if the orbital is strongly bound (as in F ·) it will act as an electrophile.

A. 2. *Elimination* $ML_n \rightarrow ML_{n-1} + L$

Elimination may simply be regarded as the reverse of the addition reaction – it will be favoured by a crowded, distorted coordination shell. Heterolytic elimination will be favoured by polar solvents stabilising the Lewis acid and Lewis base formed. Homolytic elimination (i.e. dissociation) from a non-radical species is difficult and generally requires activation (either thermal or photochemical) – elimination of a radical from a larger radical is much easier. Radical elimination is also sometimes found after an electron transfer reaction as in the case of the reaction between $[Cr(OH_2)_6]^{2+} + C_6H_5CH_2Cl$ giving $[Cr(OH_2)_5Cl]^{2+}$ and $C_6H_5CH_2$. The $C_6H_5CH_2Cl$ is reduced by the chromium (II) giving, it is believed, a chloro complex of Cr(III) and the relatively stable benzyl radical.

A. 3. *Substitution reactions* $ML_n + L' \rightarrow ML_{n-1}L' + L$

Substitution reactions are perhaps the most studied field of mechanistic inorganic chemistry; particularly popular have been nucleophilic substitution reactions at metal centres. Since most of these reactions in inorganic chemistry are fast, interest was initially concentrated on those complexes which react slowly enough to be followed by classical means – this accounts for the apparent obsession of such studies with complexes of chromium (III) and cobalt (III) (both octahedral) and square planar Pt(II) and Pd(II). The increasing availability of fast kinetic methods is expanding the field of study (see Problem 7.16, page 280).

In any substitution reaction we need to break one bond and form another – the different mechanisms involved differ in the relative timing of these two steps. The associative (A) mechanism involves an initial addition of the substituting ligand L′ followed by the loss of the replaced ligand L – in the limiting case an intermediate of increased coordination may be isolated. The dissociative (D) mechanism involves initial elimination of L followed by addition of L′; here it may be possible to find an intermediate of reduced coordination number. Finally, we may imagine an interchange mechanism where the bond formation and bond breaking steps are almost simultaneous and affect each other – this may conveniently be divided into two classes – the interchange (associative) or I_a mechanism where approach by the incoming ligand slightly precedes the weaking of the bond to the outgoing ligand, and the interchange (dissociative) or I_d mechanism, where slight bond weakening precedes substituent approach.

These mechanisms may also be classified using the Hughes Ingold system popular in organic chemistry, giving the following equivalences:

A	$S_\alpha 2$ (limiting)
I_a	$S_\alpha 2$
I_d	$S_\alpha 1$
D	$S_\alpha 1$ (limiting)

where α = N for a nucleophilic substitution
E for a electrophilic substitution
H for a homolytic (radical) substitution.

In introducing the A, I, D classification Langford and Gray distinguished I_a from I_d by the influence of the incoming ligand L in determining the activation energy and rate: in an I_d mechanism the rate will be little affected by the nature of L$'$.

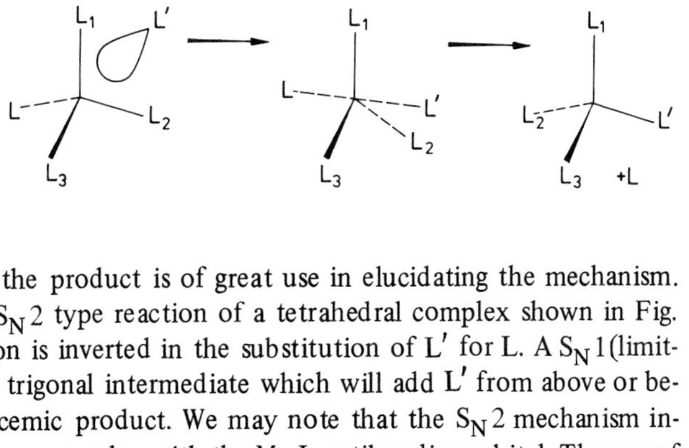

Fig. 6.10.
Inversion in a S_N2 reaction

The stereochemistry of the product is of great use in elucidating the mechanism. The classic example is the S_N2 type reaction of a tetrahedral complex shown in Fig. 6.10 where the configuration is inverted in the substitution of L$'$ for L. A S_N1(limiting) mechanism will have a trigonal intermediate which will add L$'$ from above or below the plane to give a racemic product. We may note that the S_N2 mechanism involves an approach of maximum overlap with the M–L antibonding orbital. The use of stereochemistry in inorganic mechanistic chemistry is slightly more complicated than in organic chemistry since there is usually a wider choice of mechanisms (of which several may be followed), and effects such as rearrangements of 5–coordinate intermediates must be considered. Good discussions of the subject are given by Basolo and Pearson, and Tobe.

Nucleophilic substitution is extremely common, and all of the mechanisms mentioned above are known. With the possible exception of certain low valence states (where the bonding is almost certainly highly covalent) all metal atoms in compounds are electrophilic centres, and ligand substitution involves the exchange of nucleophiles. Electrophilic substitution is less common, although proton exchange and attack by metal ions on negatively charged centres may be regarded as electrophilic attacks. It is often possible to regard nucleophilic substitution on one atom as electrophilic substitution on another – tradition favours the nucleophilic point of view. For example the reaction between $[Co(III)(NH_3)_5(OH)]^{2+}$ and N_2O_3 is thought to proceed as shown below

$$\begin{array}{c}ON(NO_2)\\ \uparrow\end{array}$$
$$[(H_3N)_5Co{-}O{-}H]^{2+} \rightarrow [(H_3N)_5CoONO]^{2+} + H^+ + NO_2^-$$

which may be regarded as nucleophilic substitution of nitrogen or electrophilic substitution of oxygen. It is also worth noting that there is no actual substitution of the cobalt, despite the apparent HO^-/NO_2^- exchange.

Radical substitution is also well known, especially for the p block elements. Given our earlier remarks about radical elimination, it is unlikely that a D mechanism will be adopted, but an associative S_H2 mechanism has been found for radical attack on organo boron compounds and inversion has been found for attack on phosphorus compounds.

If we restrict our discussion to nucleophilic substitution, there are some regular features in the mechanisms adopted. For main group elements, tetrahedral species shown all four mechanisms, but the smaller atoms favour the more dissociative reactions (carbon shows I_a I_d and D behaviour) while the heavier elements, which often show Lewis acidity, show a slight partiality for associative mechanisms. Transition metals in high oxidation states (e.g. MnO_4^-) show a tendency to associative reactions (dissociative reactions might involve the breaking of a double bond in one step), but the tetrahedral d^{10} complexes of metals in low oxidation states give dissociative reactions. Square planar 16 electron complexes give additive reactions. Five coordinate species have not been greatly studied but, with the possible exception of 18 electron transition metal species which might be expected to be dissociative, no definite preferences have been established. Octahedral complexes generally give dissociative reactions, although there is some reason to expect associative reactions in high oxidation states, especially where species with C.N. greater than 6 may be isolated (e.g. NbF_7^{2-}, $W(CH_3)_8^{2-}$, XeF_8^{2-}). There is little data on species with high coordination number (lanthanide or alkali metal ions in solution). Species with low C.N. are often Lewis acids and show associative reactions (for example linear Au(I) complexes).

A. 4. *Change in geometry without bond breaking.*

This class includes the fluxional movements we have discussed before, the inversion of pyramidal centres, and changes from tetrahedral to square-planar geometry. Since there is no bond breaking or formation, the 'reaction' may simply be regarded as a molecular vibration. Calculation of the reaction coordinate requires only the calculation of the molecular wavefunction at several geometries, leading to the potential energy surface for the transformation – this has been successfully done for the inversion of ammonia (Chap. 5). Alternatively, one may use a second order Jahn-Teller effect approach as was mentioned for PF_5. In the case of the nickel (II) tetrahedral-square planar inversion the reaction is forbidden since it involves a spin change of 2 units,[1] but this is presumably overcome by spin-orbit coupling interactions. However, the relatively slow rate ($\sim 10^6\,sec^{-1}$) by the standards of fluxional molecules, and the activation energy (45 kJ/mole) suggest that the reaction is at least beiing slowed down.

For three coordinate pyramidal species, the inversion vibration produces the other optical isomer and thus leads to racemisation (see Fig. 6.11). For nitrogen this racemi-

---

[1] It is also forbidden by orbital symmetry.

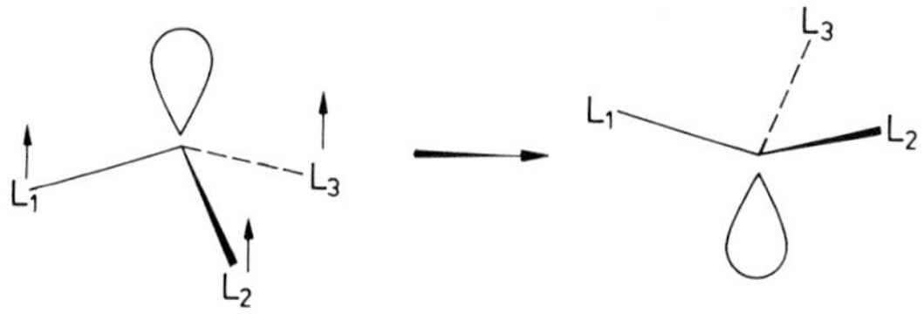

Fig. 6.11. Inversion of a three-coordinate species

sation is rapid, but phosphines are slower, and salts of the pyramidal ions R R′ R″ S^+ were resolved using classical methods by Pope. Tetrahedral carbon or silicon cannot racemise this way; the most probable route would be through a square planar intermediate, but an orbital correlation diagram shows this to be unfavorable, and that there are no low lying empty orbitals which may be mixed in by the vibration.

A. 5. *Change in ligand arrangement with bond rupture.*

In its simplest form this is an elimination reaction followed by an addition reaction leading to a new product. This may be a more favorable route to isomerisation than the distortion of the initial molecule without bond rupture. Thus, in the racemisation of $[Co(en)_3]^{3+}$ (Fig. 4.4) one Co–N bond will break, followed by rearrangement of the 5-coordinate species, and the recombination will give either enantiomer, leading to a racemisation. In a similar way *cis*-isomers may be partially converted to *trans*-isomers and so on.

For the particular case of the racemisation of metal complexes such as $[Co(en)_3]^{3+}$ there is often a choice of a bond rupture or a intramolecular twist mechanism (class A4) passing through a trigonal prismatic intermediate. Both routes have been found experimentally – dithiocarbamate ($R_2NCS_2^-$) chelates appear to be particularly attached to the intramolecular mechanism. This agrees well with the observation of the formation of trigonal prismatic complexes with certain metals by this ligand.

A. 6. *Reactions between ligands involving a change in the coordination shell.*

The proximity of two ligands coordinated to a central atom may encourage them to react with each other, and this may involve a change in coordination of the central atom: a simple example is the ligand migration reaction

$$\begin{array}{c}L\\ |\\ M - L'\end{array} \rightarrow M - L' - L$$

The central atom M may in fact encourage this coupling reaction, acting as a catalyst for an otherwise difficult reaction. Another example is the reductive elimination reaction where the M–L′ bond is also broken to give a free L–L′ molecule. Both these reactions are discussed later in the chapter.

Electron transfer reactions

Many redox reactions are simple electron transfer reactions, but as a result of the difficulties associated with the definition of oxidation state discussed in Chap. 5 some redox reactions involve no electron transfer. Thus the reaction $SO_3^{2-} + OCl^- \rightarrow SO_4^{2-} + Cl^-$ might be regarded as a nucleophilic substitution (at oxygen) of SO_3^{2-} for Cl^-; we would also class this as a redox reaction, yet it is clearly different from the reaction

$$\underset{3d^5}{Fe^{3+}(aq)} + \underset{3d^4}{Cr^{2+}(aq)} \rightarrow \underset{3d^6}{Fe^{2+}(aq)} + \underset{3d^3}{Cr^{3+}(aq)}$$

where we may definitely identify an electron transfer. Similarly, oxidative addition and reductive elimination, while they involve a redistribution of electron density, only involve electron transfer if a highly ionic model of the bonding is used. In this section we will discuss those reactions where the transfer of one or more electrons is the dominant feature. Such reactions generally involve the transfer of electrons from or into weakly bonding, antibonding or non-bonding orbitals such as transition metal d orbitals or the non-bonding s orbitals of certain main group elements (e.g. Sn(II)). Bond formation or rupture may occur in these reactions but is of secondary importance. We will discuss first those reactions where bonds are not made or broken before, during or after transfer.

B.1. *Electron transfer involving no change in coordination sphere*

This reaction consists of a simple electron jump between two species with no formation of a bond between them. An example is the oxidation of $[Fe(CN)_6]^{4-}$ by $[IrCl_6]^{2-}$:

$$[Fe(CN)_6]^{4-} + [IrCl_6]^{2-} \rightarrow [Fe(CN)_6]^{3-} + [IrCl_6]^{3-}$$

This type of reaction is referred to as an *outer sphere* electron transfer as the reductant remains in the outer coordination sphere of the oxidant (and vice versa). The electron is assumed to 'tunnel' through a potential barrier between the configuration before the reaction, and that after the reaction. Although the electron transferred is only weakly involved in chemical bonding its departure (or arrival) will have an effect on the energy of the species. Thus the removal of an electron from the t_{2g} shell of $[Fe(CN)_6]^{4-}$ would normally lead to a shortening of the Fe–CN bond (as for WF_6^-, Fig. 5.9). If the electron jump is assumed to obey the Franck-Condon principle, then the $[Fe(CN)_6]^{3-}$ produced will be in a vibrationally excited state. Similar reasoning shows that $[IrCl_6]^{3-}$ will also be formed in a vibrationally excited state, the Ir-Cl bond length of $[IrCl_6]^{2-}$ being too short for $[IrCl_6]^{3-}$. This additional vibrational energy will provide an energy barrier to the reaction, which can be partially overcome by distorting (or 'rearranging') the two complexes before the electron transfer.

The principle of microscopic reversibility which requires that the reverse reaction involves the same distortions of internuclear distances as the forward reaction leads us to conclude that the oxidant and reductant are distorted to a configuration intermediate between their structures before and after the reaction. This distortion requires an activation energy referred to as a *Franck-Condon barrier.*

Various models for calculating activation energies for outer sphere reactions have been established (notably by Marcus) allowing for ligand rearrangement, repulsion or attraction between the two reacting species, changes in solvation, and the total free energy change of the reaction. Their predictive power is fairly satisfactory, and they confirm the general view of this reaction. The initial approach of the oxidant and reductant is generally fairly rapid and electrostatic repulsion effects (e.g. between $[IrCl_6]^{2-}$ and $[Fe(CN)_6]^{4-}$) not very important, although catalysis may be important here. The rearrangement, or Franck-Condon, barrier is often quite considerable. This is particulary true when the orbital filled or emptied has an appreciable bonding or antibonding role, since the change in its occupancy will produce an appreciable change in equilibrium bond lengths, and thus an appreciable Franck-Condon barrier. Thus removal of an electron from the e_g* d levels of an octahedral complex will require a larger rearrangement energy than removal of an electron from the non-bonding t_{2g} orbitals, even though the ionisation energy of the t_{2g} shell is, in principle, higher. Large negative free energy changes for the reaction may help to reduce the importance of the Franck-Condon barrier. Finally, for the tunnelling to take place, it appears to be important that there should be some slight overlap between the donor and acceptor orbital although the actual extent of the overlap may be rather small. In the case of $[IrCl_6]^{2-}$ it will be recalled that the acceptor orbital of the t_{2g} set is partially delocalised over the chlorine ligands (from ESR results) and will thus be able to overlap weakly with reducing agents.

Outer sphere electron transfer reactions vary enormously in speed, from only slightly slower than diffusion control ($10^9 s^{-1}M^{-1}$) to rather slow ($10^{-4} s^{-1}M^{-1}$). They are particularly slow for reactions where a change in spin state and electronic configuration is involved, as for example, for the Co(III) low spin – Co(II) high spin reaction (Fig. 3.6).

B.2. *Electron transfer involving a change in coordination shell*

There is now abundant evidence for an electron transfer mechanism involving the initial formation of a complex between the oxidant and reductant in which at least one ligand is bound to both initial species. This mechanism, since it involves a merger of two coordination shells, is referred to as an *inner sphere* mechanism. The classical example, due to Taube, is the reduction of $[Co(NH_3)_5Cl]^{2+}$ by aqueous Cr^{2+}, for which the mechanism is assumed to be

$$[(H_3N)_5CoCl]^{2+} + [Cr(OH_2)_6]^{2+} \rightarrow [(H_3N)_5Co^{III}\text{–}Cl\text{–}Cr^{II}(OH_2)_5]^{4+} + H_2O \quad (1)$$

$$[(H_3N)Co^{III}\text{–}Cl\text{–}Cr^{II}(OH_2)_5]^{4+} \rightarrow [(H_3N)_5Co^{II}\text{–}Cl\text{–}Cr^{III}(OH_2)_5]^{4+} \quad (2)$$

$$[(H_3N)_5Co^{II}\text{–}Cl\text{–}Cr^{III}(OH_2)_5]^{4+} \rightarrow Co^{2+}aq + 5NH_3 + [Cl\text{–}Cr^{III}(OH_2)_5]^{2+} \quad (3)$$

Steps 1 and 3 are simple substitution steps (which may influence the rate of reaction as a whole), and step 2 is the actual electron transfer mediated by the bridging ligand (Cl). The speeds of steps 1 and 3 are clearly dependent on the susceptibility of the reactants to substitution – in this particular case, Cr^{2+} (aq) is extremely rapid in its substitution reactions and step 1 will be fast. Step 3 (decomposition of the complex) need not necessarily involve a ligand transfer: $[IrCl_6]^{2-}$ when reduced by an inner sphere mechanism, retains its coordination shell intact. In the example given above, the chloride is retained by the Cr^{3+} ions as Cr(III) complexes undergo substitution reactions slowly. Co^{2+} forms very labile complexes, and will readily give up the chloride ligand and lose its coordinated ammonia ligands.

For the initial step to be favourable, it is clear that at least one of the reactants must be substitutionally labile, and that a suitable bridging ligand must be available. Thus $[V(OH_2)_6]^{2+}$, which undergoes substitution more slowly than $[Cr(OH_2)_6]^{2+}$, will often react by an outer sphere mechanism when Cr^{2+} aq reduces via an inner sphere process.[2]

[2] The variation in rates of substitution reactions is discussed on page 273.

We may also note the lower Franck-Condon barrier for V^{2+} where the reducing electron leaves a t_{2g} orbital and not an $e_g{}^*$ orbital as for Cr^{2+}.

Ammonia and (it is currently believed) H_2O are unable to act as bridging ligands, and $[Co(NH_3)_6]^{3+}$, for example, is thought to be reduced by an outer sphere mechanism. The range of ligands which may act as bridges is considerable, including the halide ions, OH^-, NCS^-, $N_3{}^-$ and $SO_4{}^{2-}$. It has also been found that the two metal centres need not be coordinated to the same atom in the ligand; if the bridging chloride ligand in the reaction discussed above is replaced by 4-carboxyaminopyridine

N⟨◯⟩CONH$_2$ the bridging complex has the structure:

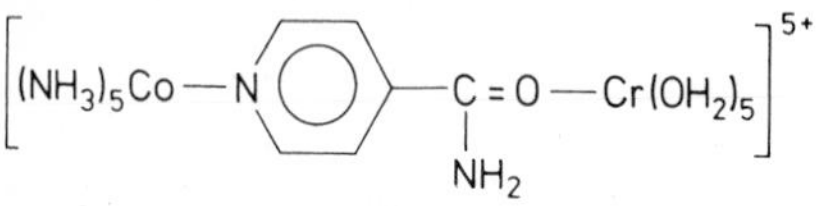

$$\left[(NH_3)_5Co-N\langle\bigcirc\rangle-\underset{\displaystyle NH_2}{\underset{|}{C}}=O-Cr(OH_2)_5\right]^{5+} \tag{4}$$

this mechanism is referred to as remote attack; attack by Cr^{2+} on the nitrogen coordinated to the cobalt would be described as adjacent attack. The reduction of $[(NH_3)_5CoSCN]^{2+}$ by Cr^{2+} appears to involve both remote attack (on N) and adjacent attack (on S). Clearly remote attack may reduce steric interactions between the two metal complexes, but it is remarkable that electron transfer over such a long bridge is still possible.

The role of the bridging ligand

Clearly the formation of a bridged complex has the advantage of bringing the reactive centres close together, but it also appears to favour the electron transfer in other ways. For a simple bridging ligand such as Cl^-, the bridging appears to increase the interaction between the donor and acceptor orbitals, and to reduce the energy barrier through which the electron has to tunnel. A simple theory is discussed by Halpern and Orgel.

The case of remote attack is somewhat better understood. It has been found that only ligands with delocalised π systems joining the two metal binding sites may act as bridging species, implying some sort of conduction through the ligand. In the case of ligands which are relatively easily reduced, it has been shown that the reaction 2 appears to take place in two steps, an initial reduction of the ligand by the reductant, followed by a reduction of the oxidant by the reduced ligand species.

In certain cases, the final dissociation step (3) is slow enough for the bridged complex to be isolated after electron transfer. A particularly interesting case occurs when oxidant and reductant are the same element, as for example for the intermediate

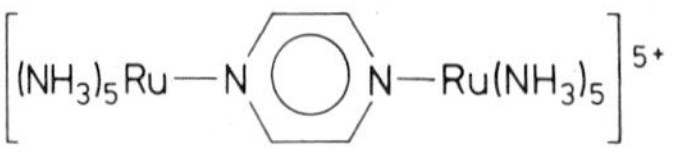

$$\left[(NH_3)_5Ru-N\langle\bigcirc\rangle N-Ru(NH_3)_5\right]^{5+}$$

since the intermediate is now a mixed valence complex (Chap. 4). We may reasonably expect the case of electron transfer between the two Ru centres to show some correlation with the mixed valence class of the complex – in this case class II. It has been possible to synthesise similar binuclear complexes with different metal atoms and measure the rate of electron transfer between the two centres.

The symmetries of the orbitals involved also play an important part in determining the rate. The rates of reduction of some pentammino (4-carboxyamino-pyridyl) complexes by Cr^{2+}aq are:

Metal	Configuration	Rate constant k $sec^{-1} M^{-1}$ at 25^oC
Cr(III)	t_{2g}^3	1.8
Co(III)	t_{2g}^6	17.4
Ru(III)	t_{2g}^5	3.8×10^5

All these reactions will pass through a bridged intermediate similar to (4). The similarity of the rates for Co(III) and Cr(III) when the total free energies of reaction differ by 60kJ/mole may be explained by assuming the initial electron transfer step involves reduction of the bridging ligand. The much greater speed for Ru(III) reduction is explained by the fact that the reducing electron enters a t_{2g} orbital and not an e_g*, and the t_{2g} may overlap and mix with the ligand π system. The reduction of Ru(III) by the reduced ligand is thus greatly accelerated. It seems reasonable to imagine that $t_{2g}-t_{2g}$ electron transfer will be most successfully mediated by a π system on the ligand, and that $e_g^*-e_g^*$ transfer will favour passing through a σ system. We might regard the single atom bridging species such as Cl^- as examples of a σ system. There is often a considerable *trans*- effect for transfers to e_g* orbitals, strong ligands *trans*- to the bridging group slowing down the reaction, presumably as a result of the greater reorganisation energy when the *trans*- ligand interacts strongly with the e_g* orbital.

Factors favouring inner sphere or outer sphere paths

Very frequently the inner sphere and outer sphere paths have roughly equal probability, and it is not uncommon for both to contribute to the reaction. Species with no bridging ligands, with low reorganisation energies, or with very slow substitution rates will tend to favour outer sphere mechanisms, while substitutionally labile species will be more attracted to inner sphere paths. An inner sphere mechanism may offer a means of reducing reorganisation or Franck-Condon barrier too large for an outer sphere mechanism to be followed readily.

Two electron electron transfer reactions

Many chemical reactions involve the transfer of more than one electron, notably those involving main group elements (e.g. Sn(II)/Sn(IV)). For gaseous atomic species two electron transfers are only slightly less probable than one electron transfers, but in solution, or in complex molecular species, the rearrangement barriers will be considerable. We would accordingly expect inner sphere reactions to be more probable, possibly involving two distinct transfers of one electron. We might regard the SO_3^{2-} + OCl^- reaction discussed earlier as being a two electron inner sphere process. Another example is the Pt(II)/Pt(IV) exchange reaction thought to pass through a bridged intermediate as shown in Fig. 6.12.

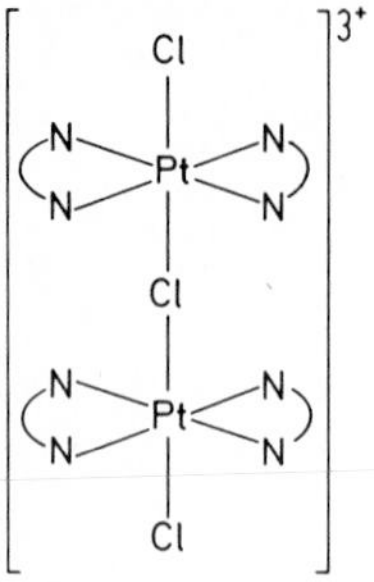

Fig. 6.12

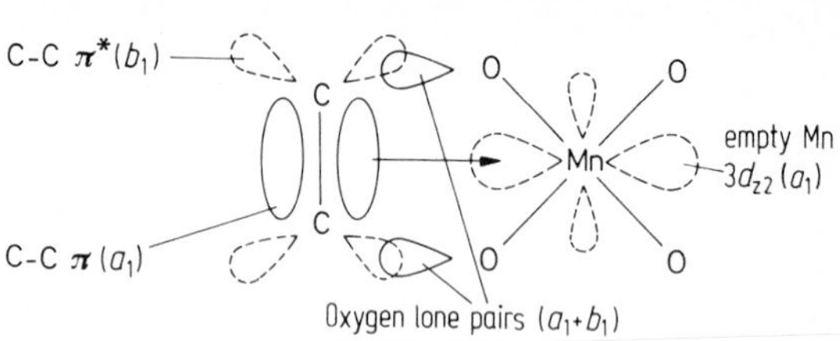

Fig. 6.13

Complementary and non-complementary electron transfers

When a reductant has a tendency to lose two electrons and an oxidant will only accept one electron (or vice versa) it is evident that a reaction between these two species will not be straightforward. Such a reaction is classified as non-complementary, as opposed to complementary reactions where the number of electrons lost by the reductant equals the number gained by the oxidant. An example of a non-complementary reaction is

$$2Fe^{2+} + Tl^{3+} \rightarrow 2Fe^{3+} + Tl^{+}$$

This reaction could proceed by a simple termolecular mechanism, but the probability of this is very low, and a two step bimolecular process will be favoured, such as

$$Fe^{2+} + Tl^{3+} \rightarrow Fe(IV) + Tl^{+}$$

$$Fe(IV) + Fe^{2+} \rightarrow 2Fe^{3+}$$

or, alternatively, a mechanism involving intermediate formation of a Tl(II) species. In both mechanisms the intermediate existence of an unstable oxidation state (Tl(II) or Fe(IV)) is necessary. It is now quite well established that unstable intermediates can be formed during non-complementary reactions, and there is evidence from trapping reactions, etc., for species such as Tl(II), Sn(III), Pt(III) and Cr(IV). Advances in synthetic chemistry have also allowed the preparation of stable compounds of supposedly unstable oxidation states such as Ir(II), Au(II) and Mn(V), which gives some justification for the invocation of unstable oxidation states as intermediates. Non-complementary reactions are often rather slow, but are very susceptible to catalysis (see later). A simple way of circumventing the problem is to form a dimeric species of a one electron reductant which may then react rapidly with a two electron oxidant – for example

$$M(II) + M(II) \rightarrow [M(II) - M(II)]$$

$$[M(II) - M(II)] + A(III) \rightarrow [M(III) - M(III)] + A(I)$$

Atom transfer reactions

Hitherto we have concentrated on electron transfer from essentially non-bonding orbitals so that bond formation and rupture have essentially secondary importance. Very many redox reactions involve species which have no obviously non-bonding orbitals, and electron transfer in these systems will clearly involve changes in the bonding. The reaction of Cr^{2+} and benzyl chloride is an example

$$[Cr(OH_2)_6]^{2+} + Cl-CH_2C_6H_5 \rightarrow Cr(OH_2)_5Cl^{2+} + CH_2C_6H_5$$

The reaction produces the benzyl radical, presumably as the result of an inner sphere complex formation followed by transfer of an electron to the C–Cl antibonding orbital. However we may regard the reaction as involving the transfer of a chlorine atom from the organic species to the metal ion. It is often possible to regard redox reactions as atom transfer reactions, and it was even proposed at one stage that reactions between aquo ions could take place in this fashion:

$$[M(III)OH]^{2+} + [H_2OM'(II)]^{2+} \rightarrow [M(II)OH_2]^{2+} + [HOM'(III)]^{2+}$$

Although this approach is currently out of favour, the atom transfer approach is often useful, especially in reactions with covalent systems. In such cases orbital symmetry arguments are useful, as in the reaction of permanganate with an alkene, Fig. 6.13. The lone pairs on two oxygen ligands transform as $a_1 + b_1$, and the b_1 M.O. can overlap and mix with the C–C π^* M.O., weakening the C–C π bond, which may be further weakened by loss of electrons from the π bonding M.O. into the empty d_{z^2} orbitals of MnO_4^-. The result is the transfer of electrons to the MnO_4^- group to give Mn(V), the rupture of the olefinic double bond and the formation of two C–O bonds. The cyclic system can then be hydrolysed to break the Mn–O bonds, leaving a diol. This is clearly a more satisfactory mechanism than a leap of electrons from the occupied π M.O. to the empty manganese d orbital. Pearson gives a similar argument for the *cis* chlorination of an alkene by $PbCl_4$.

In summary, the forms of electron transfer and redox reactions we have discussed represent a progression from simple electron transfer involving non-bonding orbitals and little change in bonding (outer sphere reactions), through systems involving weak or partial bond formation (inner sphere processes), to reactions where the redistribution involves substantial changes in occupation of bonding orbitals and thus a considerable change in bonding (as in atom transfer reactions).

C. Selected Reactions and Topics

a) Reactions Showing the Importance of Orbital Symmetry

In the general classification orbital symmetry arguments were used only incidentally, and in this section we consider three specific reactions where orbital symmetry is very important. The reactions discussed also have a practical interest as they frequently occur as intermediate steps in homogeneous catalysis by organometallic compounds. Detailed discussions may be found in the book by Pearson.

(i) Oxidative addition. The fundamental oxidative addition mechanism involves attack by a group ML_n on a molecule X–Y, where X and Y may be atoms or groups of atoms to give an addition compound L_nMXY where definite MX and MY linkages may be established, and where the X–Y bond is weakened. Some examples will clarify this:

For reaction (i) the association of oxidation with addition is evident, but for reaction (ii) oxidation of iridium implies reduction of hydrogen to hydride while (iii) requires

$PCl_3 + Cl_2 \longrightarrow PCl_5$ (i)

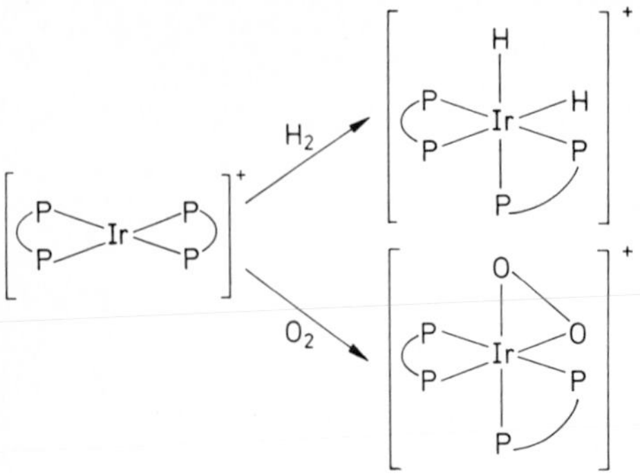

$[Ir(diphos)_2]^+ \xrightarrow{H_2} [Ir(diphos)_2H_2]^+$ (ii)

$[Ir(diphos)_2]^+ \xrightarrow{O_2} [Ir(diphos)_2O_2]^+$ (iii)

the formation of peroxide ion. For reactions (ii) and (iii) spectroscopic evidence suggests relatively little transfer of charge from metal to the new ligands, but implies a considerable redistribution of electrons within the molecular orbitals.

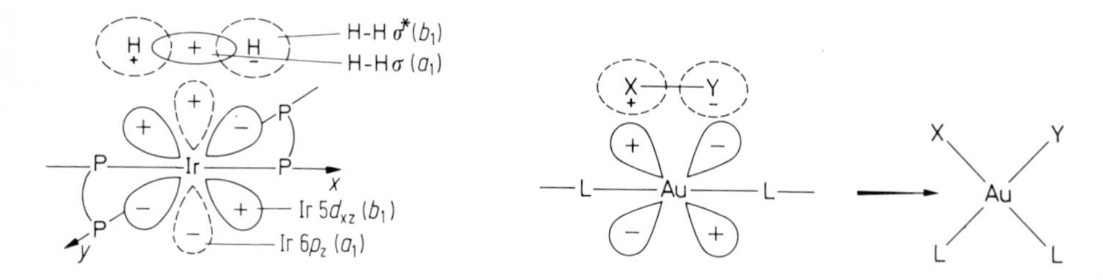

Fig. 6.14. Oxidative addition: $[Ir(diphos)_2]^+ + H_2$

Fig. 6.15. Oxidative addition to a d^{10} complex

We may examine the orbitals involved in the reaction $[Ir(diphos)_2]^+ + H_2 \rightarrow [Ir(diphos)_2H_2]^+$ in Fig. 6.14. The H–H bond will be weakened by mixing of the full d_{xz} orbitals of this d^8 planar complex (see Chap. 3) with the σ^* H_2 orbitals. A further weakening of the H–H bond might arise from the mixing of the σ H_2 M.O. with the empty iridium $6p_z$ orbital (both a_1 symmetry in the C_{2v} point group). The two iridium-hydrogen bonds have symmetry $a_1 + b_1$. A similar argument, using the π_z and $\pi_z{}^*$ M.O.s of oxygen will account for reaction (iii); this picture is in fact remarkably similar to that of the bonding of ethylene to a transition metal shown in Fig. 3.16. The O–O bond length in the oxygen adduct is greater than that of molecular oxygen, implying a transfer of charge density to the π^* M.O. or a transfer from the π bonding M.O. If the two electrons in the d_{xz} orbital are assumed to be completely lost to the ligand, and the donation from the ligand is small, the iridium may be regarded as having been oxidised to Ir(III).

Oxidative addition may occur in a similar way for linear complexes of a d^{10} transition metal (see Fig. 6.15) where the initial interaction is again between the sigma

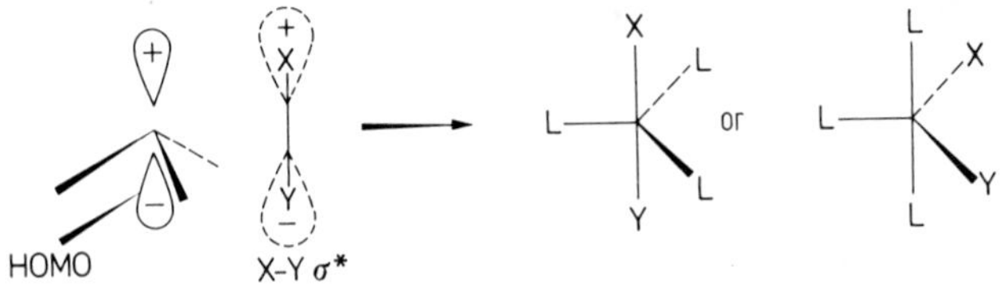

Fig. 6.16. Oxidative addition to a three-coordinate species such as PCl_3

anti-bonding orbital of XY and a filled d orbital. In this case *cis* addition is allowed, but *trans* addition would result in occupation of the σ^* d orbital which is directed at all 4 ligands (b_{1g} in Fig. 3.10). Oxidative additions of type (i) involve species with no d orbitals – the HOMO is a non-bonding hybrid of s and p orbitals, and sideways approach of the XY group is required to allow interaction with the σ^* XY orbital (Fig. 6.16) leading to an axial-axial or equatorial-equatorial adduct. It should be evident that a direct approach along the threefold axis of EL_3 will give only a repulsive interaction between the lone pair and the XY bonding M.O. In all these mechanisms, there is an interaction with the σ^* or π^* X–Y orbital, and it is consequently necessary that this orbital should be of fairly low energy – this will be the case when X and Y are electronegative and the halogen molecules, with low lying σ^* orbitals, will readily undergo oxidative addition.

There has been considerable interest in the oxidative addition reactions of Vaska's compound, *trans*- $[Ir(CO)Cl(PPh_3)_2]$ and its derivatives. Vaska's compound is a planar d^8 species similar to $[Ir(diphos)_2]^+$; it will add oxygen and hydrogen reversibly to give *cis* adducts, but is now clear that other mechanisms may also be followed. If the X–Y bond is polar, with Y the electronegative end, the addition may proceed by a nucleophilic attack by Ir on X followed by loss of Y^- giving a five coordinate intermediate $[IrX(CO)Cl(PPh_3)_2]^+$ which can later add Y^- to give an octahedral complex. If the XY bond is weak and non-polar, a radical mechanism is possible:

$$\underset{Ir(I)}{Ir(CO)Cl(PPh_3)_2} + X{-}Y \rightarrow \underset{Ir(II)}{Ir(CO)Cl(PPh_3)_2X} + Y\cdot$$

$$\underset{Ir(II)}{Ir(CO)Cl(PPh_3)_2X} + Y\cdot \rightarrow \underset{Ir(III)}{Ir(CO)Cl(PPh_3)_2XY}$$

This has been established for the addition of methyl iodide, which gives a *trans* addition product (Fig. 6.17). Which of the three mechanisms is followed depends on the substrate and on the conditions – thus HCl adds *cis* in benzene, implying a concerted re-

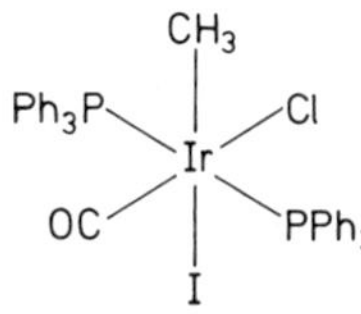

Fig. 6.17

action, but *trans* in more polar solvents, implying a two step process, probably via a heterolytic process.

The reverse of oxidative addition, *reductive elimination,* is of interest as it offers a means of forming an X–Y bond. The orbital picture of the reaction is essentially the same, with charges moving in the opposite sense.

(ii) Ligand migration. The ligand migration reaction involves the transfer of a ligand from a central atom to a second ligand bonded to the central atom. A well-known example is the methyl migration of $(CH_3)Mn(CO)_5$:

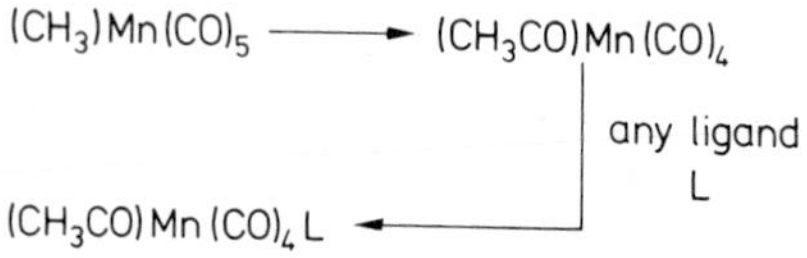

$$(CH_3)Mn(CO)_5 \longrightarrow (CH_3CO)Mn(CO)_4 \xrightarrow{\text{any ligand L}} (CH_3CO)Mn(CO)_4L$$

The methyl group migrates to the carbonyl group, giving an acyl group which is σ bonded through carbon to the manganese. In the first step of the reaction the metal coordination number is reduced by one, and the intermediate accepts a new ligand L

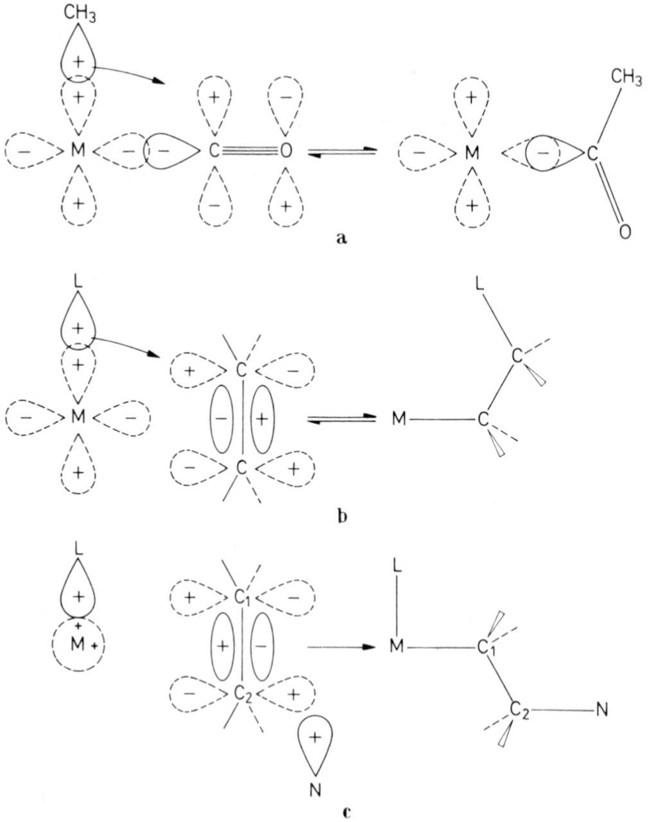

Fig. 6.18. (a) Methyl migration to a carbonyl group, **(b)** ligand migration to a coordinated alkene, **(c)** *trans*-addition to an alkene coordinated to a metal using only an s orbital

to complete its coordination shell. The reaction takes place in a concerted fashion with no loss of CH_3 radicals, and the migration is always to the group *cis* to the migrating group (see page 235). The acceptor group invariably possesses some double bonding.

Figure 6.18 a shows the orbitals involved in this process. The σ bonding CH_3 group will interact with the $d_{x^2-y^2}$ or d_{z^2} orbital of the metal, and the σ orbital wave function directed at the *cis* ligand will thus have opposite sign. As the methyl group migrates to the positive lobe of the π^* CO orbital, the negative lobe of the π^* orbital may mix with the metal d orbital.

Figure 6.18 b shows another interesting reaction, the migration of a ligand to a coordinated alkene to give a metal alkyl. In this case, vertical movement of the alkene will allow overlap of the π^* orbital with the ligand σ orbital and the metal d orbital to give a *cis* product.

In both these reactions the mixing of orbitals only occurs as a result of nuclear motion lowering the symmetry and allowing second order Jahn-Teller mixing. Needless to say, the arguments used here may equally be applied to the reverse reaction to predict the activation energy. We would expect that a transition metal ethyl might readily transfer a hydrogen atom to the metal by the reverse of the reaction in Fig. 6.18 b:

$$M-CH_2-CH_3 \longrightarrow \overset{H}{\underset{|}{M}}-(C_2H_4) \longrightarrow \overset{H}{\underset{|}{M}} + C_2H_4$$

Metal alkyls do indeed often decompose by this reaction (β elimination).

Figure 6.18 c shows the orbitals involved in the ligand migration reaction for a central atom where the s orbital is responsible for ligand binding. It will be seen that concerted attack by L on C_1 and C_2 on M is not possible, and L cannot migrate as in Fig. 6.18 b. An external nucleophile N can however attack at C_2 to give a *trans* addition product, without ligand migration; this reaction is of course equally possible in Fig. 6.18 b and is indeed found as a side reaction. For metals using s orbitals the ligand migration reaction is thus disallowed, as is its reverse, the elimination reaction.

Species using p orbitals for bonding use different atomic orbitals for ligands *cis* to each other, and suitable overlaps are possible. An example is the hydroboration reaction presumed to pass through the step

$$\begin{array}{c} CH_2{=}CH_2 \\ \vdots \\ H_2B{-}H \end{array} \longrightarrow \begin{array}{c} CH_2-CH_3 \\ | \\ BH_2 \end{array}$$

(iii) Cyclisation. Certain pericyclic reactions involving cyclic transition states may be catalysed by metals. An example, the cyclisation of two ethylene molecules, is shown in Fig. 6.19; since the two π bonds of the olefins are broken and three new bonds (one carbon-carbon and two metal-carbon bonds) are formed, it is clear that two electrons must be furnished by the metal from a non-bonding orbital – it is thus referred to as an oxidative cyclisation. Figure 6.19 b shows that if the metal moves down from the middle of the two olefins, suitable overlap between orbitals will be found with C_{2v} symmetry conserved. The occupied orbitals of the initial compound correlate in symmetry with those of the product if the metal donor orbital has a_1 symmetry and the reaction is allowed although the cyclisation of ethylene is forbidden

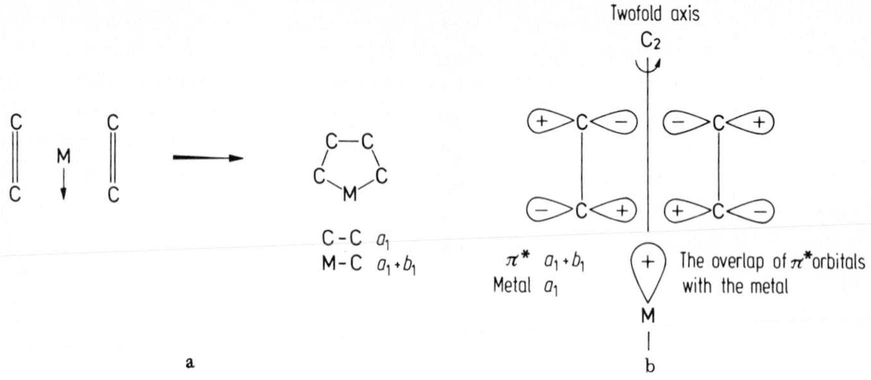

Fig. 6.19 a and b. Oxidative cyclisation

in the absence of a metal catalyst: by contrast, the allowed condensation of ethylene and butadiene (the Diels-Alder reaction) is not allowed for ethylene and butadiene bound to a metal. The catalysis of these cyclic reactions has been reviewed recently by Mango.

b) Solvent Effects

Most chemical reactions are carried out in solution, and in this section we consider the effects of the solvent on the mechanism of a reaction. A solution is a relatively condensed phase of matter and any translational movement of a molecule will involve collisions with other molecules (generally solvent molecules) in solution. Any two reactant molecules which collide will undergo collisions with surrounding solvent molecules before they can move apart, and will thus have a good chance of colliding again before separation – this phenomenon is referred to as the *solvent cage* effect, and will clearly increase the probability of a reaction (or two successive reactions) between the two molecules.

In general, however, the interaction between solute and solvent is more important than the cage effect. In water solvation energies of several hundred kJ/mole are quite common (for cations), and the rearrangement of solvent molecules will clearly affect the rate of reaction. Let us consider an example where the solute-solvent interaction is strong, the substitution of an anion X^- in the coordination sphere of a cation M^{n+} in aqueous solution. The cation will be surrounded by water molecules, and we may imagine two coordination spheres, the inner sphere comprising the coordination sphere of the metal (i.e. those water molecules bound to the metal), and the outer sphere comprising the water molecules attracted to, but not actually bound to the cation.

If the substituting anion X^- is assumed to be only weakly solvated (as is generally the case), the first step of the reaction will be the substitution of X^- in the outer coordination sphere of M^{n+}. This is generally a very fast reaction, close to the diffusion rate, and a pre-equilibrium concentration of the outer sphere complex is set up; this concentration depends on the electrostatic attractions between X^- and $M^{n+}(aq)$. The second step involves exchange of X^- and a water molecule in the inner coordination sphere to give the product, an inner sphere $[M-X]^{(n-1)+}$ complex. The fact that the rate of the second step is generally independent of the nature of X^- and is generally only slightly slower than the rate of water exchange between inner and outer coordination spheres is taken as evidence for a dissociative mechanism (water exchange will be slightly faster because of the large number of water molecules in the outer sphere). It is possible to make good order-of-magnitude calculations of ligand substitution rates using the water exchange rater constant and an estimate of the equilibrium constant for outer sphere complex formation

An example of the route followed is:

$$\text{(i)}\quad [Cr(OH_2)_6]^{3+} + Cl^- \underset{}{\overset{\text{fast}}{\rightleftharpoons}} \underset{\text{outer sphere complex}}{[Cr(OH_2)_6{\cdot}Cl]^{2+}}$$

$$\text{(ii)}\quad [Cr(OH_2)_6{\cdot}Cl]^{2+} \xrightarrow[\text{slower}]{} [Cr(OH_2)_5Cl]^{2+} + H_2O$$

For bivalent or trivalent metals step (ii) is slower than step (i), but for the alkali metals or metal ions where the charge: size ratio is small, the inner coordination sphere is not strongly bound and steps (i) and (ii) may have similar rates. In solvents less polar than water, where solvation of ionic species is weaker, ion pairs may be predominant: in this case the equilibrium step (i) will lie completely to the right.

The capacity of the solvent to stabilise or destabilise a transition state or intermediate may also be important. Formation of a polar intermediate, or even dissociation into two ions will be much more favorable in a polar solvent. In non-polar solvents there is little solvation of polar species formed, and there is frequently a considerable loss of entropy as a result of the ordering of solvent molecules by the polar intermediate. Choice of solvent may favour one of several possible mechanisms: in the oxidative addition of hydrogen halides to Vaska's compound, a polar solvent such as methanol favours the S_N2 heterolytic mechanism, giving a *trans* adduct, while the less polar benzene gives a *cis* adduct expected for the concerted addition of an HCl molecule.

Quite apart from effects of solvation on the activation energy, the solvent may actually take part in the reaction. The solvent is the species present in highest concentration and its participation is thus generally favoured. It is unfortunately difficult to detect solvent participation from rate laws as the concentration changes negligibly during the reaction. Apart from the obvious case where the solvent reacts with a product of the reaction, solvent participation without reaction has frequently been detected. In dissociation reactions a solvent molecule may move into the inner coordination sphere to replace the lost ligand – although the ligating power of the solvent molecule may be low, it will generally lower the energy of the dissociation. In the ligand migration reaction discussed above, a solvent molecule is thought to move into the vacant coordination site produced by the migration, to be replaced later by any stronger ligand present in solution or by the reverse of the ligand migration.

One of the best known examples of solvent participation is nucleophilic substitution at a four-coordinate square planar species. It was mentioned earlier that the open geometry of this system favours an associative mechanism, yet the rate law shows that two mechanisms are followed, only one being dependent on the nature of the incoming ligand. This mechanism is clearly an S_N2, A type mechanism, but the other process, dependent only on the concentration of the four-coordinate species is believed to involve the solvent displacing the outgoing ligand, followed by displacement of the solvent by the incoming ligand:

$$ML_3X + \underset{\text{(solvent)}}{S} \xrightarrow{\text{slow}} ML_3S + X$$

$$ML_3S + Y \xrightarrow{\text{fast}} ML_3Y + S$$

This proposal is supported by the observation that the solvent assisted mechanism is not found in non-coordinating solvents, only a simple ligand-dependent rate law being observed.

c) Catalysis

In this section we shall discuss various ways of accelerating chemical reactions by chemical rather than thermal or photochemical methods. By catalysis we imply the acceleration of a reaction by a relatively small amount of a species which may be recovered more or less unchanged at the end of the reaction; there are of course reactions where the 'catalyst' accelerates the reaction but remains bound to the products (necessitating a stoichiometric amount of catalyst) or where the reactants or products may themselves have a catalytic effect. Equally, there are cases where a relatively small amount of a species added to the reaction mixture will accelerate the reaction, yet be consumed in the course of the reaction: this is particulary common for species which readily yield free radicals and may thus initiate chain reactions – such species may be described as initiators or sensitizers.

We may envisage two methods of accelerating a reaction: by altering the electronic structure of the reactants so that a mechanism similar to the uncatalysed reaction is rendered more favorable, or by enabling an entirely new and faster mechanism to be followed. An example of the second case would be the solvent assisted substitution at a square planar centre discussed above.

Acid-base catalysis

Interaction of the reactants or one of the reactants with an acid or base may render the system more reactive. As an example we may consider the hydrolysis of $[Co(NH_3)_5Cl]^{2+}$ which is strongly catalysed by hydroxide ion (alkaline hydrolysis is 10^5 times faster than acid hydrolysis). The mechanism is thought to be

$$\text{(i)} \quad [Co(NH_3)_5Cl]^{2+} + OH^- \underset{\text{fast}}{\rightleftharpoons} [Co(NH_3)_5NH_2Cl]^+ + H_2O$$

$$\text{(ii)} \quad [Co(NH_3)_4NH_2Cl]^+ \xrightarrow{\text{slow}} [Co(NH_3)_4NH_2]^{2+} + Cl^-$$

$$\text{(iii)} \quad [Co(NH_3)_4NH_2]^{2+} + H_2O \xrightarrow{\text{fast}} [Co(NH_3)_5OH]^{2+}$$

The dissociative part of the reaction now involves the conjugate base $[Co(NH_3)_4NH_2Cl]^+$ of the initial complex and for this reason the mechanism is referred to as S_N1CB. One may propose various reasons for the rate determining step (step (ii)) being faster for the conjugate base than for $[Co(NH_3)_5Cl]^{2+}$: it will be easier to remove a chloride ion from an ion with charge +1 than +2; the strong NH_2^- ligand may push the chloride off; the NH_2^- ligand with two non-bonding pairs may be able to stabilise the five coordinate intermediate $[Co(NH_3)_4NH_2]^{2+}$.

Acid catalysis is shown in the aquation of *trans* $[Co(en)_2F_2]^+$, where the loss of fluoride is accelerated by the protonation of one of the fluoride ligands.

When the catalysing acid or base is the characteristic cation or anion of the solvent (in the case of water, H^+ and OH^-) the reaction is said to show specific acid or base catalysis; if any acid or base (for example CH_3CO_2H or $CH_3CO_2^-$) will catalyse the reaction, the reaction shows general acid or base catalysis.

Metal ion catalysis

Metals are almost always found in solution in cationic form, and may therefore function as Lewis acids and exhibit appropriate catalytic activity. Their catalytic functions are extremely important in biological systems where high or low pH levels are undesirable. As an example we may consider an organic carbonyl group bonded to a metal ion:

$$M^{n+} \text{----} O{=}C\begin{matrix} R \\ R \end{matrix}$$

The effect of the ion will be to draw negative charge towards itself and consequently away from the carbonyl carbon atom which will thus become more susceptible to nucleophilic attack: in aqueous solution, this may correspond to attack by water. Metal ions (e.g. Zn^{2+}) are extremely effective at hydrolysing peptide chains, where there is a possibility of chelation to the metal; similarly phosphate chains may be hydrolysed by metal ions, notably Mg^{2+}. For fuller details the reader is referred to specialist works on bio-inorganic chemistry.

The hydrolysis of $[Co(NH_3)_5Cl]^{2+}$ is accelerated by the addition of Hg^{2+} ions which may complex with the chloride ligand and encourage its dissociation. A simpler, purely electrostatic case of catalysis is the outer sphere electron exchange between MnO_4^- and $MnO_4^=$ catalysed by added cations such as Cs^+, as a result, it is thought, of a weak complex of the type $MnO_4^- \ldots\ldots Cs^+ \ldots\ldots MnO_4^=$. The electrostatic repulsion between the anions is thus reduced. Finally we may note that ligands bonded to metal ions may also undergo condensation reactions to give multidentate macrocyclic ligands – this offers a useful synthetic route to certain complicated organic molecules using the 'template' effect of the metal ion.

In all the examples given above the electrostatic or Lewis acid properties of the metal ion can be seen to accelerate a particular reaction of a coordinated ligand. It is, however equally important to notice that the relative ease of substitution of metal ions is important since the reactant molecules may bind to and the products dissociate from the metal by relatively low energy pathways. We may understand the frequency of occurence of Mg^{2+} and Zn^{2+} as Lewis acid catalysts since they have a sufficiently high charge to polarise the coordinated ligands, while undergoing substitution reactions relatively easily.

Redox catalysis

Catalysis of electron exchange (or catalysis of other reactions by electron exchange) almost invariably involves the adoption of a completely new mechanism. Catalysis is particulary common for non-complementary reactions, and species which may undergo several different redox reactions (notably transition metal compounds) are often extremely powerful catalysts. Manganese, with at least seven well characterised oxidation states, is often a source of catalysts as in the reaction

$$KClO_3 \xrightarrow{\Delta/MnO_2} KCl + 1\tfrac{1}{2}O_2$$

where considerable rearrangement of the chlorate ion will clearly be necessary to produce chloride and molecular oxygen. A more concrete example is the catalysis of a non-complementary reaction by silver ions – a two electron oxidant (such as peroxydisulphate $S_2O_8{}^{2-}$) may oxidise Ag(I) to Ag(III) in a single step; Ag(III) may then oxidise a one electron reductant by two one electron steps: Ag(III) → Ag(II), Ag(II) → Ag(I). Catalysis of simple complementary reactions may also be favoured if the catalyst redox couple allows two fast steps rather than one slow step.

Biological compounds show a whole series of catalysed redox reactions, and copper and iron containing catalysts are common. It is thought that the distorted environments sometimes found for the metal ions in these systems may result in lower rearrangement energies on electron transfer, with a consequent increase in reaction rate. The importance of catalysis in biology is understandable when it is remembered that the most common natural oxidant, molecular oxygen, is a four electron oxidant. Transition metal ions are very effective at catalysing the reactions of molecular oxygen and hydrogen peroxide, and their ability to complex both these oxidants must be helpful for this. More generally, the fact that many transition metals readily undergo one electron redox reactions offers a useful low energy path to the formation of free radicals as in reactions of the type

$$M^{2+} + HOOH \rightarrow [M(HOOH)]^{2+} \rightarrow [M(III)OH]^{2+} + HO.$$

Catalytic processes involving electron transfer can also accelerate substitution reactions. Charcoal catalyses substitution reactions of cobalt (III) and there is reason to believe that this occurs by a reduction to Co(II) which undergoes substitution much more rapidly than Co(III). Gold (III) complexes undergo substitution much more rapidly in the presence of a reducing agent able to produce even very small amounts of gold (II). A more complicated example is the chloride exchange of $PtCl_6{}^{2-}$ which was found to be catalysed by $PtCl_4{}^{2-}$: there is a Pt(II) – Pt(IV) exchange involving an intermediate similar to that shown in Fig. 6.12.

Catalysis by organometallic compounds

Some of the specific reactions discussed in illustration of orbital symmetry effects offer low energy pathways for the formation of covalent bonds such as the carbon-carbon bond formed by ligand migration in $CH_3Mn(CO)_5$. It is therefore not surprising that many organometallic compounds show considerable catalytic activity. In an excellent review of the subject, Tolman proposes that organometallic compounds react by elementary reactions involving only organometallic species with 16 or 18 valence electrons. 18 electron species may be transformed into 16 electron species by ligand dissociation, migration, reductive elimination, or oxidative cyclisation; the 16–18 electron transformation is effected by the reverse of these reactions. An explanation that this approach provides for the hydrogenation of ethylene catalysed by $Rh(PPh_3)_3Cl$ (Wilkinson's catalyst) is shown in Fig. 6.20, and it may be seen that the reaction loop involves only simple steps that we have already shown to be allowed. A fuller discussion, with some of alternative pathways, is given by Tolman.

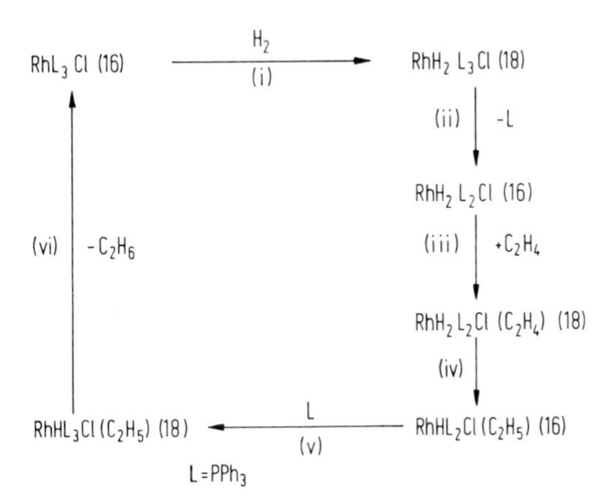

Fig. 6.20. Figures in brackets after the compounds indicate the number of valence electrons associated with the metal. Step (i) Oxidative addition, (ii) ligand dissociation, (iii) ligand addition, (iv) ligand migration, (v) ligand addition, (vi) reductive elimination

A large number of organometallic catalysts are known and their mode of action has been rationalised. Chemists are, however, still some way from designing 'tailor made' catalysts, although they are less and less surprised when they find them. In a very general sense we may see why organometallic compounds are so effective as catalysts: the variety of orbitals available (s, p, and d) eases orbital symmetry restrictions; the relatively weak bonding of organic species to the metals (metal alkyls ~150–200 kJ/mole, alkenes ~100 kJ/mole, CO ~200–250 kJ/mole) enables bonds to be formed and broken with much less expenditure of energy than the homolytic paths these non-polar reactions would otherwise follow.

Heterogeneous catalysis

Many reactions in the gas and liquid phases are catalysed by solid surfaces. In all these reactions, adsorption of one or more of the reactants on the surface is essential, and for any pronounced catalytic effect it is important that the interaction should be sufficiently strong to produce an appreciable chemical change in the adsorbed molecule.

Some of the atoms on the surface of any material must be coordinatively unsaturated, and can function as Lewis acids or bases. Some surfaces (notably SiO_2) carry acidic protons which may react with adsorbed species, while others (such as ZnO) may act as bases, and capture protons from the substrate. An alumina/silica catalyst is used for reactions of alkenes, and is thought to act by protonation of the double bond, producing carbonium ions. Many transition metals and their compounds can undergo electron transfer reactions with adsorbed species, and act as redox catalysts.

These examples differ only from those discussed earlier in that the catalyst is immobile, and fixed to a solid support, although this may be of great use in separating products and catalyst after the reaction. More specific to heterogeneous catalysis is the fact that the catalyst may be presented to the substrate or substrates in the gas phase without any interference from a solvent, and that many small molecules adsorbed on metal surfaces undergo dissociation. Thus platinum metal absorbs hydrogen to give dissociated, and therefore more reactive, hydrogen atoms. Photoelectron spectroscopy suggests that the interaction of iron metal with CO is sufficiently strong to break the C≡O triple bond. These dissociative adsorptions will clearly accelerate homolytic reactions.

d) Steric Effects and Stereochemistry

We have not hitherto discussed steric effects, due to repulsive interactions between atoms or distortion from equilibrium geometries, yet these frequently play a dominant role in determining mechanism. A model of a molecule with high coordination number such as an octahedron will often show an associative attack on the central atom to be impossible without considerable repulsion from ligands already bound to the central atom – this in itself is a good explanation for the predominantly dissociative reactions of octahedral complexes.

We may often use our knowledge of the structure of stable compounds to predict the steric possibility or impossibility of an intermediate. Thus, very few 7 coordinate complexes of transition metal ions are known, but in some cases where 7 coordinate species are known (for example Fe^{3+} in $[Fe(EDTA)OH_2]^-$) there is some evidence for associative reactions. The size of the central atom is often important: carbon is very rarely found with coordination number higher than 4, while for the larger silicon and phosphorus atoms higher C.N. are known in stable compounds. Not surprisingly, carbon shows only I_a associative substitution mechanisms, while silicon and phosphorus show A or limiting S_N2 associative mechanisms.

Bulky ligands can also affect the mechanisms followed: the classic case of tertiary butyl compounds $(CH_3)_3CX$ and their tendency to react by dissociative mechanisms is well known. An interesting example from inorganic chemistry is the d^8 iridium (I)

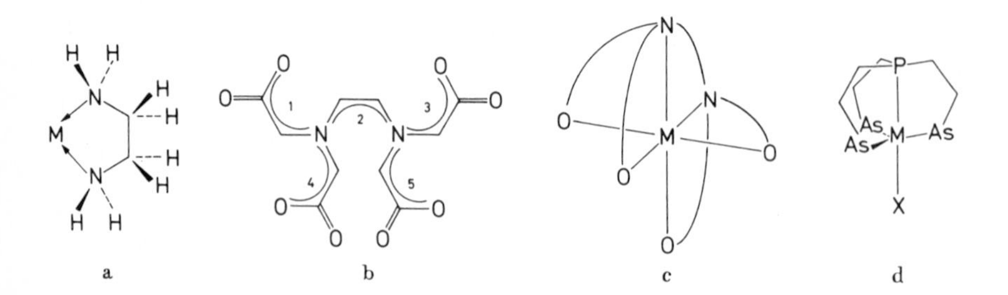

Fig. 6.21. (a) The five-membered metal ethylene diamine ring, **(b)** the five chelating rings of EDTA, **(c)** a metal-EDTA complex, **(d)** a complex of the 'tripod' ligand $P(CH_2CH_2AsR_2)_3$

complex $[Ir(PPh_2Me)_4]^+$ which differs from the chelate compound shown in Fig. 6.14 only by the replacement of a $-CH_2-CH_2-$ link by two CH_3 groups. The complex $[Ir(PPh_2Me)_4]^+$ is appreciably distorted (towards a tetrahedral structure) and does not oxidatively add oxygen or hydrogen. The choice of suitably bulky ligands can effectively prevent associative mechanisms: ethylation of dien (Table 5.3) can block the approach of ligands in substitution reactions of compounds such as $[Pt(Et_4dien)Cl]^+$, and favour a dissociative mechanism.

Steric effects are very important for chelating or multidentate systems: thus the metal-ethylenediamine linkage (Fig. 6.21) forms a stable five membered ring with little or no distortion of bond lengths or angles – dissociation of a metal-nitrogen linkage will involve some distortion of other bonds in the ring, and this provides a kinetic barrier to the dissociation reaction. This effect is most spectacularly seen in EDTA complexes where there are five five-membered rings, conferring exceptional kinetic stability on these complexes. Choice of a suitable polydentate ligand may force an unusual coordination number on a metal, rather than produce an intolerable strain in the ligand – 'tripod' ligands (Fig. 6.21 d) have been much used in this context. Biological compounds are past masters in this field, using ligand strain effects to complex metals in unusual geometries (where they may be more reactive – this is the basis of the theory of the entatic state due to Vallee and Williams), to keep possibly interacting species apart, and, by choice of suitable weakly hindering ligand groups, favour only the attack of particular species – as in hydrophobic pockets (page 125).

Stereospecificity

As we have been mainly concerned with the relationship between electronic structure and reactivity we have not specifically discussed stereochemistry in this chapter, commenting only in passing on any stereochemical predictions (such as *cis* addition for concerted oxidative addition). Consideration of the stereochemistry of products and reactants has been of great help in unravelling mechanisms: an example is the ligand migration reaction where the compound *trans* $[CH_3Mn(CO)_4L]$ gives the product $[L'Mn(CO)_3(CH_3CO)L]$ where the acetyl group is *cis* to the ligand L, and L' is *trans* to it, showing that it was the methyl group that moved. The presence of the ligand L enables the stereochemistry of reactants and products to be related, and is often referred to as a stereochemical signpost; much more complete discussions are given by Basolo and Pearson, and by Tobe.

In certain cases the steric repulsions of the ligand group can give rise to a stereospecific complex formation reaction. One of the most notable cases is the cobalt(III) complex of PDTA, propylenediaminetetraacetic acid (which may be regarded as EDTA with one hydrogen replaced by a methyl group in the $N-CH_2-CH_2-N$ chain). The ligand itself is chiral and the (–) isomer of the ligand forms only the L* $[Co(-)PDTA]^-$ complex (a species with two chiral centres) as a result of the repulsive interactions of the methyl group with other parts of the ligand in D* $[Co(-)PDTA]^-$. By contrast EDTA gives a racemic mixture with Co(III). The subject of stereospecificity has been reviewed by Sargeson.

The most famous case where electronic rather than steric effects lead to stereospecific reactions is the *trans* effect shown in associative nucleophilic substitution at square planar d^8 metal ions, most notably for Pt(II). It was observed that certain ligands exerted a strong labilising effect on the *trans* ligand to the point where substitution reactions would result in the exclusive displacement of this *trans* ligand. Thus, since chloride has a greater *trans* effect than ammonia, the reaction of $[PtCl_4]^{2-}$ with ammonia gives *cis* $[(PtCl_2(NH_3)_2]$, the two chlorides displaced being *trans* to other chlorides; the reaction of $[Pt(NH_3)_4]^{2+}$ with chloride gives *trans* $[PtCl_2(NH_3)_2]$ since the second chloride enters more readily *trans* to the first than *trans* to an ammonia ligand. A generally accepted series for the *trans* effect is

$$H_2O \sim OH^- \sim NH_3 \sim \text{amines} < Cl^- \sim Br^- < SCN^- \sim I^- \sim NO_2^- < CH_3^- < \text{sulphur + phosphorus ligands} < H^- < CO \sim CN^- \sim \text{alkenes.}$$

The *trans* effect is a kinetic effect, showing up in rate constants of particular reactions; there is also a ground state influence of the *trans* ligand on a metal ligand bond which has been extensively studied – this is best referred to as the *trans* influence, although it is sometimes confusingly called the ground state *trans* effect. Once this confusion is resolved, there remains the problem that the *trans* effect of a given ligand is dependent on the nature of the incoming ligand, although this was not immediately recognised. Two explantations have been proposed for the *trans* effect, both of which appear to play a part in determining the *trans* effect.

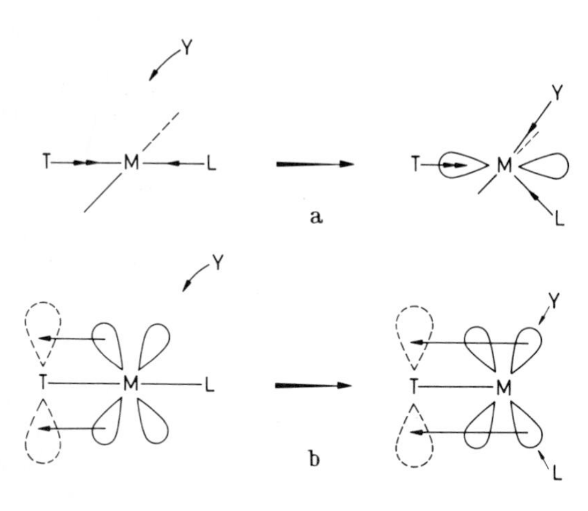

Fig. 6.22. (a) The σ *trans*-effect – The M–L bond is weakened in the initial complex, **(b)** the π *trans*-effect – π acceptance by T reduces M–Y and M–L repulsion in the intermediate or transition state

The σ bonding explanation predicts that a strong σ donor T will bind strongly to the metal, using the metal p orbital it shares with the *trans* ligand L. There will thus be a repulsion between the T–M bond and the M–L bond which will give a weakening of the M–L bond, and will favour the distortion necessary to produce the five-membered transition state of the associative substitution (Fig. 6.22 a). This effect should clearly show some correlation with the ground state *trans* influence, and indeed appears to do so.

The π bonding explanation is invoked to explain the high kinetic *trans* effect of ligands generally held to be good π acceptors (such as CO, alkenes). The ligand T accepts charge from a filled d orbital which would otherwise repel the ligand L and the incoming nucleophile Y in the 5 coordinate transition state (Fig. 6.22). The transition state is thus stabilised by the interaction, and the reaction facilitated.

These two theories are useful as a qualitative guide, but, apart from the combination of σ and π effects it is probable that other effects such as reorganisation energy in the transition state and the preference of ligands for d, s or p metal atomic orbitals, are important. Our understanding of the kinetic *trans* effect is still imperfect, as, indeed, is our understanding of most ligand-ligand interactions.

e) Solid State Reactions

The chemical reactions undergone by solids are of immense importance in industrial, geological, and many other fields. The study of these reactions is extremely complicated and a great variety of experimental rate laws are found for apparently quite similar reactions. In this section only a brief discussion of some of the important features is given, the intention being to underline the difference between solid state and fluid state reactions.

Diffusion in solids

The most obvious difference is the low mobility of atoms, ions, and molecules in solids by comparison with fluids, and this is often the determining feature in solid state reactions. It is, however, erroneous to imagine that a species trapped in a solid is immobile: SiO may be trapped in a nitrogen matrix at 20^oK, but warming to 30^oK produces rapid polymerisation to $(SiO)_n$ even though the melting point of nitrogen is 63^oK. Defects and irregularities in the crystal play a vital part in determining solid state mobility, and we discuss their effect in some detail. In a perfect, closely packed, solid it will clearly be very difficult to move an atom, ion, or molecule without encountering considerable repulsion from neighbours. If, however, the structure is not perfect (as is always the case) and there is, for example, a vacancy, some mobility is possible since an atom can diffuse into the vacancy, leaving a new vacancy behind it – this may be regarded either as atomic diffusion or as diffusion of the vacancy. We may identify three regions where mobility may be enhanced:

1) Near point defects (Chap. 3)
2) Irregularities in packing of units, such as dislocations in the crystal and the boundaries between grains
3) Surfaces.

Any crystal will have a surface, and an equilibrium distribution of defects; the presence of irregularities in the packing will depend on the history of the crystal, and rapidly grown, poorly formed, or physically maltreated crystals will show many irregularities, and possibly a defect concentration higher than the equilibrium value. There will also be a break in the regular packing of the solid at surfaces and irregularities which will give some room for atomic movement. The sudden break in regularity may also involve a sudden change in potential energy: we would expect that the surface of an ionic crystal would show a high electrostatic potential close to the ions, falling away rapidly as one moves away from the crystal. This high potential region may well be chemically active. Finally, we may note that the space available to allow motion in the solid may also give room for the formation of a new species with the rearrangement of atomic positions that this requires. This formation of a new crystallite is referred to as *nucleation;* many solid state reactions begin with a nucleation reaction at a surface or dislocation etc., and proceed by steady growth of the new crystallite.

By way of illustration we may consider the effects on diffusion of the point defects in ionic crystals discussed in Chap. 3 (Fig. 3.43). Schottky defects will produce vacancies which will be able to diffuse through the solid (Fig. 6.23). The cation vacancies of the Frenkel defect will also be mobile; the mobility of the interstitial species depends on the radii of the crystal components. Small impurities in interstitial sites are often very mobile and can diffuse rapidly through a crystal. In general, the mobility of an ion is dependent on its charge and size – monovalent ions are more mobile than divalent ions, cations are more mobile than anions (see Table 5.1). Impurities producing vacancies will also increase mobility.

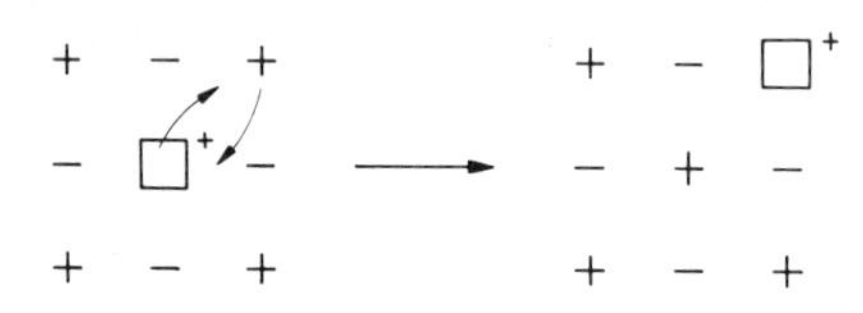

Fig. 6.23. Diffusion of a cation in an ionic solid

Electronic defects, such as the electron holes (due to the presence of Ni^{3+} ion) (Fig. 3.44) in non-stoichiometric $Ni_{1-\delta}O$ may often be quite strongly bound (a Ni^{2+}–Ni^{3+} electron exchange would have a Franck-Condon barrier as for any electron transfer). It should be borne in mind that these diffusion reactions produce a displacement of charge, the left to right movement of a cation vacancy corresponding to a right to left movement of positive charge. The build up of an appreciable polarisation in the crystal is thermodynamically unfavourable in the absence of an applied field, and ionic motion in crystals will thus preserve local charge balance. For this reason defect mobility is often greater in non-polar solids such as metals.

Some simple reactions

(i) Annealing and sintering. Annealing reactions essentially involve the return of a solid to its equilibrium state. Thus crystals which have been quenched rapidly from high temperature (and thus have too high a defect concentration) or have been subject to radiation damage, or physical maltreatment, may be annealed to reduce irregularities or defects in the structure. For alkali halides, quenched from high temperature, where the most common defects are Schottky defects, annealing will involve diffusion of valencies to the surface. The spinel (page 117) $NiAl_2O_4$ ideally has Ni^{2+} ions in octahedral sites and Al^{3+} ions in tetrahedral sites, but entropy effects give rise to increasing interchange of the two ions with increasing temperature. A sudden drop in the temperature will be followed by annealing as cations move to an equilibrium distribution.

Sintering reactions involve the growth of the crystal at one point, accompanied by the erosion of the crystal at another: the net effect is to reduce the total surface area and increase the links between the crystallites of the solid. The reaction involves diffusion of crystal components in such a way as to lower the total free energy of the system.

(ii) Reactions between two solids. This can be formidably complicated subject, and we consider here only three possible reaction paths for a reaction of the type $AO + B_2O_3 \rightarrow AB_2O_4$. The simplest case would be for AO and B_2O_3 to mix in a solid solution as a result of diffusion, followed by nucleation of the AB_2O_4 phase. In the second possibility AB_2O_4 may form quite readily at the junction, and the reaction will then be controlled by the rate at which the reactants can diffuse through the product to react with each other (Fig. 6.24). This rather bald statement does nothing to tell us how the diffusion takes place – A^{2+}, B^{3+} and O^{2-} ions are all possible migrating species, but there is also the possibility of loss of gaseous oxygen by one reactant, oxygen take up by the other coupled with electron transport through the AB_2O_4 layer; several possible mechanism's may be proposed (see Schmalzried), and all may be followed to some extent, depending on the relative mobility of the species involved. In the final path the reaction rate is controlled by the rate of reaction at the AO/B_2O_3 interface, and the diffusion together of the reactants is no longer rate determining.

Fig. 6.24. Formation of a layer of duct between two reactants

(iii) Surface oxidation of metals. We will take as a simple example the oxidation of zinc, but it must be emphasied that this is an idealised picture of a specific reaction path, and that five different rate laws (not to mention combinations of them) have been established for metal tarnishing.

We assume that the Zn metal/O_2 reaction is fast and rapidly gives a surface layer of ZnO. It was mentioned in Chap. 3 that ZnO will absorb Zn metal to give $Zn_{1+\delta}O$ with interstitial Zn^{2+} and associated trapped electrons. It is a n type semiconductor. The reaction may then proceed further by a diffusion of interstitial Zn^{2+} and its electrons into the layer of ZnO to react with oxygen at the surface to give more ZnO – this diffusion is favoured by the concentration gradient for Zn^{2+} (interstitial), falling from the metal/oxide junction to the oxide surface. Further zinc can then enter the oxide layer, and the reaction continues. (Fig. 6.25).

A similar mechanism for the oxidation of aluminium is not possible since the aluminium cannot enter the Al_2O_3 layer interstitially; diffusion of Al^{3+} ions might be possible, but the electrons necessary to reduce the molecular oxygen cannot readily be transported to the surface. The reaction thus stops after a thin layer of oxide is formed. It should however, be noted that the resistance to corrosion owes much to

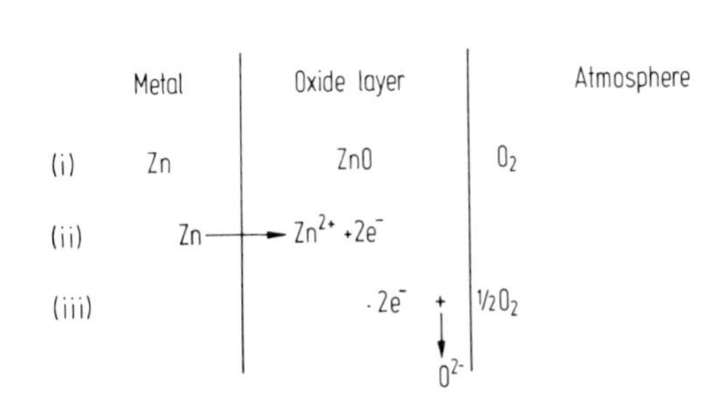

Fig. 6.25. The corrosion of zinc. (i) The initial layers, (ii) zinc metal diffuses into the oxide layer to give Zn^{2+} and trapped electrons, (iii) the electrons reduce oxygen to O^{2-} which diffuses into the oxide layer

the hard, non-porous nature of the oxide preventing further attack by oxygen at the unreacted metal.

It is hoped that this very brief survey has given the flavour if not the substance of solid state chemistry, and that it has indicated the importance of crystal imperfections in allowing mobility and nucleation of new species. So many effects can change the rate of a reaction that each reaction must be studied as a particular case, but we may see that poorly formed crystals with high surface areas and may defects will be appreciably more reactive than more perfect solids. Impurities in the solid will generally increase the defect concentration and thus reactivity, but impurities which reduce the number of vacancies, for example, may actually lower the reactivity.

f) Photochemistry

The better understanding of molecular structure and spectra resulting from the increased use of M.O. theory has led to a great deal of interest in the photochemistry of inorganic compounds. Photochemistry essentially involves two processes: the absorption of energy of the reactant system, and the chemical reaction of the excited state produced. In the vast majority of cases the absorption of radiation is not followed by chemical reaction but by other physical processes allowing the dissipation of the absorbed energy. The most important processes are:

1) Internal conversion. The excited state molecule loses energy to vibrational modes and by collisions in the fluid phase.

2) Energy transfer. The energy is lost to some other species which is simultaneously excited. This may occur by various mechanisms, and can be useful as a means of exciting a second species (the acceptor, A) which is otherwise difficult to excite. The initially excited molecule (the donor *D) is thus relaxed to the ground state by the transfer: $^*D + A \rightarrow D + {}^*A$

This is the basis of sensitization, where a strongly absorbing donor is used to excite a weak or colourless acceptor.

3) Luminescence. We may split this up into *fluorescence* and *phosphorescence.* In Fig. 6.26 the excited state produced by absorption is a vibrationally excited state of the state A_1 which will have the same spin multiplicity as A_0. This will rapidly lose vibrational energy and can follow two alternative paths: (i) it decays to the vibrational ground state of A_1 (ii) it will jump to the potential surface of a second excited state

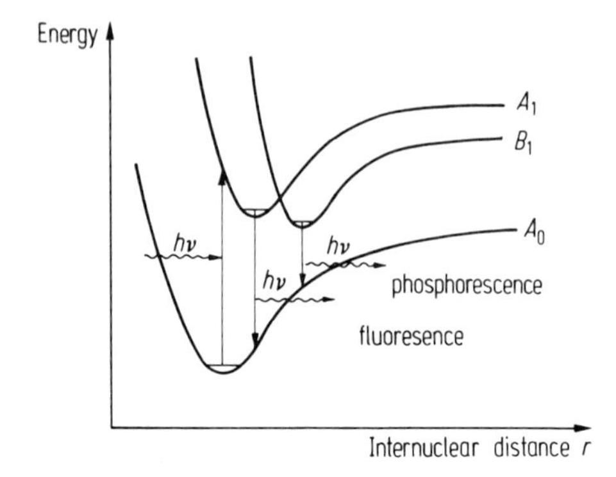

Fig. 6.26. Fluorescence and phosphorescence of excited states

B_l. In the first case, the ground state of A_l can decay by emission of radiation to a vibrationally excited state of A_o – this emission of radiation, at a lower frequency than the exciting line, is *fluorescence.* It will be favoured when the vibrational relaxation between the A_l groundstate and A_o is slow – this is most frequently the case when the vibrations available are of low energy.

The second case where the system moves to a different state, B_l, is more complicated. If B_l and A_l have the same spin multiplicity the jump will be rapid and allowed by the internal conversion mechanism, and the decay may continue by any of the mechanisms described above. If B_l and A_l have different spin multiplicity the jump will be forbidden by the spin conservation rule, but mechanisms such as spin-orbit or vibronic coupling may partially allow the jump. If B_l is a triplet and A_l a singlet, the vibrational groundstate of B_l will lie lower than that of A_l (by Hund's rule) and the decay to the lowest B_l level will clearly be favourable. If this groundstate is to decay to A_o by radiation emission the spin conservation rule $\Delta S = 0$ will be broken and the emission has a low probability; this is *phosphorescent emission.* The forbidden nature of the transition results in phosphorescent states having much longer lifetimes than fluorescent states.

Any chemical reaction of the excited state will have to compete with these processes, and therefore the irradiation must produce a very considerable increase in reactivity. In some cases, interaction of excited species with other bodies can result in deactivation with no net reaction – this is called quenching.

We must therefore turn our attention to the nature of the excited state. Even absorption at the lowest energy of the visible spectrum (10kK) corresponds to an energy of over 100kJ/mole, so we may reasonably expect to see some difference in chemical behaviour. Generally speaking, the excited state A_l will decay very rapidly to its vibrational ground state or to the vibrational ground state of the lowest excited state of the same multiplicity, and it is at this point we may expect some chemistry. The basis of photochemistry is the fact that the excited state has a different electronic structure, and therefore a different reactivity. The absorption of radiation may involve the movement of charge from on part of a molecule to another, and this can produce a completely new polarisation in the molecule. Changes of pK_a in the excited states of organic acids of over 6 units are known. Although the Franck-Condon principle 'freezes' the molecule during the absorption, the subsequent loss of vibrational energy

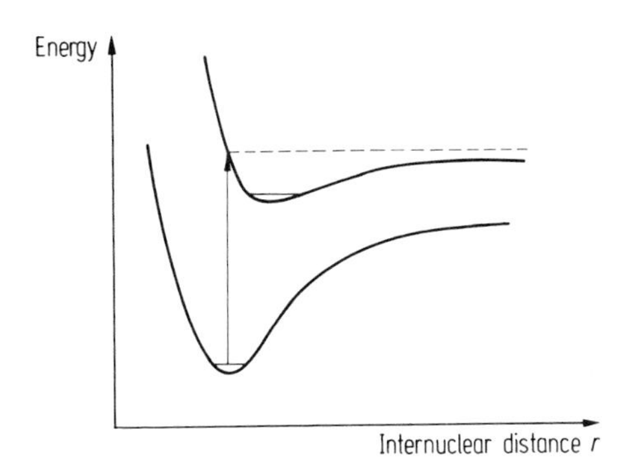

Fig. 6.27. Dissociation of an excited state

will involve a change in internuclear distances, and the vibrational ground state of the excited molecule may well have a very different geometry – this can, of course, affect symmetry controlled reactions.

With these generalities in mind we will now discuss some of the general effects of absorption, and then relate these to the absorption spectra discussed in Chap. 4.

A. Homolytic bond cleavage. Consider a halogen molecule X_2. The lowest lying empty orbital is the σ_u* and photo-excitation will occupy this orbital, consequently weakening the X–X σ bond. Unless the excitation is from the occupied σ_g orbital (Chap. 2) this will not produce a net antibonding arrangement, but since the excited state produced will have vibrational energy as well (Fig. 6.27), it may shake itself apart before it can decay to the vibrational ground state. The net result is homolytic fission of the X–X bond. The photolytic cleavage of halogens is quite well known and offers a useful way of producing radicals. A more complicated example is the photolysis of $[(OC)_5Mn{-}Mn(CO)_5]$ to give two $[Mn(CO)_5]$ radicals. The weak Mn–Mn bond is cleaved by excitation to the Mn–Mn σ* orbital, the electron coming from either the σ bonding orbital, or d orbitals involved in π bonding to CO. For photolytic bond cleavage, it is important that the antibonding orbital should be of fairly low energy, and that it should be fairly localised or the excitation will only produce a bond weakening in several bonds, and no actual cleavage. The reader may care to show for himself that this would be the case for the weak interactions of two d_{z^2} orbitals for two $3d^7$ $Mn(CO)_5$ fragments.

B. Charge distribution changes. A simple example is the carbonyl group >C = O. The lowest-lying absorption is from the non-bonding p orbital of the oxygen to the π* orbital of the CO bond, which is strongly localised on the carbon atom:

$$>C=O \rightarrow >\dot{C}-\dot{O}.$$

The carbon is now more nucleophilic, while the oxygen, having lost an electron, is now electrophilic, and the species may be regarded as a di-radical. Not surprisingly, charge transfer excited species show an enhanced activity in a similar fashion.

C. Orbital correlation. When the absorption or radiation involves highly delocalised M.O., or only a small amount of charge redistribution, we are obliged to use the more general molecular orbital apporach, starting with a consideration of the structure of the excited state. This will depend on the compound being irradiated, but there are two features of general interest.

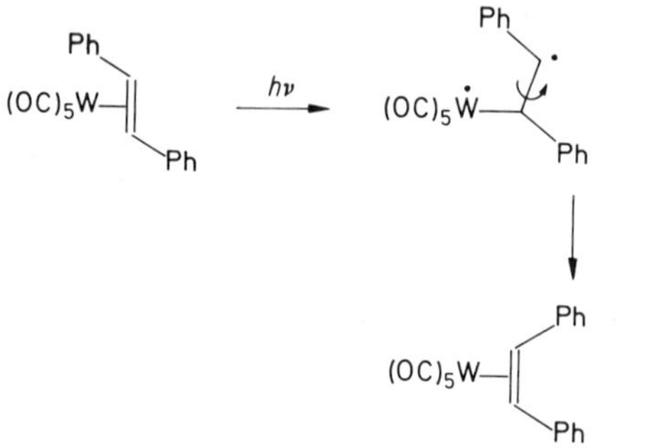

Fig. 6.28

The excited state will have a structure different from that of the ground state. This may well change the symmetry restrictions on a chemical reaction, and may also cause the approach of two reactive centres. As an example of this effect we may consider the photo-isomerisation of *trans*-stilbene coordinated to $W(CO)_5$ (Fig. 6.28). It has been suggested that a σ bonded di-radical is produced which has much freer rotation about the olefin C–C bond, and can thus isomerise. A more important effect may be seen by considering orbital correlation. A reaction will be favoured if the ground state orbitals of reactants and products correlate (Fig. 6.29 a). If the orbitals do not correlate (Fig. 6.29 b) there will clearly be an energy barrier to the reaction. If one electron is excited (Fig. 6.29 c), then as one orbital rises in energy, the other falls, and there is little change in energy on going from reactant to product and the reaction is much more favourable.

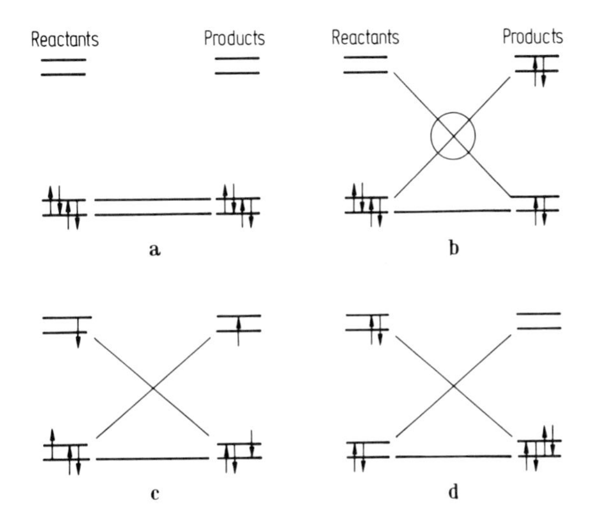

Fig. 6.29 a – d. Orbital correlations. **(a)** Allowed thermally, **(b)** forbidden thermally, **(c)** allowed photochemically, **(d)** allowed thermally

In fact this is an oversimplified view of the reaction as we should really be considering electronic *states* and not individual electrons. Reaction (6.29 b) can in fact occur as a result of mixing with the highly excited state (6.29 d) near the crossing point (circled in Fig. 6.29 b) giving the product in its electronic ground state but this process will have a high activation energy. The one electron excited state (6.29 c) will move more readily along the reaction coordinate to give the products, and will mix more readily with state (d). The reaction is thus downhill all the way to the product ground state. A much more detailed discussion is given by Pearson.

The foregoing arguments may be generalised to give the useful rule that reactions disfavoured in the ground state will be favoured in the excited state, and vice versa.

We will now discuss some examples of the photochemical activity resulting from the electronic absorptions discussed in Chap. 4.

d–d transitions

As an example of some of the photophysical effects discussed at the beginning of this section we may consider the tetrahedral $3d^5$ $MnCl_4{}^{2-}$ ion. In the solid state (but not in solution) this almost colourless compound gives a strong green luminescence in ultra-violet light. The lowest energy transition of this $3d^5$ high spin species is an $e^2t^*{}_2{}^3 \rightarrow e^3t^*{}_2{}^2$ transition which is spin forbidden (see Fig. 3.9). The emission of light is thus due to phosphorescence. The movement of an electron to the e (non-bonding) orbitals from the $t_2{}^*$ orbitals would be expected to produce a shortering of Mn–Cl bond lengths in the excited state. In solution, the interactions with the solvent rapidly deactivate the excited state but, in the solid, the energy is lost to metal-ligand and lattice vibrations which are less efficient.

Apart from cases where a change in spin state is involved, most d–d transitions involve promotions to antibonding d-orbitals, and, not surprisingly can have an effect on metal-ligand substitution rates. A set of rules, initially due to Adamson, but modified by later workers enables the prediction of the photochemical behaviour of octahedral complexes. The first rule states, that, in a non-symmetric octahedral complex, the axis with the weakest ligand field will be labilised. This is reasonable since an electron localised on this axis (for example in a d_{z^2} orbital) will experience the least antibonding effect; even if the initial excitation were to the $d_{x^2-y^2}$ orbital at slightly higher energy, internal conversion would very rapidly give the lowest excited state, with the electron in the d_{z^2} orbital. Thus in $[Co(NH_3)_5Cl]^{2+}$, the NH_3–Co–Cl axis will be labilised as Cl is a weaker donor than NH_3, and this axis has the weakest overall interaction with the metal.

The second rule determines which ligand on the axis will be more weakened. In the case of σ bonding only, the ligand with the greater overlap with the metal σ^* (i.e. d_{z^2}) orbital will be most affected by the excitation; in cases where the overlaps are similar, the ligand with the highest (i.e. least negative) donor orbital energy will be lost. Generally speaking, this implies that the stronger ligand of the two will be lost (in the case of $[Co(NH_3)_5Cl]^{2+}$, this is the *trans* NH_3 group). π bonding ligands complicate matters, as the electron excited comes from the π bonding (or antibonding) t_{2g} set. If the ligand is a π acceptor, the t_{2g} is π bonding and excitation to $e_g{}^*$ will weaken the metal ligand π bond, favouring dissociation; if the ligand is a π donor, the t_{2g} is π antibonding, and $t_{2g} \rightarrow e_g{}^*$ excitation will strengthen the metal-ligand π bond.

These rules are reasonably well obeyed, and explanations can be found for the exceptions. The fact that the stronger ligand on the labilised axis is lost is often of considerable preparative usefulness since thermal reactions generally result in the weaker ligand being lost. Even when stereospecificity is not sought, photochemical activation can allow reactions to take place at temperatures where undesirable side reactions are very slow. Another use is the production of unstable species as a result of ligand dissociation after a d–d transition. Thus:

$$Fe(CO)_5 \rightarrow Fe(CO)_4 + CO$$

The 16 electron $Fe(CO)_4$ species will very readily undergo oxidative addition, followed possibly by ligand migration etc., and can thus act as a catalyst. Absorption of energy by a d–d transition can also produce isomerisation. This *cis-trans* isomerisation of square Pt(II) complexes has been observed on photolysis. Both dissociative and intramolecular (passing through a tetrahedral intermediate) mechanisms are known.

Charge transfer processes

Ligand to metal transfer corresponds to a metal-ligand redox reaction, and, if the absorbed energy is not lost by non-chemical processes, the products of this redox reaction will be found. Thus, irradiation of $[Co(NH_3)_5Br]^{2+}$ in the ligand to metal charge transfer region produces Co^{2+}, free ammonia and Br_2 as a result of:

$$[(H_3N)_5Co(III)Br^-]^{2+} \rightarrow [(H_3N)_5Co(II)Br]^{2+} \rightarrow Co^{2+} + 5NH_3 + \frac{1}{2}Br_2$$

After the charge transfer, a bromine radical leaves the Co(II) complex, which then decomposes to give the aquated Co^{2+} ion, as cobalt(II) complexes are very labile. This process may also be regarded as homolytic cleavage of the Co–Br bond. Similarly metal oxalates may be photolysed to give reduced metal ions and CO_2. When the metal-ligand bond is too strong to be broken, a very reactive radical may be formed. The photochemistry of the uranyl ion $[UO_2]^{2+}$ is very complex, but it seems certain that one of the primary processes is ligand to metal (5f) transfer, giving a very electrophilic oxygen radical which is capable of hydrogen abstraction from solvents to give a uranium (V) species and an oxidised radical in solution. Irradiation of cerium (IV) in the charge transfer region gives rise to an oxidation of water (or any other suitable species in solution) – this reaction is very slow in the absence of light.

Metal to ligand transfer can also give some interesting reactions: $[Fe(CN)_6]^{4-}$ on photolysis gives $[Fe(CN)_6]^{3-}$ and solvated electrons above 28kK – at energies lower than this, photosubstitution occurs as a result of d–d excitation. Many other examples are known but related substitution reactions (as in the $[W(CN)_8]^{4-}$ / $[W(CN)_8]^{3-}$ couple) can complicate the system. One of the most interesting complexes in this field is $[Ru(bipy)_3]^{2+}$, where bipy is the chelating aromatic ligand (2,2'dipyridyl). This complex has a relatively long lived metal-to-ligand charge transfer state which phosphoresces even in solution. The reactions of the excited state may thus be followed by the change in phosphorescence. It has been found that it can act as an excited state electron donor (i.e. reducing agent) or electron acceptor. The related complex $[Ru(bipy)_2(CN)_2]$ shows a considerable change in acid-base properties on irradiation. Interestingly enough, the complex $[Ru(bipy)_3]^{3+}$ shows chemiluminescence on reduction (e.g. by BH_4^-) showing the Ru(II) complex to be formed in an excited state.

Solid state photochemistry

This is a field of immense importance in photography and radiation detection. We discuss two examples, one of great practical importance which is incompletely understood, and another of potential practical use.

Photography. Very small crystallites of silver bromide (or an iodide and bromide mixture) form, on exposure to visible light, a latent image which, on treatment with a suitable reducing agent, can be developed to give a visible amount of metallic silver where light struck the emulsion of crystallites. This is the basis of conventional photography. It should come as no surprise to find that the defect structure of the solid is of fundamental importance in explaining the process in all the explanations advanced so far. We give here a simple explanation, but this is still a controversial subject.

Silver bromide is virtually ionic: the conduction band is thus empty and localised on silver ions, and the full valence band is full and localised on bromide ions. Silver bromide has a relatively high instrinsic defect concentration of cation vacancies and interstitial silver ions (Frenkel defects),and the interstitial ions are quite mobile. The primary photographic process involves excitation of an electron from the valence band to the conduction band. The hole in the valence band can diffuse rapidly from the electron in the conduction band (it must do this to avoid recombination) and is thought to be 'trapped' by a bromide ion at a dislocation or on the surface, forming a bromine atom. Meanwhile the electron in the conduction band can diffuse to the surface and be trapped by a silver ion. By consideration of the defect equilibria maintained in the crystal, and remembering that silver ions may move interstitially it is possible to explain the build up of nuclei of metallic silver at the surface or at dislocations. On addition of a mild reducing agent (the developer) reduction proceeds much faster at the metallic silver nuclei, and thus causes a build up of metallic silver at areas of high light intensity.

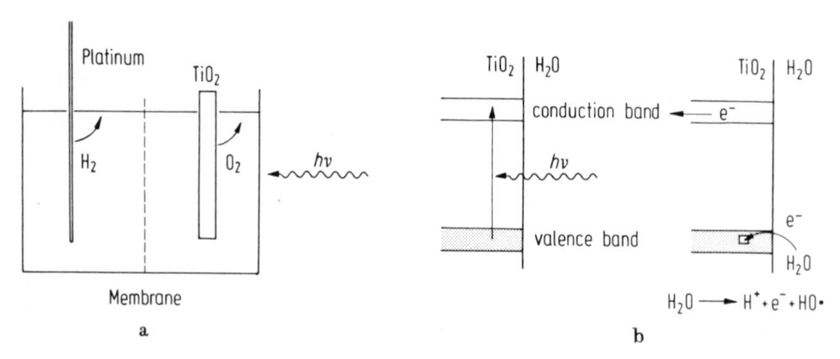

Fig. 6.30 a and b. Light assited electrolysis. **(a)** The cell, **(b)** the reactions at the TiO_2 electrode

Light assisted electrolysis. There is currently considerable interest in the use of solar energy to decompose water into oxygen and hydrogen. Irradiation of TiO_2 (an n-type semiconductor) with energies higher than the band gap (24.4 kK) gives liberation of O_2 in a cell such as that shown in Fig. 6.30. Hydrogen is liberated at the non-irradiated Pt electrode. The explanation is thought to be as follows: absorption excites an electron into the conduction band and leaves an electron hole in the valence band; this hole can abstract an electron from water at the surface of the electrode, and the electron in the conduction band can move round the electrical circuit to reduce hydrogen ions at the platinum electrode.

Bibliography

General treatments of mechanistic inorganic chemistry

Basolo, F., Pearson, R.G.: Mechanisms of inorganic reactions, 2nd Ed. London, New York, Sydney: J. Wiley 1967
Hague, D.N.: Fast reactions. Experimental details with some discussion of inorganic mechanisms. London, New York, Sydney, Toronto: J. Wiley 1971
Pearson, R.G., Ellgen, P.C.: Mechanisms of inorganic reactions in solution: in Physical chemistry, an advanced treatise, Vol. II, H. Eyring (ed.). London, New York: Academic Press 1975
Purcell, K.F.; Kotz, J.C.: Inorganic chemistry. Philadelphia, London, Toronto: W.B. Saunders 1977
Sykes, A.G.: Kinetics of inorganic reactions. London: Pergamon 1966
Tobe, M.L.: Inorganic reaction mechanisms. London: Nelson 1972
Tobe, M.L. (ed.): MTP International Reviews of Science, Inorganic Chemistry, Vol. 9, Reaction mechanisms in inorganic chemistry. Series 1 1972, series 2 1974. London: Butterworths. Reviews of several of the subjects discussed in this chapter.
Wilkins, R.G.: The study of kinetics and mechanism of reactions of transition metal complexes. Boston: Allyn and Bacon 1974

More specific references

Adamson, A.W., Fleischauer, P.D.: Concepts in inorganic photochemistry. London, New York, Sydney: J. Wiley 1975
Arnold, D.R., Baird, N.C., Bolton, J.R., Brand, J.C.D., Jacobs, P.W.M., de Mayo, P., Ware, W.R.: Photochemistry, an introduction. London, New York: Academic Press 1974
Balzani, V., Carassiti, V.: Photochemistry of coordination compounds. London, New York: Academic Press 1970
Burwell, R.L. Jnr.: Heterogeneous catalysis, in Survey of progress in chemistry, A.F. Scott (ed.) *8,* 1 (1977). New York, San Francisco, London: Academic Press
Edwards, J.O., Pearson, R.G.: Scales of nucleophilicity. J. Amer. Chem. Soc. *84,* 16 (1962)
Eichhorn, G.I. (ed.): Inorganic biochemistry, Vols. I and II. Amsterdam, London, New York: Elsevier 1973
Haim, A.: Inner sphere electron transfer. Acc. Chem. Res. *8,* 264 (1975)
Halpern, J., Orgel, L.E.: Redox reactions. Disc. Faraday Soc. *29,* 32 (1960)
Heck, R.F.: Organotransition metal chemistry. A mechanistic approach. New York, San Francisco, London: Academic Press 1974
Hughes, M.N.: The inorganic chemistry of biological processes. London, New York, Sydney, Toronto: J. Wiley 1972
Langford, C.H., Gray, H.B.: Ligand substitution processes. New York: Benjamin 1965
Mango, F.D.: Transition metal catalysis of pericyclic reactions. Coord. Chem. Revs. *15,* 109 (1975). See also: Mango, F.D., Schachtschneider, J.H.: J. Amer. Chem. Soc. *93,* 1123 (1971)
Marcus, R.A.: Chemical and electrochemical electron transfer theory. Ann. Rev. Phys. Chem. *15* 155 (1964)
Pearson, R.G.: Symmetry rules for chemical reactions. London, New York, Sydney: J. Wiley 1976. A very thorough treatment; more elementary introductions are given by the same author in Acc. Chem. Res. *4,* 152 (1971) and Chem. in Britian *12,* 160 (1976)
Sargeson, A.M.: Conformation of coordinated chelates, in Transition metal chemistry, Vol. 3, R.L. Carlin (ed.). New York: Dekker 1966
Sargeson, A.M., Buckingham, D.A.: Conformation analysis and steric effects in metal chelates, in Topics in stereochemistry, N.L. Allinger, E.L. Elliel (eds.) *6,* 219. London, New York, Sydney: J. Wiley 1971
Schmalzreid, H.: Solid state reactions. Translated by A.D. Pelton. Weinheim/Bergstr.: Verlag Chemie; London, New York: Academic Press 1974
Taube, H.: Electron transfer reactions of complex ions in solution. London, New York: Academic Press 1970
Taube, H.: Electron transfer through bridging ligands. Ber. Buns. Phys. Chem. *76,* 964 (1972)
Tolman, C.A.: Organometallic catalysis. Chem. Soc. Revs. *1,* 337 (1972)
Wehry, E.L.: Review of the photochemistry of transition metal complexes. Quart. Revs. *21,* 213 (1967)
Wrighton, M.S.: Good introductory review of inorganic photochemistry. Topics in current chemistry *65,* 37 (1976)
Zink, J.I.: Stereochemical predictions in transition metal photochemistry. J. Amer. Chem. Soc. *96,* 4464 (1974)

Problems

1. For the following species, decide which molecular orbitals (occupied or unoccupied) will determine the reactivity, and predict some of the possible reactions:

F_2, $P(C_6H_5)_3$, $[SnCl_3]^-$, BF_3, $[MnO_4]^-$, Cl^-, CO_2, NO_2, SO_2, $[Cr(OH_2)_6]^{2+}$, B_2H_6, H_2O.

(It may be useful to refer to Chap. 2 and 3)

2. Why, as general rule, are coloured molecules more reactive than colourless ones?

3. Consider the attack of HO^- on CO_2. Would you expect the O=C=O unit to bend as the HO^- group approaches? (See Fig. 3.3)

4. (i) The out of plane bending vibration of BF_3 has symmetry A_2''. Which orbitals will be mixed by excitation of this vibration? (See Fig. 3.4) Can you see any parallels between this vibration and the addition of a Lewis base to BF_3?

(ii) We may assume that a planar species is intermediate in the inversion vibration (symmetry A_1) of NH_3. Draw a correlation diagram for the M.O. of the normal, planar, and inverted forms, showing only those orbitals which are mixed by the vibration.

5. Classify the following reactions:

(i) $NH_3 + H^+ \rightarrow NH_4{}^+$

(ii) $[Au(PPh_3)_2]^+ + PPh_3 \rightarrow [Au(PPh_3)_3]^+$

(iii) *trans* $[Co(en)_2Cl_2]^+ \rightarrow$ *cis* $[Co(en)_2Cl_2]^+$

(iv) $Ni^{2+}(aq.) + 6NH_3 \rightarrow [Ni(NH_3)_6]^{2+}$

(v) $[MnO_4]^- + HCOO^- \rightarrow [MnO_4]^{3-} + H^+ + CO_2$

(vi) $Ce^{4+}(aq.) + [W(CN)_8]^{4-} \rightarrow Ce^{3+}(aq.) + [W(CN)_8]^{3-}$

(vii) $Cu^{2+}(aq.) + 2I^- \rightarrow CuI + 1/2\, I_2$

6. Draw orbital correlation diagrams for substitution of L into a simple complex ML, assuming (i) a S_N2 and (ii) a S_N1 mechanism, including the orbitals of any intermediates formed. Ignore any π bonding.

7. What type of reaction is involved in the reactions between:

(i) aryl halides and $AlCl_3$

(ii) chloride ion and exposed metal surfaces (e.g. in sea water)

8. $[Co(en)_2(^{18}OH)_2]^+$ reacts with acetylacetone (acacH) to give $[Co(en)_2acac]^{2+}$, with the labelled oxygen atoms still bound to the cobalt atom. Propose an explanation. (acacH = $CH_3COCH_2COCH_3$)

9. What is the origin of the general observation that one electron transfers are faster than two electron transfers?

10. An oxidant is often classed as an outer-sphere oxidant if it is reduced more rapidly by V^{2+}(aq.) than Cr^{2+}(aq.). What is the basis of this distinction?

11. Many Cr(III) complexes are synthesised using Cr(VI) as a starting material – why?

12. The tendency to undergo oxidative addition decreases in the series:

Os(O) (extremely reactive) > Ir (I) > Pt(II) > Au(III) (adds only F_2) Comment.

13. Explain why transition metal methyls are much more stable than other alkyls, given that this effect is not seen for alkyl magnesium compounds.

14. Rhodium(I) and iridium(I) square planar complexes which are capable of losing at least one ligand by dissociation are known to catalyse the reaction $RCOCl \rightarrow RCl + CO$. If the first step of the cycle is an oxidative addition, propose a mechanism for the whole reaction.

15. A crystal of molecular iodine is often added to reaction mixtures as a catalyst. What is its probable catalytic action?

16. Polymerisation of alkenes may be catalysed by transition metal compounds. If the catalytic species is a metal alkyl with one vacant coordination site, propose a catalytic mechanism. Will non-transition metals act as catalysts?

17. Lithium or sodium perchlorate is often added to reaction mixtures in kinetic experiments in order to maintain constant ionic strength. Unfortunately, it is sometimes found that the rate is not independent of which of the two salts is used. Propose an explanation.

18. Why is MnO_2 a good catalyst for the decomposition of H_2O_2?

19. The ligand migration reaction of $CH_3Mn(CO)_5$ is independent of added ligand in coordinating solvents, but, in non-coordinating solvents, the rate is dependent on the amount of the incoming ligand L present in solution. Propose an explanation.

20. Why does heating destroy transistors and influence photographic films?

21. Why does X-irradiation darken sodium chloride crystals?

22. (i) Irradiation of $[Cr(NH_3)_5Cl]^{2+}$ in the ligand field region results in loss of NH_3. On irraditation in the charge transfer region. Cl^- is lost preferentially. Propose an explantation.

(ii) A low-lying doublet state of Cr(III) involves the pairing of two electrons in the t_{2g} shell. Would you expect this state to have a chemistry greatly different from that of the groundstate?

23. Reactions in the atmosphere frequently follow free radical (homolytic) mechanisms, but heterolytic mechanisms are more common for reactions in the hydrosphere and lithosphere. Discuss the various factors responsible for this.

7. Descriptive Chemistry

Chemistry is the study of the elements and their behaviour, not the study of chemical theory. In this chapter we shall endeavour to use the various theoretical points discussed in previous chapters in an approach to descriptive chemistry.

The LCAO M.O. approach suggests very strongly that we may usefully begin by considering the atomic orbitals available for bond formation. From our discussion in Chap. 2, it should be evident that two features will be important: the orbital energies and the ability to overlap with other orbitals. If two interacting orbitals are energetically far apart, we expect an ionic bond to be formed, the bonding M.O. being localised essentially on the anion and stabilized by the electrostatic interactions between ions. If the orbitals are close in energy, overlap and covalence become important.

A. Orbital Energy

Orbital energy is thus of great importance – how shall we measure it? Ionisation energies will give too negative an orbital energy as they allow for the decreased electron repulsion in the cation, while electron affinities are not readily available and give too positive an energy as a result of increased electron repulsion. Furthermore, we would like to know the energies of all the valence shell A.O. – for example, for nitrogen 2s and 2p orbitals. It is tempting to take the one electron ionisation energies measured by photoelectron spectra, but as these are not available for all elements, we will take advantage of the relatively satisfactory state of atomic structure calculations and use the calculated one electron energies for neutral atoms given by Herman and Skillman.

The 2s and 2p orbital energies for the series Li–Ne are given in Table 7.1 and plotted in Fig. 7.1. It will be seen that the orbital energies drop smoothly, the 2s much faster than the 2p as a result of their greater penetration (Chap. 1). The ionisation energies follow the orbital energies more or less closely. We conclude that the occupied 2s shell will become less and less interested in chemical bonding. We may now refer to Table 3.2 where the photoelectron ionisation energies for simple EH_n molecules were given. The bonding M.O.s all have slightly lower energy than the original atomic orbitals, as one would expect. The ls orbital energy for hydrogen marked in Fig. 7.1 leads one to foresee a polar bond (with hydrogen negative) for Li, Be and B, a non-polar for carbon and nitrogen with 2s and 2p participation and polar bonds with hydrogen positive for oxygen and fluorine, the 2s orbital playing less and less part in the bonding.

Table 7.1. Calculated orbital energies and experimental ionisation energies (in kJ)

Shell	Li	Be	B	C	N	O	F	Ne
2s	530	789	1213	1693	2226	2814	3459	4159
2p			643	866	1109	1367	1641	1931
Ionisation energy	520	899	800	1086	1402	1313	1680	2080

1 eV = 96.5 kJ

Calculated data from F. Herman and S. Skillmann 'Atomic Structure Calculations', New York: Prentice-Hall 1963

Atomic orbitals are sensitive to the charge and configuration of the parent atom: as the charge on the atom increases the orbitals drop in energy. It appears that the weakly penetrating d and f orbitals are most affected by this, and many transition metals with groundstate $3d^n\ 4s^2$ give monopositive $3d^{n+l}$ ions; the lanthanides are found as $4f^n$ Ln^{3+} ions with the 6s orbitals vacant. By contrast, gain of electron density increases the orbital energy. Even a charge of configuration can produce dramatic effects. In Fig. 7.2, the 3d orbital energy increases smoothly across the series Sc–Zn, apart from Cr and Cu which have configuration $3d^5 4s^l$ and $3d^{10}4s^l$ respectively instead of the usual $3d^n4s^2$. The increased electron repulsion in the 3d shell produces a sharp drop in orbital energy. Now let us consider $Cr(CO)_6$ which has a molecular structure similar to $W(CO)_6$ (Fig. 3.8 b). In this compound all 6 valence electrons of the chromium will be in the t_{2g} subshell of the 3d orbitals – the 3d orbitals will thus be of relatively high energy and thus more likely to interact with the π^* orbitals of CO. Conversely, in a compound such as WCl_6 where the metal will probably carry a positive charge, the d orbitals will be of very low energy, and more able to accept charge from the ligand π orbitals.

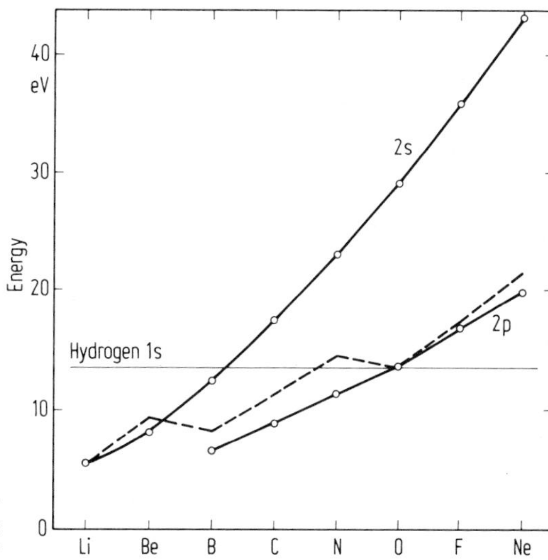

Fig. 7.1. Calculated orbital binding energies from Li to Ne. The dotted line is the experimental first ionisation energy

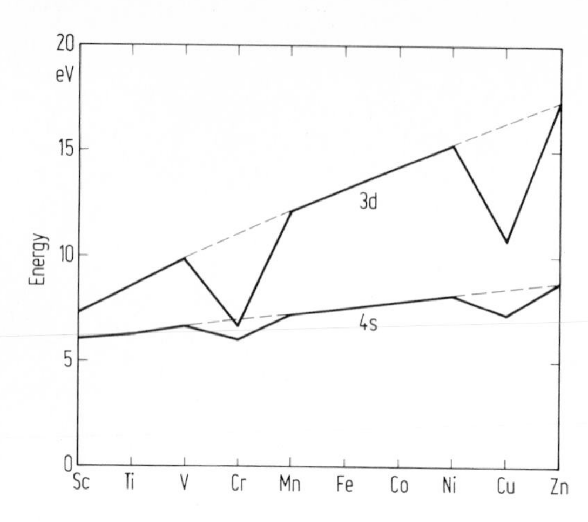

Fig. 7.2. Calculated valence orbital binding energies for Sc–Zn. The dips are due to the configuration $3d^{n+1}4s^1$ rather than $3d^n4s^2$

Let us consider now how orbital energies vary from element to element. Figure 7.1 shows that on crossing the periodic table a given *nl* shell drops in energy as a result of the poor shielding of the electrons in the same *nl* shell (see Chap. 1). Figure 7.2 shows that for the first row transition metals, the 3d orbitals drop sharply in energy but that the 4s drop relatively slowly – this is also found for the outermost s orbitals of the other transition metals and lanthanides and actinides. The change in 4s energy is however sufficiently great for Zn to have the 4s orbital more strongly bound than calcium.

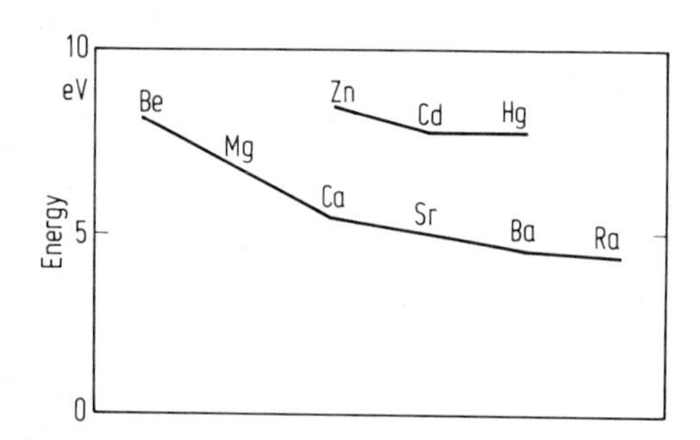

Fig. 7.3. The calculated binding energy for the valence orbital in Group IIA and IIB

Figure 7.3 shows the change in orbital energy for the metals of group IIA and IIB. Both show that the valence electron binding energy decreases as one descends a group, and this is a general trend for s and p block elements.[1] The higher binding energies for Zn, Cd and Hg result from the drop in energy of the outer s orbital on crossing the transition series – the so called d-block contraction. For transition metals this change in *IE* and orbital energy on descending a group is much less marked. Figure 7.4 shows the energy variations for group VB. Again there is a fall, but the discontinuity at As is due to the d block contraction, the filled 3d shell inadequately compensating for the increased nuclear charge – the effect is more marked for the 4s orbital which penetrates the shielding d electrons more than the 4p.

What of the sensitivity of orbital energy to change in orbital occupation? We might measure the sensitivity to positive ionisation by the differential ionisation energy men-

[1] See also Fig. 1.6 (page 33) for ionisation energies.

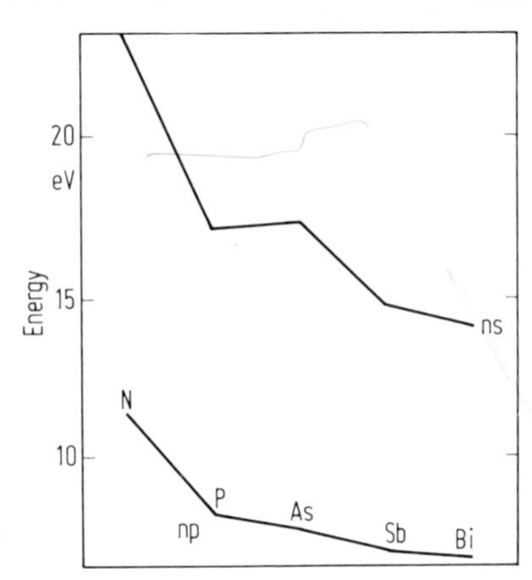

Fig. 7.4. Calculated binding energies for the valence orbitals of group VB

tioned in Chap. 5 where the zth ionisation energy is given by $I(z) = a_o + a_1 z + a_2 z^2$. The coefficients a_1 and a_2 give an indication of the resistance to ionisation with increasing charge. It is found that a_1 and a_2 increase on filling of a given *nl* subshell, and drop on descending a group. The rare earths and transition metals show low a_o values but high a_1 and a_2 values. In particular, shells where a given *l* value is first found (i.e. 2p, 3d, 4f) show particularly high a_1 values and may be thought to show the greatest variation of orbital energy with charge.

The addition of electron density to a subshell will clearly depend strongly on electron repulsion parameters. Interelectronic repulsion will increase as the size of the orbitals involved decreases. It increases by a factor of 2 on crossing a series such as the transition metals: it will be noted that the dip of the 3d levels at copper is appreciably greater than at chromium (Fig. 7.2). The first subshell of a given *l* value (2p, 3d, 4f) shows these electron repulsion effects most strongly (see Table 4.3 for d and f shells). This is no doubt the reason why the electron affinities of B, N, O and F are lower than their second row congeners, even though the electron binding energies of the first row elements are higher. We may now reexamine the problem of the lanthanide fluorides discussed in Chap. 2 (page 61) where the partially occupied 4f shell frequently lies lower than the ligand orbitals – transfer form the ligand to metal would involve too great an increase in enectron repulsion in the 4f shell.

B. Overlap

While it is generally agreed that overlap between orbitals is important, it is difficult to assess it quantitatively: no direct physical measurement gives a value for the overlap integral. Theoretical calculation requires good atomic wave functions and a knowledge of the distance between the overlapping atoms. Given this, the discussion here will be qualitative. We may consider separately the effect of the angular and radial properties of atomic orbitals on the overlap integral.

s orbitals are spherically symmetric and have no preferred direction of bond formation. Overlap will be maximised by the highest possible coordination number consistent with reasonable bond lengths (too high a C.N. will produce ligand-ligand repulsions). p orbitals have maximum density along one axis, and are thus strongly directional: compounds where only the p orbitals are involved in bond formation tend to have near 90° bond angles. If σ bonding only is considered, a p orbital can overlap strongly with 2 ligands at the most, and elements using only p orbitals tend to show lower coordination numbers. However, the strong directional character of the p orbital results in empty p orbitals showing strong Lewis acidity (and occupied p orbitals Lewis basicity) and this effect is often dominant in structural chemistry.

d and f orbitals have much less directional character and consequently tend towards high coordination number. A case where the directional nature of the d orbitals is important is the square planar geometry where the $d_{x^2-y^2}$ orbital is particularly strongly overlapped, but, generally speaking, it is easier to predict the stereochemistry of transitions metal compounds by steric rather than overlap considerations.

Radial effects are somewhat more subtle (or, more sceptically, speculative), but we may make one or two general remarks and use our common sense to see when they are valid. The approach of two atoms is ultimately checked by core-core electron repulsion and so the overlap of atomic orbitals depends on their projection outside the core valence obitals. Thus the valence orbital in Fig. 7.5 a will be able to overlap strongly but that in Fig. 7.5 b will be mainly buried inside the core and will overlap poorly. As the principal quantum number increases, the number of radial nodes increases and this leads to an increasing diffuseness (Fig. 7.5 c).

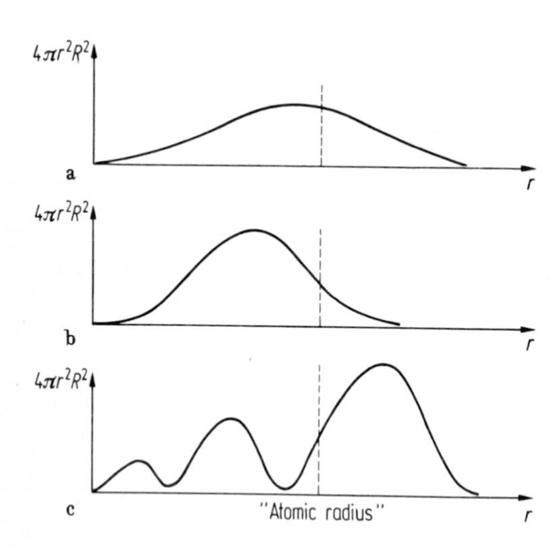

Fig. 7.5 a – c. The dependence of overlap on radial function

For light elements (Li–Ne) the atomic core is very small and the valence orbitals project enough to give good overlap. As the atomic number increases, increasingly large proportions of the valence orbitals are in the lobes inside the core, and overlap integrals appear to drop for s and p valence shells on descending the periodic table. By contrast, 3d and 4f orbitals in the valence shell appear to be quite compact, as shown by their high electron repulsion parameters, and, in the many compounds of the $3d^n$ and $4f^n$ elements where the metal carries a positive charge, they appear to 'shrink' into the core; atomic spectra show an increase in electron repulsion parameters as the charge on the metal atom increases. It was mentioned above that the first shell of a given l value has much higher electron repulsion, and it is thus not surprising that a lower energy may be obtained by atomic orbitals 'shrinking' as a result of less electron repulsion than by overlap with ligands. In the 4d, 5d and 5f valence shells, electron repulsion is lower, and the orbitals are more diffuse (type c in Fig. 7.5); contraction is less important and the more covalent nature of the heavy transition metals may reasonably be associated with increased overlap (orbital energies are quite similar).

Thus, for s and p valence orbitals, overlap decreases on descending a group of the periodic table while the opposite is true for the transition metals. This hypothesis explains the observation that the heat of atomisation of the main group metals falls with increasing atomic number, while the opposite trend is observed for the transition metals and (as far as the data go) the lanthanide/actinide family.

C. Main Group Elements

For the rest of this chapter, we will take a stroll around the Periodic Table, looking for illustrations of some of the points made earlier in the book and examining critically some of the explanations proferred for chemical behaviour. It should be borne in mind that only a highly selective treatment is given, but it may help to whet the reader's appetite.

Group IA elements have only one valence electron which is relatively weakly held and their chemistry is predominantly cationic. The increase in ionic radius on descending the series has a considerable effect – thus lithium burns in air to give Li_2O, sodium to give Na_2O_2 and potassium to give KO_2, larger cations being unable to stabilise highly charged oxygen species with a high lattice energy. Separations of alkali metals are generally based on ionic size discrimination.

The alkali metal cations are typical hard Lewis acids and complex with oxygen donors. Not surprisingly, lithium, with the most negative s orbital energy and the lowest core repulsion effects, complexes most strongly, and in aqueous solution the heat of hydration of the cations drops down the series. An interesting recent development is the syntheses of several macrocyclic polyethers, where by varying the size of the cycle or cycles and the distance between the oxygen atoms, it is possible to tailor ligands to specific cations. The interaction with these ligands can be very strong – with the 2, 2, 2, cryptate ligand L (Fig. 7.6) in a suitably inert solvent, it has been possible to prepare $(NaL)^+ Na^-$ (solv) from metallic sodium. This is one of the few chemical manifestations of the positive electron affinities of the alkali metals, and shows the relatively low electron repulsion in the valence ns^2 shell.

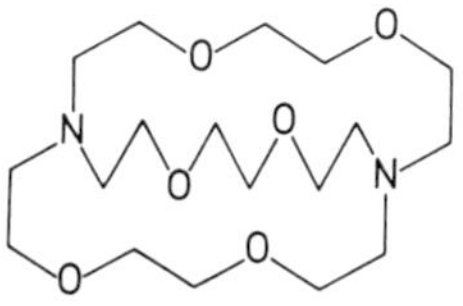

Fig. 7.6

Lithium often shows a rather different chemical behaviour from the other alkali metals, but as it has the most strongly bound valence orbital and, with its small core, the greatest tendency to overlap with ligand atoms, a deviation towards more covalent character is not surprising. Whereas most of its chemistry may be described as polarised ionic, it is in compounds with less electronegative elements that lithium shows its most interesting behaviour. Thus, solid methyl-lithium is a tetramer, in which each lithium atom is bonded to three methyl groups. Lithium forms many stable molecular organometallic compounds, many of which associate to give polymers – the other alkali metals only give organometallic compounds which may reasonably be assumed to be ionic such as $Na^+ \ C_5H_5^-$. Lithium thus shows a much greater similarity to hydrogen than other members of the group, although it differs from hydrogen by its lower orbital energy and tendency to higher coordination number, as in CH_3Li.

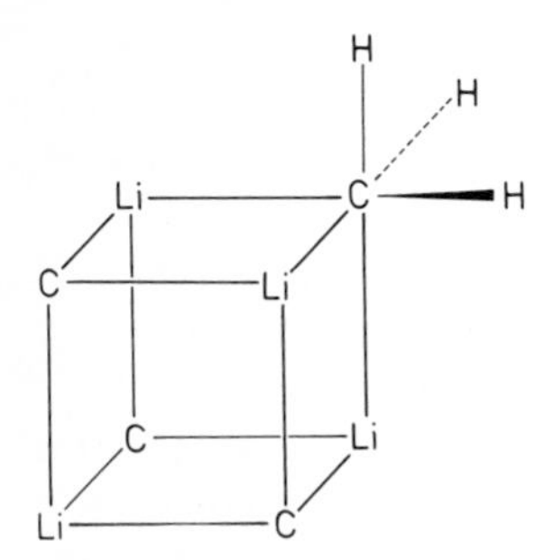

Fig. 7.7. The structure of $(CH_3Li)_4$

As expected the orbital energies of the *Group IIA elements* are more negative but for the heavier elements this does not result in the chemical behaviour being anything other than ionic. Compounds of the two lightest elements (Be and Mg) do however show considerable covalent character, and the availability of low lying, empty p orbitals causes Be and Mg to show some Lewis acid behaviour, and the organometallic compounds of magnesium (such as Grignard reagents) form adducts with ethers. The high Lewis acidity of beryllium results in the formation of many bridged, polymeric species, and berylium hydrides and alkyls show electron deficient structures with bridging hydrogen or methyl groups.

The heavier elements show little or no Lewis acid character. Their greater size results in relatively low solvation energies and anhydrous salts are often formed on crystallisation from solution. Salts of these elements with di- or tri-valent anions are generally insoluble and they are frequently found in geology and biology as phosphates, carbonates and sulphates. Although the sum of the first two ionisation potentials is quite high, an estimate of the heat of formation of the +1 oxidation state, assuming the M^+ ion to be of similar size to the preceding alkali cation, shows that the equilibrium

$$2MX(s) \rightleftharpoons M(s) + MX_2$$

lies well to the right. Their chemistry is predominantly ionic and as M^{2+} they are able to stabilise such species as the nitride (N^{3-}) and the carbide $(C_2)^{2-}$ ions as ionic solids which, however, hydrolyse readily. As a result of its greater ionic size, barium is able to stabilise a peroxide BaO_2, and the formation of this (by heating BaO in air), followed by hydrolysis to $Ba(OH)_2 + H_2O_2$ was at one time used as a source of H_2O_2.

For Ca, Sr, Ba and Ra, we may ask if the lowest lying empty orbitals are the *np* or the $(n-1)$d. The discovery that gaseous CaF_2, SrF_2 and BaF_2 are non-linear suggests that the d orbitals can play some part in the chemistry of these elements. In Fig. 2.5 it was shown that the magnesium $3p_z$ orbital could overlap with the ligand σ_u orbital; Fig. 7.8 shows that bending of the molecule will allow overlap between the σ_u and an empty d_{xz} orbital, without affecting the σ_g/metal s orbital overlap.

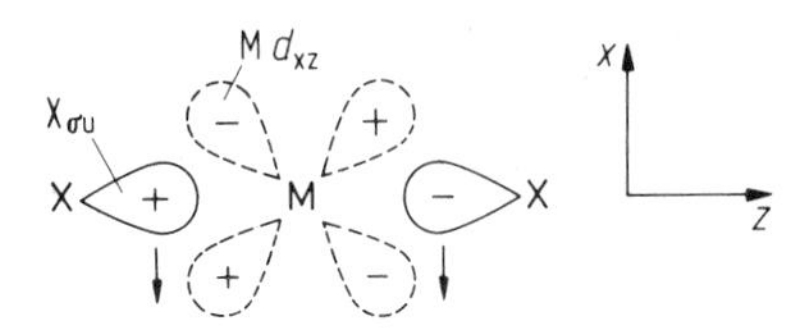

Fig. 7.8. Bending of MX_2 gives overlap between d_{xz} and σ_u

The Group IIIB elements are the first to have two occupied shells in the valence shell, and the participation or non-participation of the s subshell determines the oxidation state. For boron and aluminium both s and p subshells are invariably involved in bonding, giving the oxidation state three, but the heavier elements show increasing stability of the oxidation state +1. The mean orbital energies of s and p orbitals in boron and aluminium are quite negative and the bonding in their compounds is covalent. The structure of aluminium compounds may often be discussed conveniently in terms of an ionic model but there is usually need to invoke considerable polarisation (i.e. covalence) to explain the chemical behaviour. Thus the aqueous $[Al(OH_2)_6]^{3+}$ ion has been used as an acid since the time of the alchemists as a result of its extensive hydrolysis to $[Al(OH_2)_5OH]^{2+}$ etc. All the elements have more valence orbitals than valence electrons available and this leads to strong Lewis acid properties, especially for boron and aluminium compounds such as BF_3 and $AlCl_3$. In the absence of suitable Lewis bases, dimeric species such as Al_2Cl_6 (whith two bridging chlorine atoms) or electron deficient species (such as Al_2Me_6 and the boron hydrides) are found.

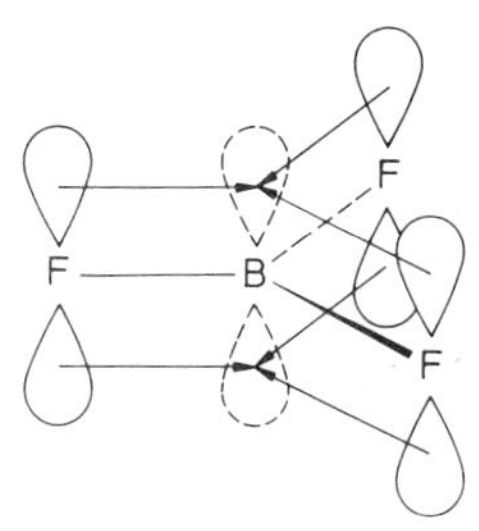

Fig. 7.9. p_π–p_π bonding in BF_3

The very high bond energies found for the B–O and B–F bonds merit comment. As mentioned in the introduction, overlap between atoms of first row elements is favoured by the small atomic core. There is good reason to believe that this is particularly important for p_π–p_π overlap, and that, when fluorine and oxygen are bonded to three-coordinate, planar boron, there is considerable π donation to the empty boron 2p orbital (Fig. 7.9), with consequent stabilisation of the B–O or B–F bond. In support of this we may note the lengthening of the B–F bond on addition of a Lewis base (e.g. F^-) to BF_3 (see page 186) and the high out-of-plane bending force constant of BF_3 when compared with BCl_3; in both cases distortion of the planar structure will destroy this π bonding.

The heavier elements (Ga, In, Tl) all have a full $(n-1)$d shell beneath the valence shell, and show a d-block contraction when compared with Ca, Sr and Ba. Their chemistry is predominantly covalent, and they all form stable, molecular organometallic species. The +1 oxidation state becomes increasingly stable, and, as orbital energies vary little, this is presumably due to the decreasing overlap of the valence s subshell with ligands. Lewis acidity is also much less marked, and thallium forms some stable, linear, two coordinate species such as $[TlCl_2]^+$. The heavier elements show distinct class B behaviour, and are often resistant to hydrolysis, $[(CH_3)_2Tl]^+$ being stable in aqueous solution.

The +2 oxidation state is not stable, and compounds such as $GaCl_2$ are mixed valence compounds ($Ga^+[GaCl_4^-]$). The chemistry of the +1 oxidation state involves only the weakly bonded p electron and is predominantly ionic. Thallium (I) is the stable form in aqueous solution, although Tl(III) may readily be stabilised by complexation; Tl^+ shows properties intermediate between those of K^+ and Ag^+, both of similar ionic radius – TlCl is rather insoluble, but TlOH is soluble and a strong base.

The fluorides, oxides and chlorides of the metallic elements show close packed structures with octahedral coordination (except for $GaCl_3$) as might be expected from an ionic description. The bromides and iodides are more frequently dimerised molecular solids (although TlI_3 is $Tl^+ I_3^-$). The nitrides are hard stable compounds with the wurtzite structure (i.e. closely related to diamond); boron nitride is isoelectronic with diamond and has similar physical properties, but is subject to hydrolysis. It may also be found in a graphitic modification. With the heavier group VB elements (P, As, Sb, Bi), Al, Ga, In give semiconducting solids, similar to elemental silicon and germanium, which have been much studied for their electrical properties.

With the *Group IVB elements* we see a continuation of the usual trends in orbital energy and bond strength on crossing or descending the period table. The +2 oxidation state, where the s electrons are not involved in chemical bonding, is found frequently for tin, and is the most common oxidation state for lead. For carbon and silicon the +2 oxidation state may be found in transient species but is too reactive to be isolated.

Carbon shows much greater differences from silicon than boron from aluminium, and it is interesting to look at these: apart from the ready formation of C–C chains it is noticeable that carbon readily forms stable double or triple bonds with other first row elements, whereas even double bonds are unknown for other group IVB elements – this may be explained by the strong p_π–p_π overlap for first row elements mentioned for boron. Carbon is also unusual in the first row elements in forming a strong covalent single bond to itself, as in diamond, although this is scarcely surprising in that the number of valence orbitals is equal to the number of valence electrons. The quite negative orbital energies of carbon militate against a cationic chemistry, whilst the low electron affinity disfavours an anionic form (although $(C_2)^{2-}$, isoelectronic with N_2, is known in CaC_2). The Si–Si bond in elemental silicon is quite strong (226 kJ/mole cf. 356.1 for diamond), but an extensive silane chemistry is limited by the fact that the hydrogen-hydrogen bond is 25 percent stronger than the Si–H bond, and that the kinetics of decomposition are relatively fast.

In its molecular compounds, carbon almost invariably has a coordination number of 4 or less; the heavier elements show an increasing tendency to higher coordination number with increasing atomic weight and in the +IV oxidation state Lewis acidity is common for tetrahedral compounds. Many supposedly tetrahedral tin compounds have been found to be 5 or 6 coordinate polymers in the solid state. This readiness to expand the coordination shell is presumably the reason for the much faster substitution reactions of the heavier elements.

The acid base properties of the divalent state are more subtle. A simple compound such as molecular $SnCl_2$ has four valence orbitals and six electrons in its valence shell: it is therefore to be expected that it can add another pair (i.e. function as a Lewis acid) to give species such as $SnCl_3^-$; $SnCl_2$ is thus a Lewis acid. The adduct $SnCl_3^-$ has a lone pair of electrons which are, in principle, in the 5s subshell; the geometry of the $SnCl_3^-$ ion shows that the Cl–Sn–Cl bond angles are greater than 90° and there must be some slight s–p hybridisation, resulting in the lone pair having directional character and consequent Lewis base properties. Many metal–$SnCl_3$ complexes are known. Thus the lone pair in $SnCl_3^-$ shows some activity, and as a result of its directional nature is sometimes referred to as a stereochemical lone pair. Many Sn(II) compounds show distorted structures implying the presence of a stereochemical lone pair.

Two of differences between carbon and the heavier elements have attracted some theoretical interest: the Lewis acidity of tetrahedral species, and the strength of Si–O and Si–F single bonds in comparison with the C–O and C–F bonds. It has been proposed that the heavier elements have low lying empty d orbitals which are capable of accepting charge, either from Lewis bases, or, by π acceptance (as for transition metals) from occupied π orbitals of ligands. In such a way, octahedral complexes will be stabilised by donation into s, p, and $d_{x^2-y^2}$ and d_{z^2} orbitals. Compounds with oxygen and fluorine will have enhanced bond strength if partial double bond character results from donation into empty d orbitals from the occupied p_π orbitals of O and F – this is referred to as p_π–d_π bonding. For heavier halogens and chalcogens the p_π overlap will be smaller, and this effect will be less marked.

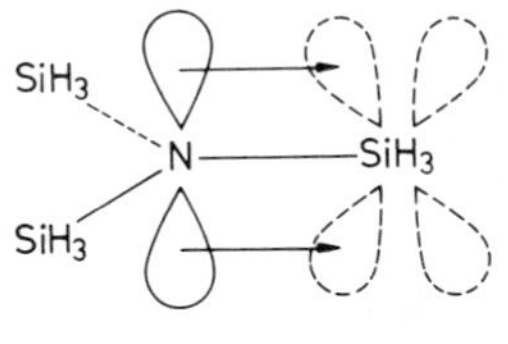

Fig. 7.10. p_π–d_π bonding between the nitrogen lone pair and empty silicon d-orbitals postulared for planar $(SiH_3)_3N$

The availability of the d orbitals is clearly a convenient distinction between carbon and the other elements, but the implication of d orbitals needs to be examined closely. Atomic spectra show that the d orbitals are of much higher energy than the occupied p orbitals but that their energetic participation is most favourable for the group IVB and VB elements, increased occupation of the p orbitals screening the d orbitals for the np^4 and np^5 elements. Furthermore, when the group IVB or V element is bonded to electronegative ligands, the positive charge on the central atom will favour a drop in d-orbital energy. The participation of d orbitals can thus be justified to some extent. The experimental data in favour of d orbital participation include the high Si–O and

Si–F bond energies, the apparent π acceptor properties of tertiary phosphines and arsines, and the feeble basicities of compounds such as $(SiH_3)_2O$ and $(SiH_3)_3N$. The latter compound has planar Si–N bonds (Fig. 7.10) although trimethylamine is pyramidal. The analogous phosphorus compound, $(SiH_3)_3P$, is pyramidal as a result, it is claimed, of weaker overlap of the phosphorus $3p_z$ orbital with the silicon $3d_\pi$ orbitals.

Against these arguments may be set the fact that the d orbitals are still relatively high in energy to show much electron affinity, as was noticed for the group IIA elements. Other arguments may be proposed to explain some of the effects discussed above, and convincing spectroscopic evidence of p_π–d_π interactions is lacking. LCAO M.O. calculations suggest the involvement of d orbitals, but this is more a reflection of some of the defects of LCAO M.O. theory; X–α calculations show no central atom d character in the molecular orbitals. In the author's opinion, the role of d orbitals in explaining main group chemistry can be over-emphasized; this results from a historical tendency to describe octahedral molecules as involving only 2 centre 2 electron bonds with sp^3d^2 hybridisation rather than delocalised multicentre bonds. Once introduced, the simple rationalisations that d orbitals allow makes their retention attractive and discourages the search for other explanations.

In their solid state chemistry, the Group IVB elements are quite similar, the exceptional element being lead, which is the only element not to have a diamond structure in the elemental state. As might be expected, carbon forms some salt like carbides with electropositive elements, but otherwise a wide range of stable compounds are found with many elements. The unusual stoichiometries of these compounds suggest the bonding to be described best in terms of band structures. The non-stoichiometric, interstitial carbides were mentioned in Chap. 3. Just as binary compounds of Group IIIB and VB elements have structures based on the diamond structure, so do compounds of Group IVB and VIB such as SnS and GeSe have structures similar to group VB elements (the two examples cited have structures related to black phosphorus).

The *Group VB elements,* with configuration s^2p^3 have one more electron than the number of valence orbitals (ignoring any d orbital participation for the moment). They are thus electron rich in the sense discussed in Chap. 3. The oxidation state 3 is common for all elements, implying a certain shyness on the part of the s electrons, but the structures of nitrogen and phosphorus compounds suggest at least partial hybridisation of the s orbitals as bond angles are greater than the 90° expected for pure p orbital bonding. The pyramidal 3-coordinate compounds of oxidation state III show Lewis basicity (e.g. ammonia, triphenylphosphine). Although the valence orbitals are strongly bound there is no true anionic chemistry, and the description of nitrides as ionic is only justified for very electropositive elements.

As might be expected, nitrogen shows a strong tendency to form double bonds with other first row elements (and itself, as in N_2, N_2O and N_3^-). Single bonds with species with full p_π shells are often weak, presumably as a result of p_π–p_π repulsion as in NCl_3. By contrast, the heavier elements form much stronger bonds with ligands with full p_π shells (such as the halogens) – this increase in bond strength on passing from first to second row elements is quite general and has been attributed to the possibility of p_π–d_π bonding in the second and later rows enhancing the strength of formally single bonds. Phosphorus also forms a strong double bond with oxygen in the +V valence state, as in $POCl_3$ and P_4O_{10}. The short P=O bond length and low polarity of the bond in comparison with amine oxides (R_3NO) suggest very strongly that there is some sort of back bonding to the phosphorus from the oxygen. Phosphorus in fact forms both double bonds with oxygen and –P–O–P– catenated bonds in contrast to silicon which gives only catenated –Si–O–Si– bonds. Hydrolysis of P–O–P bonds in adenosine triphosphate is an important source of energy in biology.

The oxidation state V is shown by all the elements: for nitrogen it is found only with oxygen bound to the nitrogen (e.g. in R_3NO).[2] Nitrates are thermodynamically strong oxidising agents, but are kinetically stable; the +V state is very stable for phosphorus but decreases in stability on descending the group and bismuthates are strong oxidising agents. Oxidative addition to P(III) compounds was mentioned in Chap. 6; many trivalent compounds of this group are quite strong reducing agents. This is sometimes referred to as a tendency to undergo valency expansion in accordance with the idea that a compound with valency or coordination number greater than 4 must use d orbitals. The penta-coordinate complexes of the heavier elements show strong Lewis acidity, and SbF_5 is one of the strongest Lewis acids known.

In aqueous solution all the elements are found strongly associated with oxygen (except, of course, for NH_4^+). Nitrogen is found as NO_2^- and NO_3^-, both of which are weak bases. The other elements show a much stronger tendency to hydrolysis. Phosphorus (V) is only found as PO_4^{3-} in very basic solution, and the basicity of phosphate salts is the origin of the use of phosphate buffers. H_3PO_3 is only a dibasic acid with the structure ($HPO(OH)_2$), the hydrogen bonded to the phosphorus not being acidic. Arsenic is similar, but the acids are weaker and $As(OH)_3$ is tribasic (cf. H_3PO_3). Antimony and bismuth show some cationic chemistry, but they are almost invariably associated with oxygen as in SbO^+ and $(Bi_6O_6)^{6+}$. Both elements complex readily with anions such as $C_2O_4^{2-}$, SO_4^{2-} and the halides to give anionic complexes or insoluble products. In the +III oxidation state, these complexes often show irregular coordination geometry as a result of the stereochemical activity of the lone pair (see VSEPR theory).

In their compounds with elements of similar electronegativity the series P–Bi often show their electron rich nature, forming chains and rings rather than close packed solids. A typical example is phosphorus sulphur chemistry: five sulphides are known, all of which may be regarded as P_4 tetrahedra with sulphur atoms added either apically or across P–P bonds. S_4N_4 and As_4N_4 were mentioned in Chap. 3. Phosphorus and nitrogen form a series of cyclic compounds of the type $(PNX_2)_n$ where the two halogen (X) atoms are bound to the phosphorus. These cyclo-phosphazenes have attracted some interest as quasiaromatic systems. $(PNCl_2)_3$ has a planar structure with 6 equal P–N distances, much shorter than the expected P–N single bond distance (1.56 Å instead of 1.78 Å). It is difficult to explain this without resort to $d_\pi - p_\pi$ delocalised bonding. A closer analogue to benzene is borazine ($B_3N_3H_6$) which is isostructural (and isoelectronic) with benzene, but much more reactive as a result of the greater polarisation of the bonds.

For the *Group VIB elements,* the energy of the p orbitals has dropped to a point where a definite anionic chemistry can be found; however, the addition of two electrons to the p subshell produces a large increase in electron repulsion energy and the anions are only stabilised in ionic lattices. All the anions are hydrolysed in aqueous

---

[2] NF_4^+ is a rather reactive exception to this.

solution to give XH^- or H_2X species. The covalent bond energies in H_2X drop steadily down the group, and H_2Te is endothermic. As for other elements, the bond energies for sulphur bonded to possible π donors (F, Cl, etc.) are much higher than for oxygen – presumably due to p_π–d_π interactions.

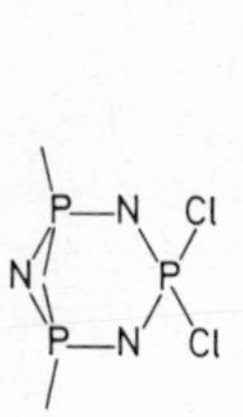

Fig. 7.11. A cyclo-phosphazene

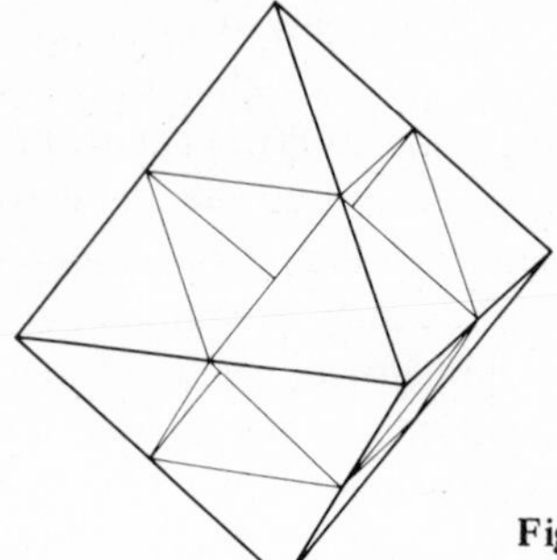

Fig. 7.12. Six edge-sharing NbO_6 octahedra in $(Nb_6O_{19})^{8-}$

Oxygen alone shows only the oxidation state 2, although for formal reasons compounds with O–O bonds may be classed as having different oxidation state – the ionic compound $O_2^+ \ PtF_6^-$ is clearly also an exception. The other elements show oxidation states –2, +2, +4 and +6, together with odd oxidation states due to element-element bonds (as in $H_2S_2O_6$ and $[Te_4]^{2+}$). None of the elements shows any simple cationic chemistry. As usual the stability of the highest oxidation state decreases down the series, while the lower oxidation states possess lone pairs which may show Lewis base properties. However the strongly bound lone pairs are much weaker donors than those of elements in earlier periods. The lone pairs show a strong stereochemical effect for sulphur and selenium, and chain and ring structures typical of electron rich elements are common.

In its covalent chemistry oxygen readily forms strong bonds with less electronegative elements, especially when there is a possibility of some π bonding (e.g. p_π–p_π in B–O; p_π–d_π in Si–O). By contrast, with itself and the halogens it forms weak bonds where there is appreciable p_π–p_π repulsion (or occupation of π^* orbitals). The weakness of the oxygen-oxygen bonds results in a high reactivity of peroxide and superoxide species and these are frequently proposed as intermediates in the reactions of molecular oxygen. O_2 itself is a triplet molecule with one electron in each π^* orbital in its ground state. The excited $^1\Delta_g$ state, with one π^* orbital doubly occupied and the other empty, has a different reactivity and can, for example, add to butadiene in a Diels-Alder typer reaction. The differences between $^1\Delta_g$ (singlet) and $^3\Sigma_g^-$ (ground state) O_2 are interesting examples of orbital symmetry effects.

The ionic chemistry of oxygen in the solid state is a vast and important subject. Oxygen has the capacity to stabilise many high oxidation states of metals, which may be explained in an ionic model by the small size and high charge of the oxide ion leading to high lattice energies, or in a M.O. approach by the σ and π donor properties of the oxide ion. Structurally, there is an enormous variety, with species which may be regarded as arrangements of metal ions in close packed oxide ion lattices (e.g. perovskites, spinels), layer structures such as PbO and SnO, and many species which may be regarded as built up from vertex, edge or face sharing polyhedra with oxygen at the vertices and metal atoms at the centres; many minerals fall into the final class – thus montmorillonite, an important clay mineral, has the basic structure of a sandwich of MO_6 octahedra between two layers of SiO_4 tetrahedra sharing vertices with the octahedra. Another example is given in Fig. 7.12.

The positive oxidation state chemistries of sulphur, selenium and tellurium are quite different from oxygen. The most stable positive oxidation state of sulphur in solution is S(VI) – all other sulphur oxidation states may be oxidised to sulphate, although in acid solution, H_2S will reduce sulphate to sulphur. As might be expected for species where there is a fair amount of covalent bond formation, the redox reactions are frequently slow. For the heavier elements the +VI oxidation state is very oxidising, and even the +IV oxidation state shows oxidising powers. The tendency to higher co-ordination number for the heavier elements is shown by the non-existence of tetrahedral $TeO_4{}^{2-}$ (telluric acid is found as $Te(OH)_6$), and Te(VI) is also thought to form $TeF_7{}^-$ and $TeF_8{}^{2-}$ anions.

Although experimental evidence suggests that in the higher oxidation states the heavier elements all carry positive charge, there is no simple cationic chemistry, and in aqueous solution they are found as oxy-anions. In very strong acids polymeric cations such as $[Te_4]^{2+}$ are formed on solution of the elements. $[Te_4]^{2+}$ is a planar square species and is sometimes described as being aromatic since there are 6 electrons available for π bonding. Interestingly enough, if the 6s electrons are assumed to be inert, the 14 valence electrons would imply a *nido* octahedral cluster according to the cluster theory discussed earlier.

In compounds with elements of similar electronegativity the heavy elements form long chains and rings, showing electron rich behaviour. The structures of elemental sulphur are varied, but the normal form (rhombic sulphur) contains an S_8 ring. At high temperatures S_2 is formed, and may be stabilised by complexing to a transition metal. The mixed molecule SO is extremely unstable. Selenium does form Se_8 but is more stable in the form of long helical chains – the same structure is found for Te. The ring compound S_4N_4 was shown in Fig. 3.31 – interestingly enough, this compound also exists as a superconducting polymer $(SN)_x$.

With very electropositive metals, S, Se, and Te form ionic salts which dissolve with hydrolysis in water. With other metals a wide range of compounds are found, but their physical properties (metallic lustre, semi-conduction), chemical properties (inertness, resistance to hydrolysis), unusual structures, and surprising or non-existent stoichiometry (e.g. Fe_7S_8) show these to be band structure compounds.

The halogens (Group VIIB) form one of the earliest recognised families of the periodic table. The valence shell is formally $ns^2\ np^5$ but in the more usual oxidation states there is little evidence of much s orbital involvement. The very negative orbital energy results in quite respectable electron affinities and the halide anions X^- are stable and show little Lewis basicity, although fluoride ion has a sufficiently strong affinity for the proton to be a weak base [3]. With sufficiently strong Lewis acids the halides will give complexes (e.g. $SbF_6{}^-$) and will often act as bridging ligands to give polymerised species e.g. NbF_5 is a tetrameric with bridging fluorine atoms.

---

[3] It appears that the 'undissociated' acid HF is in fact $[H_3O^+F^-]$ in aqueous solution.

With electropositive elements in low oxidation states the halogens give close packed ionic compounds, but as the electronegativity or oxidation state of the non-halogen increases, there is a tendency to low C.N. of the halogen and to formation of discrete molecular species – a classic example is $PbCl_2$, an almost ionic layer structure, compared with $PbCl_4$ (discrete $PbCl_4$ tetrahedra).

All the elements are found as diatomic molecules. For iodine (and to a lesser extent bromine) there is some tendency for intermolecular forces to result in much closer intermolecular distances than would be expected from van der Waals radii. The bond energies in the halogen molecules are all low: this confers on the halogens both thermodynamic (ready formation of more stable compounds) and kinetic reactivity (ready formation of free radicals). The fluorine-fluorine bond in F_2 is both longer and weaker than would be expected from extrapolation from the other halogens: this has been attributed to the considerable bond weakening effect of the doubly occupied $\pi_g{}^*$ M.O. as a result of strong p_π–p_π interaction, leading to a F–F bond weaker than Cl–Cl. An alternative explanation, based on the fact that the electron affinity of F is lower than Cl, attributes the weakness to very strong destabilising electron repulsion effects in F_2.

Fluorine is universally accepted to be the most electronegative element and shares with oxygen the ability to stabilise the highest oxidation states of elements: this may be explained by the ionic model (ease of formation of F^- and small anionic radius giving high lattice energies to balance high ionisation energies); by a M.O. model (strong σ and π donor properties of F^-); or by simple electronegativity arguments on the strength of the bonds formed. The electronegativity of fluorine is often used to produce electronegative groups in organic chemistry – the polarisation effects of the CF_3 group and other fluorocarbons can appreciably affect reactivity. CF_3COOH is 10^4 times stronger as an acid than CH_3COOH.

The other halogens all form compounds in positive oxidation states, the most important being the oxygen compounds (in particular the oxy-acids) and the interhalogen compounds. The hypohalous acids HOX are known for Cl and Br but are thermodynamically unstable to disproportionation into X^- and $XO_3{}^-$, although this reaction is slow for Cl. The acids cannot be isolated as pure compounds. More stable are the anions of the halic acids ($ClO_3{}^-$, $BrO_3{}^-$ and $IO_3{}^-$) which may all be isolated as pure salts. They are strong oxidising agents and have pyramidal structures as predicted by VSEPR theory, although for $IO_3{}^-$ the Mössbauer spectra imply the 5s orbital to be relatively inert. The perhalic anions $ClO_4{}^-$, $BrO_4{}^-$ and $IO_4{}^-$ are all tetrahedral, and the parent acids are very strongly dissociated. Although perchloric acid is thermodynamically a very strong oxidising agent, it has a high kinetic stability. Under controlled conditions it is very useful for the degradation of organic matter, but the very favourable thermodynamics can lead to an explosive reaction if care is not taken.

The perbromate ion has an interesting history: it was only prepared in 1969 after many explanations of its non-existence had been proposed. It is a stronger oxidising agent than $IO_4{}^-$ and $ClO_4{}^-$; it is quite common in the 4p series that the higher oxidation states are less stable than their 3p or 5p counterparts, as shown by the non existence of $AsCl_5$, the oxidising power of SeO_2 etc. This has been tentatively explained as being due to a weakening of d_π–p_π bonding when 4d orbitals are involved, possibly as a result of the radial node in the 4d orbitals. For the 5p series this effect is thought to be absent possibly as a result of 4f participation; whatever the explanation, the effect is quite definite. The periodate ion is a stable species with strong oxidising properties. As with most heavy elements it shows a tendency to increase its co-ordination number, and octahedral species such as H_5IO_6 are common.

It is convenient at this point to review the subject of d_π–p_π bonding. The remarkably stable series SiO_4^{4-}, PO_4^{3-}, SO_4^{2-}, and ClO_4^- all show short E–O bond lengths. The electronegativity of oxygen leads one to expect the central atom to have a slight positive charge which will lower the energy of the 3d orbitals. However, it is known that oxide ion can act as a π donor, and it is difficult to see the Cl–O bond in ClO_4^- being highly polar as chlorine is also an electronegative element; furthermore, if the bond were polar one might expect the $O^{\delta-}$ atoms to show rather more Lewis basicity than they actually do show. These features certainly give very strong support to the idea of appreciable d_π–p_π bonding as a stabilising influence. One would expect a particularly strong effect when (i) the atom with the d orbitals probably carries a positive charge, and (ii) the p_π ligand is known to be a good π donor. This will of course be particularly common for oxygen and fluorine complexes.

Halogens form much stronger bonds with non-halogens than with themselves, and interhalogens, although stable with respect to decomposition into the constituent elements, are very reactive: the higher fluorides of bromine and iodine are often used as fluorinating agents. Their structures are described well by VSEPR theory. Iodine in the +I oxidation state may quite easily be stabilised by complexation as in $[ICl_2]^-$ which is stable in concentrated HCl. In strong acids the paramagnetic I_2^+ cation is formed dimerising to the diamagnetic species I_4^{2+}. Iodine forms several weakly catenated cations and anions.

The *noble gases* and their chemistry present an interesting example of the failure of chemical theory, or, perhaps, its too ready acceptance. According to two electron, two centre bonding theory the noble gases will always form compounds with electrons in antibonding orbitals unless they are promoted to the empty d orbitals: this was ruled out by the high promotion energy which would be difficult to overcome by ionisation (as for e.g. sulphur) as the ionisation energies are too high.

The octet theory, which held that non-transition elements were stable when they obtained the electron configuration of a noble gas (either by formation of ions, or by electron sharing) also implied that a noble gas would be unlikely to form compounds. The elements were thus labelled as inert and left with their clathrate compounds, where the gas atoms are held in well-defined positions in lattices by essentially intermolecular forces. However, examination of thermodynamic properties suggested (i) that PtF_6, which forms $O_2^+\ PtF_6^-$ with oxygen, should also form $Xe^+\ PtF_6^-$ and (ii) compounds

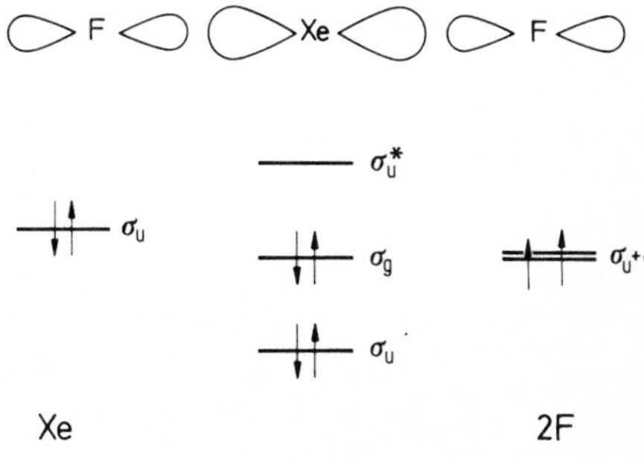

F Xe F

σ_u^*

σ_u σ_g $\sigma_u + \sigma_g$

σ_u

Xe 2F

Fig. 7.13. 3 centre 4 electron bonding in a XeF_2 unit

of xenon with fluorine should be stable. Both these predictions were verified in 1969, although the Xe/PtF_6 reaction turned out to be less simple than thought at first.

Direct reaction of F_2 and Xe under suitable conditions leads to XeF_2, XeF_4 or XeF_6. The simplest explanation of these compounds was a three centre 4 electron model. Figure 7.13 shows that two M.O. are occupied and consideration of the orbital energies of the two elements implies that the σ_g orbital may be non-bonding (or even, if it interacts with the Xe6s orbital, slightly antibonding), but that it will be strongly bound. The structure of XeF_6 is not regular, implying that the Xe6s orbital has a stereochemical effect on the structure (see page 175). The bound fluorines are slightly basic towards strong fluoride acceptors such as SbF_5. XeF_6 will accept fluoride ions to give XeF_7^- and XeF_8^{2-} salts. The fluorides hydrolyse readily to give oxygen complexes; thus XeF_6 gives XeO_4^{2-} derived from the trioxide XeO_3. Ozone oxidation of this (or disproportionation) gives the perxenate ion XeO_6^{4-} which may be precipitated as stable salts. Mixed fluoro-oxo complexes are known. Not surprisingly, these compounds are all powerful oxidising agents and are sometimes explosive.

Noble gas chemistry is at the moment more less limited to xenon fluoro- and oxo-chemistry. Xenon chlorides have been found but appear to be highly unstable. Krypton forms KrF_2 and one or two other less well characterised species. Even so, it is unwise to regard the lighter gases as totally inert: field ion microscopists are frequently embarrassed by formation of noble gas complexes of tungsten and other heavy cations in the gas phase, and many other cationic species are known in the gas phase.

D. The Transition Elements

The changes in chemical behaviour encountered on crossing the series of transition metals are much less dramatic than those found on crossing the p block. For the first row transition metals there are a large number of essentially ionic compounds of similar formula (e.g. MX_2) differing only in the number of essentially non-bonding 3d electrons possessed by the metal ion. This regularity of stoichiometry is much less common for the 4d and 5d series which tend to form compounds whose stoichiometry reflects the number of their valence electrons.

The calculated d orbital binding energies rise smoothly from about 5 eV (~500 kJ) at the beginning of the series to about 17 eV at the end (see Fig. 7.2). The change is about the same for all three series.

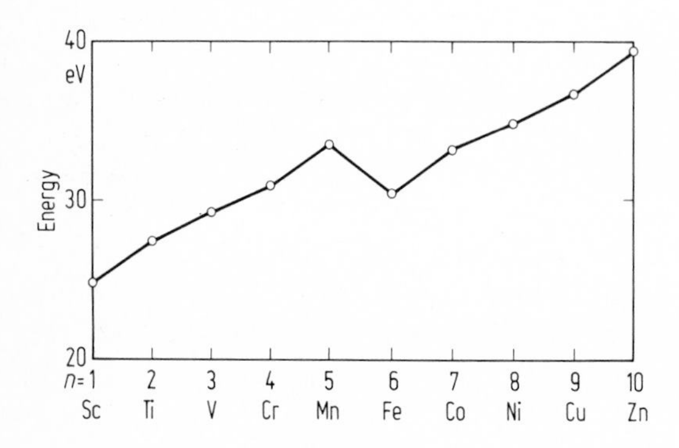

Fig. 7.14. The third ionisation energies of the 3d transition elements ($M^{2+} \rightarrow M^{3+} + e^-$, $3d^n \rightarrow 3d^{n-1}$)

For the ionic chemistry of the 3d series it is interesting to follow the behaviour of the ionisation energy. The plot of the 3rd ionisation energy (i.e. for $3d^n \rightarrow 3d^{n-1} + e^-$) for the 3d series shows a saw-tooth form (Fig. 7.14) with a sharp drop at iron. This is due to the increased electron repulsion in the Fe^{2+} ($3d^6$) ion where one electron must be paired with another: ionisation therefore requires less energy. It can be shown that the third ionisation energy is the dominant factor in determining the M^{3+}/M^{2+} electrode potential, and the stability of Mn^{2+} to oxidation when compared with Fe^{2+} is thus explained. The 4d and 5d series show a similar discontinuity in the third ionisation energy, but the drop is smaller as electron repulsion effects are much smaller in the more diffuse 4d and 5d orbitals (see Table 4.3). The 4d and 5d orbitals consequently have a greater electron affinity, and a more pronounced tendency to act as π acceptors. This must be taken with overlap effects in considering the greater covalence of the 4d and 5d series. The size of corresponding elements in the 4d and 5d series are almost identical. This is due to the „*lanthanide contraction*", the poor screening of the 4f orbitals resulting in the 5d series being smaller than expected.

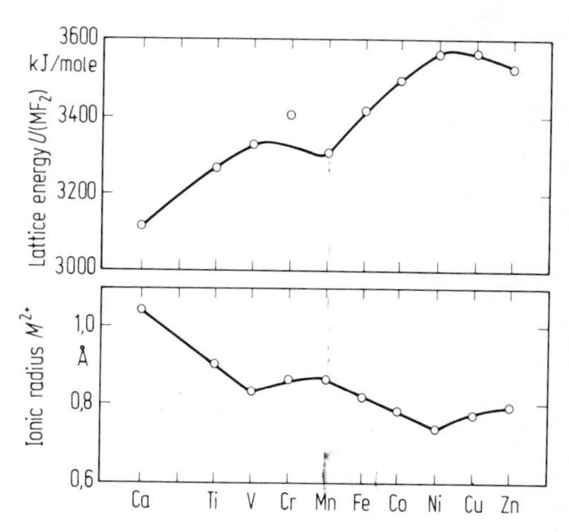

Fig. 7.15. The experimental lattice energies for the series of fluorides MF_2 (above) and ionic radii for M^{2+} (below)

An ionic approach

If we stay for a moment with the 3d series and the ionic model, it is instructive to consider the lattice energies of the fluorides. In Fig. 7.15 are plotted the lattice energy of MF_2 and the crystallographic ionic radius of M^{2+} in MO It will be noticed that the lattice energy rises and then falls slightly to a minimum at Mn^{2+}, then rises and falls again in the second half of the series. The ionic radius shows a similar variation, falling to V^{2+}, rising to Mn^{2+}, then falling to Ni^{2+} and rising slightly to Zn^{2+}. This variation in lattice energy may be directly related to the electronic structure of the salts. They all have octahedral coordination of the metal and the high spin configuration of the d electrons. As will be recalled from Chap. 3, the e_g d orbitals will be σ antibonding. The fall in lattice energy for MF_2 occurs when a d electron enters an e_g* orbital – for high spin octahedral M^{2+} ions this will be for Cr^{2+}, Mn^{2+}, Cu^{2+} and Zn^{2+}. The plotted value for Cr^{2+} is rather higher than one would hope, but this is possibly due to the distorted structure of CrF_2.[4] In summary, the lattice energy is lowered by the occupa-

---

[4] A Jahn-Teller distortion, page 139.

tion of an antibonding d orbital. The behaviour of the ionic radius may be explained similarly: the increasing nuclear charge, and the poor screening of the 3d orbitals by each other results in a general decrease in radius on crossing the 3d series. However, when a new electron enters the σ^* e_g orbitals there is a repulsion of the oxide ions giving an apparent increase in ionic radius, which falls again when the electrons start pairing in the t_{2g} shell. It should be noted that the variations in lattice energy due to these σ^* interactions amount to about 300 kJ at most – this is a small fraction (for fluorides about 10%) of the total lattice energy, showing the essentially non-bonding role of the d orbitals, and the general success of the ionic model for such compounds.

Figure 7.16 shows a similar plot for the MF_3 and M^{3+} – this time the minimum in the lattice energy is found for Fe^{3+} ($3d^5$). The deviations due to occupation of the e_g* orbitals are greater as one would expect from a greater interaction with the ligands in the higher valence state.

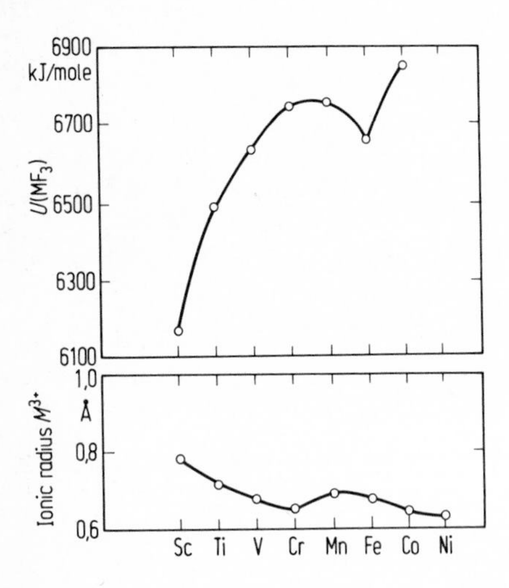

Fig. 7.16. A similar plot for MF_3

Crystal field theory

It is perhaps appropriate at this point to discuss the crystal field theory originally used to explain the spectra and magnetism of transition metal compounds. The compounds initially studied were ionic crystals (such as the fluorides) and the splitting was attributed to the differing electrostatic repulsion by the anionic ligands of the e_g and t_{2g} d orbitals. The e_g (d_{z^2} and $d_{x^2-y^2}$) orbitals point directly at the ligands in octahedral symmetry and are thus raised in energy, while the t_{2g} subset have regions of high electronic density directed between the ligands and suffer less repulsion. This pure ionic treatment has advantages in that the d orbitals may be regarded as pure atomic orbitals subjected to a perturbing electrostatic field, and the calculation of spectral and magnetic properties merely involves application of perturbation theory.

This model rapidly encountered problems with ligands which would be expected to bond covalently and which gave much too great a splitting of the d orbitals for an ionic calculation. The group theoretical treatment used by the model was so successful qualitatively that, rather than abandon the whole structure, the crystal field theory was modified to allow for covalence (by such tricks as the orbital reduction factor, page 153) and was re-baptised ligand field theory. It is now rather difficult to draw the line between ligand field theory and the M.O. approach used in Chap. 3, and any discussion of the electronic structure of transition metal compounds is liable to be classed as ligand field theory.

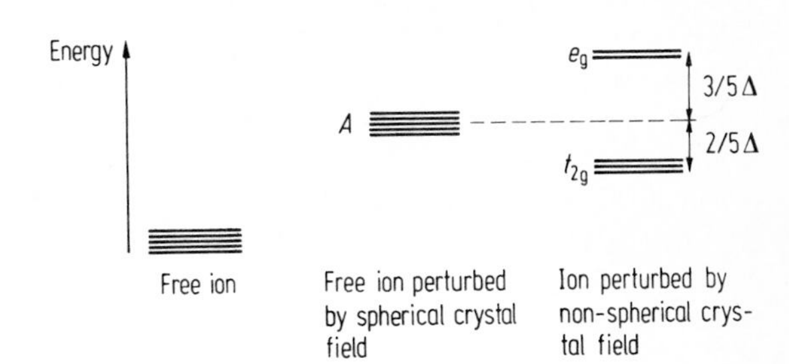

Fig. 7.17. The splitting of d-orbitals in the crystal field model

The simplicity of the electrostatic approach to d orbital splitting has favoured the retention of the crystal field model for teaching purposes. The merging of crystal field theory with M.O. theory gives little trouble except for the concept of crystal field stabilisation energy (CFSE). The crystal field approach uses an electrostatic perturbing potential which is split into two parts: one is a spherically symmetric component which raises the d orbitals as a result of electrostatic repulsion but does not split them, giving the levels marked A in Fig. 7.17, and the second part which splits the d orbitals without moving the centre of gravity. In octahedral symmetry the second potential raises the e_g by $3/5\Delta$ and lowers the t_{2g} by $2/5\Delta$. It is thus common to speak of the t_{2g} orbitals being stabilised but this is only with respect to the hypothetical species A. The crystal field stabilisation energy (CFSE) is the total energy of the d electrons with respect to the level A; as shown in Fig. 7.18 a, it is at a maximum for d^3 and d^8 in high spin d^n systems, and d^6 in low spin systems. When this variation is added to the steady increase in lattice energy to be expected from a M^{2+} ion shrinking across the series from d^0 to d^{10} as a result of steadily increasing nuclear charge (7.18 b) the resulting curve has the same form as the experimental variation in lattice energies (7.18 c). Subtraction of curve (a) using spectroscopically determined Δ values does in fact give a fairly smooth curve similar to 7.18 b.

The M.O. approach assumes the t_{2g} orbitals to be more or less non-bonding in the absence of strong π bonding, but that there will be a drop in stability from the expected smooth curve when an electron enters the antibonding e_g orbitals – the final result is the same (Fig. 7.18 d). From a qualitative point of view, the M.O. theory assumes that the experimental lattice energy will rise as a result of the drop in ionic radius and the increasing electronegativity of the 4s and 4p orbitals, giving a covalent contribution to the lattice energy. This is well borne out by the Lewis acid properties of the d^{10} Zn^{2+} ion which must be due to the 4s and 4p orbitals.

Since the calculation of CFSEs is so straightforward, they have often been used to explain chemical behaviour, notably in the field of kinetics (see below) and coordination number and site preference. Thus it may readily be shown that d^3 and d^8 high spin ions will have a distinct preference for octahedral rather than tetrahedral coordination, and the ordering of cations in spinels, where both octahedral and tetrahedral sites are available, agrees with the theory. There are, however, other effects which may determine coordination number, and it has been argued that the spinels are an unusually favourable case.

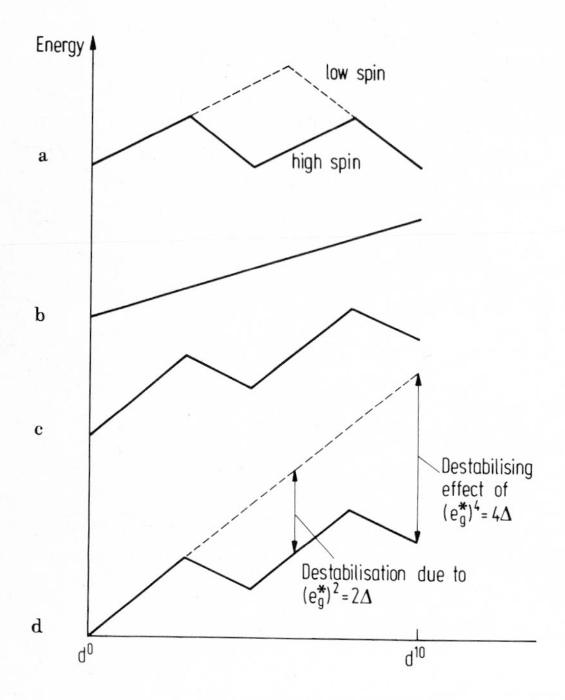

Fig. 7.18 a – d. Crystal field and M.O. predictions of stability

The 4d and 5d series

For the 4d and 5d series the interaction of d orbitals with the ligands is much greater (see Table 4.3) and there is no series of compounds of identical stoichiometry (such as MO or MF_3) in which the effects of varying d orbital population may be studied. The heavier transition metals prefer stoichiometries where there are no d electrons in antibonding orbitals. The higher d orbital splittings and the lower inter-electron repulsions (and thus lower pairing energies) favour low spin complexes. The diffuseness of the d orbitals also results in a high degree of covalence in the metal-ligand bonds: this is well shown up in the n.q.r. data for the complexes of the type $[MCl_6]^{n-}$ which show extensive metal chlorine σ and π bonding and where the metal only carries a charge of about 0.6 units even in the +IV oxidation state. The extent of d orbital covalence in these elements is also shown by the clustered structures adopted by some of the earlier members of the series in their low valence compounds (Fig. 7.19).

In solution simple aquo-ions are very uncommon for the 4d and 5d series and show a strong tendency to hydrolyse to hydroxy complexes, or, for higher oxidation states, to oxo-species which often show very short M–O bond lengths. With anionic ligands (such as halides) stable complexes are formed. The differences between the 3d and the 4d and 5d series are presumably due only to the change in radial properties of the d orbital wavefunctions as the calculated orbital energies and experimental ionisation potentials are very similar.

In the first part of the transition block the highest oxidation states of the 3d, 4d and 5d elements do not show the usual pronounced difference between the 3d series and the others. This is not unexpected for these highly covalent systems where the d orbitals are frequently empty. The stability of the highest oxidation state rises on descending a group, and falls on crossing the series: thus $CrO_4{}^{2-}$ is less oxidising than $MnO_4{}^-$, but more so than $MoO_4{}^{2-}$ and $WO_4{}^{2-}$. The unoccupied d orbitals are strong π acceptors and the M–O bond lengths in $MO_4{}^{n-}$ anions are frequently very short. In the unusual complex $[OsO_3N]^-$ there appears to be an Os–N triple bond, the nitrogen atom showing little basicity. As mentioned earlier, fluorine is also effective at stabilising high oxidation states, usually giving octahedral species. The stability of other species depends on the redox potentials involved, many high oxidation states oxidising their ligands.

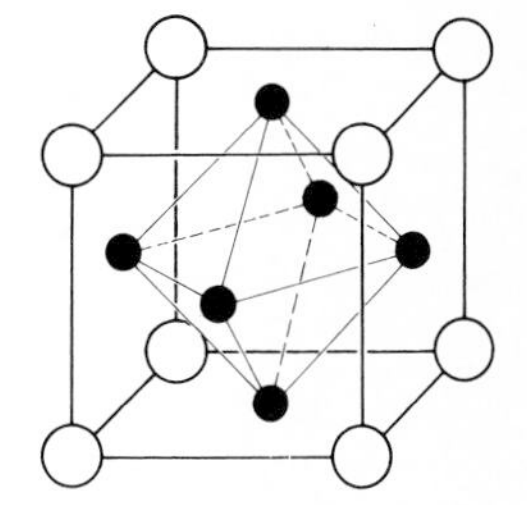

Fig. 7.19. The structure of $(Mo_6Cl_8)^{4+}$ Chlorine atoms at the corners, and molybdenum atoms at the face centres of a cube

In the second half of the series maximum oxidation numbers decrease, presumably as a result of tighter binding of the d orbitals. The heavier metals show higher maximum oxidation states than the 3d series. The covalent character of the heavier metals is well shown by the groups, Ni, Pd, Pt and Cu, Ag, Au: Ni^{2+} may be oxidised to Ni^{3+} (with suitable complexing agents present) but only gives the +IV state with fluoride or oxide ligands, while Pt and Pd give the diamagnetic +IV state without difficulty, but the M(III) state is found only as an intermediate. For copper the +II state is most stable in aqueous solution, but is increasingly less so for Ag and Au; Au is found almost exclusively as Au(I) or Au(III). In the second half of the series, where it is possible that antibonding orbitals may be filled, strong ligands can affect redox equilibria considerably.

Acid base properties

All metal cations are Lewis acids and have already been discussed as such. It is interesting to investigate the acid-base properties of complexes and molecular compounds of the transition metals. In the first part of the transition block where there are many d orbitals unoccupied, Lewis acidity is common, especially for species with low coordination number: thus $TiCl_4$ and $NbCl_5$ are good Lewis acids, the latter existing as a halogen bridged dimer.

The oxy anions (such as $WO_4{}^{2-}$) are stable in alkaline solution, but on acidification frequently polymerise to give complicated species (the poly-acids) with octahedral coordination of the metal and shared oxygen atoms (see Fig. 7.12). Coordination numbers of 7 and 8 are known in the first half of the series (especially for the heavier elements). $FeCl_3$ shows Lewis acid behaviour, and its gaseous form is Fe_2Cl_6; for this high spin $3d^5$ compound, the acceptor orbital is presumably a mixture of 4s and 4p orbitals.

Lewis base behaviour is very rare for central metal atoms: square planar d^8 Ir(I) compounds form weak complexes with Lewis acids such as SO_2 and BF_3, and also appear to act as nucleophiles in one of the mechanisms for oxidative addition. The carbonyl hydrides such as $HMn(CO)_5$ and $HCo(CO)_4$ can dissociate to give $[Mn(CO)_5]^-$ and $[Co(CO)_4]^-$ anions which may thus be regarded as Lewis bases.

Organometallic chemistry

Although there is no sharp difference in electronic structure between classical complexes and organometallic species, two features are peculiar to transition metal chemistry: the frequent stabilisation of low oxidation states by π accepting ligands, and the similarity in electronegativity of ligand and central atom. The metal-ligand interaction is essentially covalent and paramagnetic species are much less common. Titanium tetramethyl is extremely unstable, but may be stabilised slightly by the addition of a base such as pyridine; octahedral Ti(III) species with Ti–C σ bonds are important in polymerisation catalysis. Thermodynamic data for $Ta(CH_3)_5$ and $W(CH_3)_6$ show the metal-carbon bond to be reasonably strong. Chromium shows some facility in the formation of metal-carbon bonds, and inner sphere reduction of organic halides by Cr^{2+} can give Cr–C bonds; also known are the intriguing complexes $[Cr(CH_3)_6]^{3-}$ and $[Cr(CH_2SiMe_3)_4]$. In the middle of the series σ bonded organic species are found most frequently in complexes with π accepting ligands. The later elements show an greater tendency to give σ bonded species: methyl cobalt species may be made in aqueous solution, and platinum forms quite stable methyl species even as Pt(IV). The heavier metals form more stable species than the 3d series. The stability of alkyls of d^{10} species (such as the $[CH_3Hg]^+$ cation) suggests that the increasing role of s and p orbitals at the end of the series may be important.

The 18 electron rule was introduced on page 85 as an empiral result which could be justified. The wealth of preparative organometallic chemistry undertaken since its formulation has enabled a critical assessment to be made. The rule is most successful in the middle of the table and when π accepting ligands are present. For the earlier transition metals complexation of a π donor ligand appears to be favoured over addition of a large number of ligands (TiF_6^{2-} is a 12 electron complex compound by normal counting methods); it might be argued that the orbital energies are too high for all 9 metal orbitals to play an important part in bonding. At the end of the series. complexes with less than 18 electrons are again common: 16 e^- species are well known for d^8 metal ions, $MeAuPPh_3$ is a 14 electron species. This has been explained by a relatively slow drop in energy of the p orbitals leading to only limited chemical activity; alternatively, it could be repulsion from occupied tightly bound d orbitals.

The 18 electron rule is useful as a guide, but does not exclude the possibility of other species. $[Cr(C_6H_6)_2]$ and $[CH_3Ni(PMe_3)_4]^+$ are stable species but quite readily give $[Cr(C_6H_6)_2]^+$ and $[Ni(PMe_3)_4]^+$, both with 17 e^-. $[Cr(CH_3)_6]^{3-}$ with 15 electrons may seem odd, but it is isoelectronic with $[Cr(NH_3)_6]^{3+}$. The many exceptions to the 18 electron rule suggest that the proposition that all organometallic reactions pass through species with 16 or 18 electrons should be qualified: while there may be favourable paths involving such species, other mechanisms, particularly radical mechanisms involving 17 electron species cannot be ruled out a *priori.*

Reactivity of transition metal complexes [5]

The most studied reactions of complexes are the substitution reactions, and for the simple high spin aquo-ions of the 3d series ($[M(OH_2)_6]^{n+}$) it has been possible to study the changes in reactivity across the series. The characteristic water exchange rate is shown in Fig. 7.20 – it will be seen that minima in the exchange rate occur at the d^3 and d^8 configurations – immediately after these configurations electrons enter the e_g* orbitals and the exchange rate increases. It will be recalled that this water exchange reaction is dissociative, and simple MO calculations using the angular overlap model to calculate the destabilisation on going from $[M(OH_2)_6]^{n+}$ to $[M(OH_2)_5]^{n+}$ show good agreement with the rate constants if the bonding due to s and p orbitals is taken into account: this will increasingly hinder dissociation on crossing the series, and this causes a rise to the maxima at d^3 and d^8 followed by the drop on occupation of the e_g* orbitals. Crystal field theory gives qualitatively similar results but a somewhat inferior fit to the experimental data.

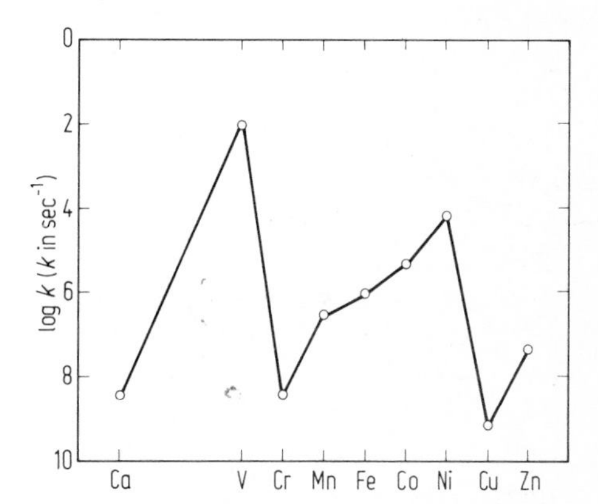

Fig. 7.20. The logarithms of the characteristic water exchange rate for M^{2+}. Note the stability of d^3 (V^{2+}) and d^8(Ni^{2+})

The slow substitution reactions of octahedral d^3 species (in particular chromium (III) have been of great importance in the elucidation of complex chemistry as has the high kinetic stability of the low spin d^6 species (in particular cobalt (III)) which may also be predicted from consideration of the d orbital occupation. In more general terms, +3 oxidation state species exchange much slower than +2 species – thus even V^{2+} ($3d^3$) and Ni^{2+} ($3d^8$) undergo relatively fast reactions, and isolation of a wide variety of complexes as for Cr(III) and Co(III) is not possible. The heavier elements generally substitute more slowly, and for exceptionally stable species such as Pt(IV) and Ir(III) (both $5d^6$ low spin) redox catalysis may be essential to obtain useful yields.

---

[5] See also Problem 16, page 280.

E. The Lanthanides

The lanthanides form the most closely related family of elements, to the point where their separation was for a long time extremely difficult. Their chemistry is essentially that of class A (i.e. hard) Ln^{3+} cations, and their differences mainly derive from the decrease in ionic size as the 4f shell fills up. The 4f shell drops very sharply from being an empty orbital subshell with no apparent chemical activity (in lanthanum) to being a tightly bound non-bonding orbital in cerium (in cerium and praseodymium it is actually just possible to remove one f electron). The 4f shell drops in energy across the series but screens the 5d and 6s orbitals very effectively, their energy scarcely changing from La to Lu.

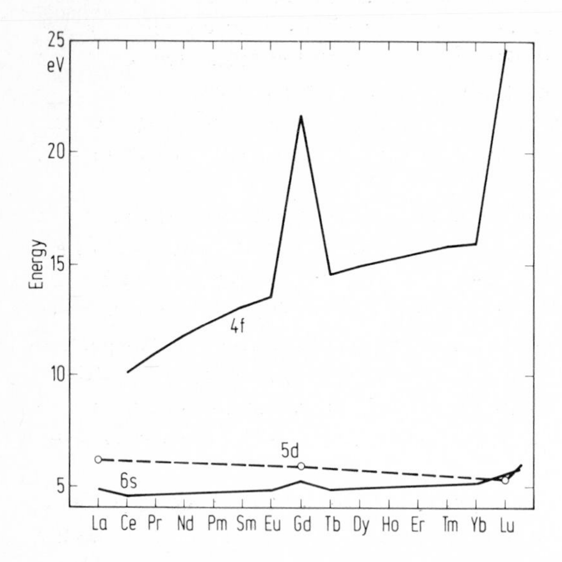

Fig. 7.21. The calculated orbital binding energies for the series La–Lu. The difference in 4f energy of Gd ($[Xe]4f^7 5d^1 6s^2$) and Eu ($[Xe]4f^7 6s^2$) shows the importance of electron repulsion effects

Interelectron repulsion is very great in the 4f shell and this is shown up by the discontinuity in the ground state configurations at gadolinium ($[Xe]4f^7 5d^1 6s^2$ rather than the expected $[Xe]4f^8 6s^2$). The spin-pairing energy of the 4f subshell is sufficiently great to favour placing an electron in the 5d orbital rather than give a $4f^8$ configuration. Similarly, the high electron repulsion in Tb^{3+} ($4f^8$) allows the formation of Tb(IV) ($4f^7$) in very oxidising conditions. The third ionisation energies of Gd and Tb are thought to be below that of Eu as a result of this.

The normal configuration of the lanthanides is $4f^{n+1} 6s^2$, and divalent ions of most of the elements may be trapped in CaF_2, but the shrinkage upon loss of one 4f electron produces such an increase in lattice energy of the trivalent state that the M^{2+} state is very reducing. Stable divalent compounds are known for samarium, europium and ytterbium – all elements where the third ionisation energy is quite high.

Evidence for the non-bonding nature of the 4f orbitals comes from the very small ligand field splittings (Table 4.3), and the crystal field theory is satisfactory for lanthanide ions. The irregularity in the third ionisation energy appears to be counterbalanced by an opposite and almost equal irregularity in the heat of vaporisation of the metals which results in a smooth drop in the standard electrode potential M/M^{3+} across the series, this, coupled with the steady drop of r (M^{3+}) across the series (there is a slight cusp at gadolinium) explains the absence from the aqueous and solution chemistry of the irregularities found in the atomic data.

The lanthanide ions are fairly large (between 1.06 and 0.848 Å) and, despite their +3 charge, do not exert a strong polarising effect on ligands, acting as class A (hard) Lewis acids. The aquo-ions undergo hydrolysis to give acid solutions of $[Ln(OH_2)_nOH]^{2+}$, and raising the pH gives precipitates of aquated hydroxides. The hydrolysis is more pronounced for the heavier elements as would be expected from ionic size considerations. The change in ionic size also produces a change in the number of water molecules coordinated to the metal ion: 9 for the early lanthanides and 8 for the later, with Sm^{3+} and Eu^{3+} apparently existing in both forms. High coordination numbers are common for the lanthanides, and octahedral complexes relatively rare – a nice example is twelve coordinate Ce^{3+} in $[Ce(NO_3)_6]^{3-}$ with the nitrate ion acting as a bidentate ligand. All other things being equal, the heavier lanthanides may show a slightly lower coordination number than the lighter, as a result of their size. The variability of C.N. may be taken as further evidence for the absence of any directed covalence.

The stability constants of lathanide complexes have attracted much attention, but the subject is more complicated than would first appear. As class A complexes, the entropy effect is often dominating in determining ΔG, and this explains the low affinity for neutral ligands (such as NH_3). The change in structure of the aquo ions on crossing the series also affects stability constants if measured with respect to the aquo ion. It now seems well established that there is an irregular variation of chemical affinity across the series, which may be represented as four dips in a smooth curve from f^0 to f^{14}, symmetric about f^7. This so-called double-double effect has yet to receive a really satisfactory explanation.

In their solid state chemistry the 4f orbitals maintain their aloofness, and appear to take no part in bonding even in the metallic elements, all of which will react readily with water and oxygen. The highly coloured metallic compounds with less electronegative elements such as sulphur generally contain metal M^{3+} ions with excess electrons in conduction bands i.e. MS is $M^{3+}\ S^{2-}\ e^-$. The dichlorides of Ce and Gd also have this structure, as do many of the hydrides. The organometallic compounds $Ln(C_5H_5)_3$ are ionic, but there is some evidence for a more covalent interaction in the compounds with cyclo-octatetraene (COT), and sandwich $[M(C_8H_8)_2]^-$ species may be identified.

F. The Actinides

The actinides have probably the most complicated electronic structures of all the elements. At the beginning of the series 5f, 6d and 7s orbitals all have roughly equal energies, and the $5f^n\ 7s^2$ configuration for the atomic groundstate is first found for Pu ($5f^6\ 7s^2$), earlier elements having partial occupation of the 6d shell. Herman-Skillmann calculations for the actinides suggest that the 5f orbitals are actually more tightly bound than the 4f orbitals of the lanthanides in the latter half of the series, and there is some spectroscopic evidence to support this, but up to neptunium it is clear that the 6d orbitals may be chemically important. A second difference from the lanthanides is the greater interaction of the 5f orbitals with ligands – this is shown by electronic spectra and is predicted from the calculated radial distributions of the wave functions.

As a result of these properties, the actinides show a much more varied oxidation state chemistry than the lanthanides. At the beginning of the series there is a strong tendency to use all the valence electrons for bonding – Th^{3+} is unstable (if it exists at all), and the most stable oxidation states for Th, Pa and U are IV, V and VI respectively. Although Np(VII) and Pu(VII) are known, the most stable oxidation state drops to III at americium, and remains so until nobelium. The +II oxidation state is known for Am (Am^{2+} is a $5f^7$ ion) and is actually quite stable for the later members, being much more stable than the +III state for nobelium. It seems quite probable that more oxidation states will be prepared, and it perhaps is best to limit discussion to remarking that the most stable oxidation state rises to +VI at uranium then falls to +III for americium. The oxidation state III remains the most stable until nobelium.

It is difficult to evaluate the role of the 5f orbitals in early actinide chemistry. Species with more or less non-bonding electrons in 5f-orbitals may be identified. The structures shown by actinide compounds are sufficiently complex and different from those of the transition metals and lanthanides to suggest some effect from the 5f orbitals. Thus the MO_2^{n+} (n = 1, or 2) ions known for U, Np, Pu and Am are linear; for Np, Pu, and Am there are non-bonding electrons in the 5f orbitals, but for UO_2^{2+} there are no non-bonding electrons. The d^0 species VO_2^+ and MoO_2^{2+} both have bent geometry so that the strong π donor oxygen ligands may donate charge into different metal d_π orbitals; if the oxide ligands in UO_2^{2+} act as π donors they must donate into the same $6d_\pi$ orbitals and compete with each other, unless the 5f orbitals are intervening. Interestingly enough, such structural data as is available suggests that the U–O bond lengths depend quite strongly on the ligands weakly coordinated perpendicular to the O–U–O axis.

The solid state chemistry of the actinides is complicated. Not surprisingly, a distinct preference for high coordination number is found (for example 12 for $U(BH_4)_4$ and 16 for $U(C_5H_5)_3Cl$), and, as a result of the many oxidation states available, non-stoichiometry is very common. U_2F_9 is a class IIIA mixed valence compound containing U(IV) and U(V). In aqueous solution the actinides are quite hydrolysing, giving acid solutions, but in contrast to the tungstates and molybdates, uranates $M^I{}_2UO_4$ are not soluble in alkaline solution, consisting of polymerised UO_4 species with high coordination number for uranium. The many redox equilibria may also complicate matters – plutonium may be found in solution as an equilibrium mixture of Pu(III), (IV), (V) and (VI).

The early actinides have a more highly developed organometallic chemistry than the lanthanides; apart from the cyclo-octatetraene species mentioned earlier, there are several species with definite metal-carbon σ bonds, such as $Li_2(U(CH_2SiMe_3)_6)$. Actinide chemistry almost certainly has many surprises in store for those who are prepared to face the considerable health hazards that the transneptunium elements present.

G. Conclusions

In this very brief survey of the periodic table we were concerned mainly with the gross differences in chemical behaviour that may be related to atomic properties of the element. In studying a particular system one is generally more interested in slight differences between closely related compounds, and it is then necessary to take into account the considerable delocalisation of bonding in inorganic compounds. The influence of next-nearest neighbour and more distant atoms in inorganic chemistry is still an under-developed subject when compared with inductive and 'resonance' or mesomeric effects in organic chemistry where the effect is almost invariably transmitted by a carbon-carbon chain. By way of illustration only we discuss some of the effects observed.

Substitution of a carbonyl group in a $M(CO)_6$ molecule by a weakly π accepting ligand produces a drop in CO stretching frequencies which may be attributed to an increased π acceptance by the remaining carbonyls in an attempt to restore a balance: this has been used as a means of ordering the π acceptor powers of different ligands. Similarly, the chlorine in the linear complexes L–Au–Cl becomes more ionic as the σ donor strength of L increases. This is clearly a ground state *trans* influence, and similar studies have been made on square planar Pt(II) and Pd(II) complexes: one of the more interesting results, from n.m.r. coupling constants, suggests that the Pt–P bond in *cis* and *trans* $(Pt(PR_3)_2X_2)$ complexes has a greater 6s character when the two phosphorus ligands are *cis* rather than *trans* to each other. There is now a fair amount of evidence to suggest that 'hybridisation' of bonds can vary according to the tastes of the ligand, and that ligand strengths are affected by the other ligands present. Bent has proposed as a general rule that, for two different ligands, the bond to the less electronegative has a higher s character. Clearly this will have consequences for the reactivity of coordinated ligands and for the symmetry control of reactions – although little understood at the moment, these delocalisation effects may yet prove to be important for the design and improvement of catalysts.

Non periodicity

We have followed the practice of most contemporary inorganic chemistry books in discussing the chemistry of an element in terms of its position in the periodic table. There are however many similarities (e.g. in orbital energy, ionic radius, electronegativity) between elements not closely related by the periodic system. Since Moseley's work on X–ray spectra and the development of the Aufbau principle, the periodic table has acquired such a dominating position in chemical ideology that the non-periodic similarities are sometimes overlooked. It suffices to read a little of the history of the development of the periodic table to realise that differences or similarities were not always as clear cut as they now seem. Even if we accept the periodic table as it is, there are regular similarities between elements of different groups, the best known being the diagonal rule, pointing out the similarity between an element and that diagonally south-east of it in the table. Thus lithium and magnesium show many similarities (insoluble fluorides, high hydration energies, stable organometallics), Be and Al (strong hydrolysis in solution), B and Si (oxygen chemistry dominant) and so on; this may be regarded as the natural result of increasing electronegativity on moving right, and decreasing electronegativity on moving down from an element on the periodic table.

Less systematically we may note the similarities between K^+, Tl^+ and Ag^+, or the series Ca^{2+} – Mn^{2+} – Zn^{2+}, Mn^{2+} showing little redox chemistry in normal aqueous solution. For large scale contempt of the periodic table we may cite geochemistry and qualitative analysis for cations: the geochemist classifies elements differently (see page 201) and in minerals, substitution at cation sites of Al^{3+} for Si^{4+}, Fe^{3+} for Al^{3+}, Fe^{2+} for Ca^{2+}, Ni^{2+} for Mg^{2+} etc. are common. The classical group separation method (Table 7.2) is also apparently unrelated to chemical periodicity.

The non-periodic similarities mentioned may all be explained on closer examination by the concepts discussed in earlier chapters, such as similar ionic radius. class A and class B character, etc. It is perhaps fitting to end this chapter by the above examples of the failure of simple periodic arguments, and by underlining the importance of the critical application of theory to inorganic chemistry. Theories are man-made and will only be useful as aide-memoires and means of prediction if they are applied with common sense.

Table 7.2. The classical separation of elements for qualitative analysis

Group	Ions	
I	Pb^{2+}, Hg_2^{2+}, Ag^+	Precipitated as chlorides
II	Hg^{2+}, Pb^{2+}, Bi^{3+}, Cu^{2+}	Precipitated by H_2S
	Cd^{2+}, Sn^{2+}, Sb^{3+}, As^{3+}	
III	Fe^{3+}, Cr^{3+}, Al^{3+}, (Mn^{2+})	Precipitated by NH_3 as hydroxides
IV	Co^{2+}, Ni^{2+}, Mn^{2+}, Zn^{2+}	Precipitated as sulphides by HS^-
V	Ba^{2+}, Sr^{2+}, Ca^{2+}	Precipitated as carbonates
VI	Mg^{2+}, Na^+, K^+	Soluble residue

Bibliography

General texts on inorganic chemistry

Bailar, J.C., Emeleus, H.J., Nyholm, R.S., Trotman-Dickenson, A.F. (eds.): Comprehensive inorganic chemistry, 5 vols. Oxford: Pergamon 1973. A useful reference work for descriptive chemistry.

Cotton, F.A., Wilkinson, G.: Advanced inorganic chemistry, 3rd Edn. New York, London, Sydney, Toronto: J. Wiley 1972. An extremely useful source for the hard facts of inorganic chemistry.

Gmelin Handbook of Inorganic Chemistry. Berlin, Heidelberg, New York: Springer. Supplements to the early volumes appear regularly.

Huheey, J.E.: Inorganic chemistry. New York, Evanston, London, San Francisco: Harper and Row 1972. A readable, if selective, discussion of many areas of inorganic chemistry.

Phillips; C.S.G., Williams, R.J.P.: Inorganic chemistry. 2 vols. London: Oxford University Press 1965. An advanced treatment giving a good 'feel' for the subject.

Purcell, K.F., Kotz, J.C.: Inorganic chemistry. Philadelphia, London, Toronto: W.B. Saunders 1977. A fairly advanced text, particularly strong on reactivity and its relation to structure.

A small selection of references relevant to the subject matter of this chapter

Bent, H.A.: Bent's rule. Chem. Revs. *61,* 275 (1961)
Burdett, J.K.: J. Chem. Soc. Dalton *1976,* 1725. A M.O. treatment of the variation of chemical properties on crossing the 3d series.
Dye, J.L.: Preparation of Na^-. J. Chem. Ed. *54,* 332 (1977)
Johnson, D.A.: J. Chem. Soc. A *1969,* 1525. Thermodynamic discussion of the variation in the third ionisation energy of the lanthanides.
Lehn, J-M.: Cryptate ligands. Acc. Chem. Res. *11,* 49 (1978)
Mitchell, K.A.: The use of outer d orbitals in chemical bonding. Chem. Revs. *69,* 137 (1969)
Mulliken, R.S.: Overlap and chemical bonding. J. Amer. Chem. Soc. *72,* 4493 (1950)
Parish, R.V.: The metallic elements. London, New York: Longman 1977. A readable introduction to the descriptive chemistry of the metals.

Problems

1. Would you expect K^- and Li^- to be more or less stable than Na^-?

2. Discuss the possible reasons for the great difference in speed of the hydrolyses of $SiCl_4$ and CCl_4.

3. Discuss the possible reasons for the planarity of the S_3N skeleton of $(CF_3S)_3N$. (See C.J. Marsden, L.S. Bartell, J. Chem. Soc. Dalton *1977* 1582)

4. Suggest a possible explanation for the fact that H_2SO_4 is one of the strongest dibasic acids known.

5. Compare the isoelectronic series I(I), Xe(II); Te(II), I(III), Xe(IV); Sb(III), Te(IV), I(V), Xe(VI). Do xenon compounds belonging to these series show any exceptional features?

6. Would you expect Ga to form electron deficient compounds?

7. Consider the stereochemistry of iodine in positive oxidation states using (i) the criterion of overlap with the 5p orbitals, and (ii) the VSEPR model. Compare your conclusions for the two approaches.

8. The oxidising power of the ClO_4^-/Cl^- couple is almost as great as that of the MnO_4^-/Mn^{2+} couple in acid solution, but the kinetics of oxidation by perchlorate are very slow. Discuss this in terms of the electronic structures of the XO_4^- ions.

9. Calculate the crystal field stabilisation energies for octahedral and tetrahedral co-ordination of d^3, d^7 (high spin), and d^8 ions. Do your results explain the ease of preparation of $[CoCl_4]^{2-}$ and the difficulty of preparing $[NiCl_4]^{2-}$? (Assume that $\Delta_{tet} = 1/2\ \Delta_{oct}$.)

10. The differences between the three transition series are least pronounced for high oxidation state chemistry and for low oxidation state, organometallic compounds. Why?

11. Cu(III) may only be stabilised by F^-, oxidising ligands such as periodate IO_6^{5-} and tellurate TeO_6^{6-}, and strong ligands such as chelating phosphines and amino-acids such as glycyl-glycyl-glycine. Comment on the various factors responsible for the stabilisation.

12. The variation of stability constant on change of metal cation is shown below for two typical ligands. The same general order of complexing power of cations is found for a wide variety of ligands, and is often referred to as the Irving-Williams series. What is the theoretical basis of this series?

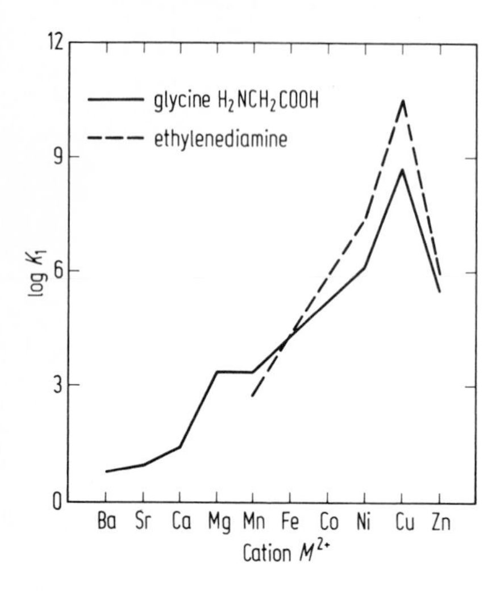

13. (i) Comment on the differences between the complexes of the alkaline earths and those of the M^{2+} ions of the 3d transition series.

(ii) The alkali metals form no ammine complexes analogous to those of the transition metals, yet their electrode potentials in liquid ammonia are different from those in water. Comment.

14. Discuss the increase in Brønsted acid strength in the series:

$$HMn(CO)_5 < H_2Fe(CO)_4 < HCo(CO)_4$$

15. CO is a very weak Lewis base for p-block elements, and no stable lanthanide carbonyl is known. Comment.

16. The figure below summarises the characteristic water exchange rate constants for several metal ions, and may be taken as an indication of the rate at which ligand substitution takes place for different metals. Comment on the variation in rate constants.

	-6 to -4	-4 to -2	-2 to 0	0 to +2	+2 to +4	+4 to +6	+6 to +8	+8 to +10
M^+								Alkalis
M^{2+}					Be	Mg		CaSrBa
M^{3+}							Lanthanides	
M^{2+}					V	Ni Co	FeMn	Cr Cu
M^{3+}	Cr		Co		Fe V	Ti	Mn	
M^{3+}				Al	Ga	In		
M^{2+}							Zn	Cd Hg

$\log_{10}$ k for water exchange of metal cations

17. Compare the solution chemistries of the rare earths and the alkaline earths.

18. Why are most of the commonly used Lewis acids and bases in the p-block of the periodic table?

19. Comment on the following enthalpies of hydration (all in kJ/mole)

Na^+	−405	Ni^{2+}	−2106
Mg^{2+}	−1922	Zn^{2+}	−2045
Al^{3+}	−4660	Cr^{3+}	−6118
Ca^{2+}	−1593	Fe^{3+}	−4378
Mn^{2+}	−1846		

20. Ruminate fearlessly on the role in biological systems of
   (i) trace quantitites of transition metal ions
   (ii) the class B metals
   (iii) relatively high concentration of alkali metal cations.

8. Physical and Spectroscopic Methods

In the foregoing chapters we have frequently used data from physical or spectroscopic methods, and the use of these techniques is central to modern inorganic chemistry. Their development has, however, been somewhat haphazard, often being governed by the availability (and price) of commercial instruments, and often seeming to follow the dictates of fashion. In some cases the claims made for a new technique have proved over-optimistic. In several cases the interpretation of the data obtained has turned out to be more complicated than initially supposed, and it is frequently desirable to seek specialist knowledge before drawing too many conclusions.

In view of this, this chapter will limit itself to a description of the general principle of a method, the origins and type of chemical information obtainable, the conditions of measurement and possible pitfalls. No detailed description of the interpretation of results will be given, but it will, I hope, be possible to foresee if a given measurement is likely to give useful information. For further details, the reader is referred to the specialised texts in the bibliography. The length of text devoted to each technique is more a measure of its familiarity or complexity than its relative importance to the inorganic chemist. Electronic absorption spectroscopy and magnetochemistry were discussed in Chap. 4 and will not be further treated. I have also excluded, on rather arbitrary grounds, any discussion of the measurement of electric dipole moments, molecular beam spectroscopy, and thermochemical and electrochemical measurements – these subjects are generally treated in physical chemistry texts.

A. Diffraction Methods

The most important diffraction technique for contemporary inorganic chemistry is single crystal X–ray crystallography. The dominating position of this technique may be gauged by the fact that, at the time of writing, it is quite common to find 50 percent or more of the papers in an issue of a journal consisting of or containing an X–ray crystal structure. However, the ready availability of the data, and its general completeness and unambiguity should not blind us to its defects, nor to the possibilities of other diffraction methods.

The scattering of X-rays is due to the electrons of the atoms in the crystals. X-ray diffraction is allowed at angles corresponding to Bragg reflection from the planes of the crystal. Analysis of these angles will give the dimensions of the unit cell. The variation of the intensities of the allowed reflections is related to the electron distribution within the unit cell – and this may readily be related to the atomic positions in the cell. There are however certain problems – the scattering power of an atom increases rapidly with the atomic number Z; in the presence of heavy atoms, the scattering by light atoms may be difficult to detect. Thus hydrogen atoms are rarely 'seen' directly – for example, in organic species their positions may be deduced from the stereochemistry of the carbon atom and the known carbon-hydrogen bond length. This is clearly an assumption. More seriously, other light atoms of considerable chemical interest may be incorrectly positioned in the presence of a heavy atom – this was in fact the case for initial studies of UO_2^{2+} and TeO_2.

The X–ray diffraction method sees a time average of molecular motion, and if there is free rotation in the crystal of a small group, this can present problems; it is also necessary to allow for vibrational effects in structure determinations. Since all the above difficulties need to be taken into account in the solution of a crystal structure, the result, at least in the finer details, is not always wholly unambiguous. For this reason, the non-crystallographer should be chary of discussing small differences in bond lengths, etc., unless he has some idea of the assumptions made in solving the structure.

The improvement in computing hardware and software has speeded up and simplified the practical side of crystal structure determination to a point where one chemist has claimed that it is easier and quicker to perform a crystal structure analysis than obtain a full elemental analysis. While it is true that X–ray crystallography can sometimes supplement or correct an elemental analysis, one is still left with the often considerable difficulty of obtaining good crystals. An interesting series of examples of possible difficulties is given by Ibers. The improvements in experimental technique have now enabled some information to be obtained on the actual distribution of valence electrons in compounds (most of the X–ray scattering is due to core electrons), and some examples are given by Coppens. However, this is at the extreme of experimental precision, and the results are by no means universally accepted.

X–ray diffraction of materials other than single crystals is also very useful. Crystalline powders give diffraction patterns which are useful fingerprints, and from which, in relatively symmetric cases, cell dimensions may be deduced. Glasses and liquids give diffraction patterns, but, as there is no extended ordering of atoms, only an average radial distribution around an atom may be seen. It has been possible to show that the first coordination shell in a liquid or glass is roughly constant, and to follow changes on addition of a ligand or changing solvent. Although difficult to interpret, these studies are of interest as they give the only relatively direct evidence of structure in solution.

Neutron and electron diffraction

The attraction of using neutrons and electrons for diffraction studies is that the scattering power is a less sensitive function of atomic number Z. Neutrons are scattered by the atomic nucleus and the scattering power varies irregularly across the periodic table. It is thus possible to distinguish elements of similar atomic number (difficult with X–rays) if their neutron scattering powers are different. More importantly, it is possible to distinguish hydrogen atoms quite easily. Neutron diffraction has been used extensively for the study of hydrogen bonding. The scattering of neutrons is a weak effect and relatively large crystals (1 mm^3) are necessary. As neutrons possess magnetic moments they may also be scattered by magnetic fields in solids, and this property has been studied extensively by physicists, as has the inelastic scattering of neutrons by lattice vibrations.

Electrons are very strongly scattered by atoms: at low electron energies the scattering is mainly by the valence electrons; at higher energies the nuclear potential becomes important. Electrons are however readily absorbed, and it is necessary to use either gaseous samples or study only the surface of solids. The lack of ordering in the gaseous phase limits the information that may be gained to a radial distribution curve from which atom-atom distances may be calculated; this will frequently be sufficient to deduce the molecular structure. Scattering by heavy elements rather than lighter ones is less of a problem for electrons than X–rays, although it does increase with atomic number.

B. Electron Spectroscopy

Electromagnetic radiation of a sufficiently high energy can expel an electron from an atom or a molecule. It the energy of the incident radiation is known and the kinetic energy of the expelled electron can be measured, it is then possible to calculate the binding energy of the electron in the atom or molecule. This is the basis of electron spectroscopy. We may divide the subject according to the means used to eject the electron.

(i) Ultraviolet photoelectron spectroscopy (U.P.S.). The electrons are ejected by ultraviolet photons with energy up to 40 eV – the usual sources are helium lamps emitting either spectra of the excited He atom or of the He^+ ion. The electrons ejected from the sample are thus fairly weakly bound (i.e. valence electrons) and the pattern of the spectra reflects the pattern of the M.O. As mentioned in Chap. 3, methane gives two signals corresponding to the t_2 and a_1 molecular orbitals at 14 and 23 eV. The electron binding energy as measured by U.P.S. is often identified with the one electron energy via Koopman's theorem (page 27). However Koopmans' theorem is not infallible (see the 'third revolution in ligand field theory', page 61) and the order of M.O. determined by U.P.S. does not always agree with calculations or electronic absorption spectra. The general agreement is however very satisfactory and is a strong argument in favour of the M.O. approach to chemical bonding.

The ejection of the electron is a rapid process and the internuclear distances in the molecule do not change during the transition – the ion produced is thus in a excited vibrational state (see Fig. 5.9). As a consequence, the binding energy is greater than the adiabatic ionisation energy, and vibrational fine structure may be seen in favourable cases (the resolution of U.P.S. is of the order 0.01 eV). This may allow the calculation of the adiabatic ionisation energy, as well as the classification of the ionised M.O. as bonding, non-bonding or antibonding depending on the vibrational frequency of the ion. The interpretation of the U.P.S. spectra of relatively simple molecules can become quite complicated – for example the spectrum of benzene was a subject of controversy for some years. From a practical point of view, the low energy of the ionising radiation results in the electron being ejected with little energy, and U.P.S. is generally restricted to the study of gases or surfaces where there is less chance of the electron being absorbed.

(ii) X–ray photoelectron spectroscopy (X.P.S.). The use of X–rays enables more tightly bound 'core' electrons to be ejected; the X–ray energy is typically 10^3 eV (from a Mg or Al source). Since the inner shells are tightly bound and take little part in chemical bonding, their binding energies and ease of ionisation by X–rays are roughly constant irrespective of the chemical environment of the atom. The initial interest in X.P.S. was as an analytical technique permitting the detection of an element (from the known energy of a given inner shell) and an estimate of its concentration (from the intensity of the emitted electrons). For this reason X.P.S. is also referred to as Electron Spectroscopy for Chemical Analysis (E.S.C.A.).

It was however found that the binding energy of an inner orbital varied by as much as ±5eV depending on the chemical environment of the parent atom. Three origins are proposed for this rather large variation in energy of a chemically 'inert' orbital:

i) The charge on the parent atom. If the parent atom carries a high positive charge it will clearly be harder to remove an electron. This is supported by the fact that the chemical shift of an X.P.S. signal often correlates well with the oxidation state of the atom – however this relationship is not universal.

ii) Madelung energy. In an ionic (or polarised covalent) system the electron ejected will have its energy affected by the electrostatic field of the surrounding ions. This electrostatic field will act in the opposite direction to the atomic charge effect.

iii) Relaxation effects. When an electron is expelled, the electron repulsion due to it will disappear and the other electrons in that atom will move closer to the nucleus. This 'contraction' is referred to as relaxation, and appears to take place very rapidly so that the energy of the ejected electron is affected by the extent of relaxation, which is itself sensitive to chemical environment – there is thus a third effect on the chemical shift. It is, in fact, another example of the failure of Koopmans' theorem.

These three effects are of approximately equal importance, and the interpretation of chemical shifts in X.P.S. is quite complicated – the best results at the moment come from X–α calculations which do not use Koopmans' theorem. It is better to regard the X.P.S. shift as a fingerprint than as a source of chemical information.

The resolution of X.P.S. is quite low (~0.5 eV) but splitting of signals and satellite signals are often observed – doublets are observed for ionisations from orbitals with orbital angular momentum $l > 0$ corresponding to the spin-orbit coupling of the remaining unpaired electron with the orbital angular momentum. If the atom ionised already possesses an incomplete shell (e.g. a $3d^9$ ion) then the production of a second incomplete shell gives rise to several possible states of different energies, and the X.P.S. spectrum consists of several overlapping lines. As an example we may consider the ejection of a 2p electron from a Cu^{2+} ion giving the configuration ($2p^5$ $3d^9$) with many possible Russell-Saunders terms. Finally we may mention that the photo-ionisation of an electron may be accompanied by an electron transfer from another part of the molecule, again leading to a well developed satellite structure.

From the practical point of view, X.P.S. may be used to study gases, surfaces or solids, although the ready absorption of electrons hinders the study of solids in bulk. When studying solids, it is important to remember that the sample can decompose in the radiation, and that the continuous loss of electrons can build up a positive potential on samples of non-conducting solids, giving an artifically low kinetic energy to the ejected electron.

(iv) Auger electron spectroscopy (A.E.S.) The Auger effect is observed after the formation of a hole in an inner electron shell. An electron in a higher shell will drop down to the vacant inner orbital and the energy released will be dissipated either as a photon of X–ray radiation or by the ejection of a second electron, an Auger electron. Auger spectroscopy involves the measurement of the kinetic energy of this electron. As with X.P.S. a chemical shift may be detected, and the Auger chemical shift is sometimes more sensitive than the X.P.S. shift. The technique is used essentially for the study of surfaces. The 'hole' in the inner shell may be produced either by X–irradiation (as in X.P.S.) or by electron bombardment. The final state lacking two electrons can yield information not readily obtainable by any other experimental method.

C. X–Ray Spectroscopy

Some of the uses of X–ray spectroscopy were discussed in Chapter 3 in the section on solids. X–ray spectroscopy is essentially a solid state technique and the interpretation of the spectra must take possible band structure and delocalisation into account. X–ray absorption spectroscopy is generally restricted to thin films, but X–ray emission spectroscopy of almost any solid is possible, and the technique is frequently used for qualitative and semi-quantitative analysis (e.g. in mineralogy). Reviews of the subject are given by Bonnelle (in the book edited by Hill and Day) and Urch.

D. Vibrational Spectroscopy

The study of molecular vibrations has been of considerable importance in inorganic chemistry, both for the determination of molecular structure and for the study of that elusive quantity 'bond strength'. As discussed on page 21, molecular vibrations may be classified using the symmetry of the molecule, and this relationship is the origin of the use of vibrational spectroscopy for the determination of molecular structure (see for example Nakamoto; Cotton; Jaffe and Orchin). Group theory also allows the prediction of the infra-red or Raman activity of a given band; one may also assign 'totally symmetric' vibrations in the Raman spectrum from the fact that the scattered light will retain at least a certain amount of the polarisation of the incident exciting line.

Each normal coordinate of vibration generally involves motion of several atoms in the molecule; in certain cases, it is however possible to 'localise' the vibration on one particular chemical bond – a good example is $Cr(CO)_6$. The carbonyl stretching vibrations may reasonably be regarded as 'localised' on the C≡O triple bond. The vibrations of the six different carbonyl groups are however coupled so that each carbonyl group does not vibrate independently – there are six normal modes (one triply degenerate, one doubly degenerate and one non-degenerate). Only the triply degenerate mode is observable in the infra-red. The coupling is sufficient to produce three distinct vibrational energies, but these are quite close to each other (2000, 2026, 2118 cm^{-1}), and at much higher energy than any metal-carbon stretching or bending vibrations. We may therefore talk of a carbonyl stretching region of the vibrational spectrum. In the case of CO_2 there is naturally much stronger interaction between the two CO bonds, and the antisymmetric and symmetric stretching frequencies are 1330 and 2349.3 cm^{-1} respectively. For a vibration to be localised it is generally necessary for its frequency to be some distance from those of surrounding bonds. This frequently occurs for double or triple bonds (high force constant) or bonds with low reduced masses (most frequently E–H bonds).

The use of vibrational spectroscopy to determine bond strengths is not without difficulties. If the frequency alone is studied one must be sure that the vibration is really localised on the bond in question. If, as a result of a more thorough analysis a force constant is extracted from the experimental data, one should be wary of comparing small differences between the force constants of very different species. The classic example is furnished by the halogens:

	Force constant mdyne/Å	Dissociation energy kJ/mole
F_2	4.46	147.7
Cl_2	3.19	243.9

The fluorine-fluorine bond, although 'tighter', is easier to break. None the less, there is a general monotonic relationship between bond dissociation energy or bond length and force constant – double bonds always have higher force constants than single bonds. The evolution of the stretching frequency of a given species does give a useful guide to changes in the bonding – as for example in metal carbonyls and nitrosyls, or the dioxygen species:

	O_2^+	O_2	O_2^-	O_2^{2-}
ν (cm^{-1})	1876	1556	1145	770

Introductions to the extraction of force constants are given by Nakamoto and Gans. Complete analysis of a vibrational spectrum is by no means a trivial procedure: it is normally necessary to confirm trial solutions by studies of isotopically labelled systems where the effects of the change in reduced mass may be followed. Isotope labelling studies have shown many assignments to be incorrect. One should also be careful in the interpretation of the vibrational spectra of solids, where interactions between adjacent molecules and vibrations of the lattice as a whole may be important (see Sherwood).

Practical details

Both infra-red and Raman spectroscopy may be used to study gas, liquid and solid samples. Since most compounds are opaque in the infra-red, absorption by solvents, cells etc. can be a problem, and infra-red spectroscopy of aqueous solutions is well nigh impossible. The great advantages of infra-red spectroscopy are the ready availability of commercial instruments and the speed of measurement. Unfortunately many infra-red active vibrations of interest to inorganic chemists are of low energy (< 600 cm^{-1}) as a result of high reduced masses: this poses some problems, and below about 200 cm^{-1} special instrumentation is necessary (e.g. Fourier analysis of interference).

Raman spectroscopy has been revolutionised by the introduction of the laser as light source. The amount of sample required has been reduced to a few microlitres of solution, and it is possible to detect very weak signals by slow scanning. All three phases may be studied, and there is no necessity to use NaCl or other costly materials for cells. Many solvents (such as water) give weak Raman spectra, and aqueous solutions may be studied easily. A particular advantage for the inorganic chemist is the fact that low frequency (< 200 cm^{-1}) vibrations are easily studied.

The very high light intensities of laser sources enable the observation of several complicated effects. A particularly important case occurs when the sample has an absorption band in the visible close to the exciting line. The high light intensity can produce photodecomposition of the sample, or other effects such as resonance fluorescence. The 'resonance' effect can however be exploited as in *Resonance Raman Spectroscopy,* where the Raman spectra of vibrations coupled to the electronic absorption are enhanced by several orders of magnitude, with a consequent increase in sensitivity. This has been used in the study of biological media (Spiro): for example, Raman spectra of vibrations close to the strongly absorbing π system of the porphyrin in hemoglobin are greatly intensified by the resonance Raman effect; it is thus possible to study the iron-porphyrin system alone without the spectrum being obscured by the many other molecular vibrations.

E. Microwave Spectroscopy

In its simplest form, microwave spectroscopy measures the transitions between the pure rotational energy levels of a molecule. All studies are made in the gas phase, and the molecule must possess a permanent dipole moment. Calculation of the moments of inertia, and, where necessary, those of isotopically substituted species, can lead to the determination of accurate bond lengths and angles. However, the spectra are very complicated for molecules of low symmetry, and it is necessary to study Stark splittings of the spectra to obtain the maximum amount of information. Nuclear quadruple couplings in the molecule (see later) also produce splittings in the spectrum, and the interpretation of the spectra is generally a specialist affair. Microwave spectra have however given a great deal of useful information, especially in the determination of the geometry of small volatile species such as SO_2 and ozone.

F. Nuclear Magnetic Resonance Spectroscopy

N.m.r. is an extremely familiar technique, especially in organic chemistry. Its application to inorganic chemistry, particularly to nuclei other than the proton, is less familiar and here we will mention some of the more specifically inorganic aspects. Any nucleus with a spin greater than zero has a nuclear magnetic moment, and, in principle, a very high proportion of the elements may be studied by n.m.r. In fact low magnetic moments, quadrupole effects, electronic paramagnetism and low natural isotopic abundance can all limit the application of n.m.r. – Harris has reviewed the amenability to n.m.r. of elements in the periodic table.

The two parameters generally associated with an n.m.r. spectrum are the chemical shift and the spin-spin coupling constants. The chemical shift gives a measure of the screening (or reinforcement) by the surrounding electrons of the magnetic field at the nucleus studied. For the proton, this may occasionally be related to the electron density around the nucleus, but if the magnetic field (generally of the order of 20000 gauss or 2 Tesla) can mix excited states into the ground state wavefunction of the molecule, the resulting change in the local magnetic fields can produce enormous chemical shifts, which, generally speaking, may not easily be related to molecular structure. The high proton chemical shifts of metal hydrides are caused by this effect and not necessarily by a negative charge on the proton. The chemical shift is also affected by changes in the magnetic susceptibility of the solvent or addition of paramagnetic solutes – this latter giving a useful method of measuring paramagnetic susceptibilities in solution. For protons the range of chemical shifts is fairly small (10 ppm for *simple* organic compounds) but for heavier elements with low lying excited states the shifts can be enormous (hundreds of ppm for ^{31}P, 14000 ppm for ^{59}Co).

Splitting of n.m.r. signals arises from the interaction of non-magnetically equivalent nuclei. In the solid state the interaction is due to direct dipole-dipole interactions and this may be used to estimate the distance between the interacting nuclei. In solution, the direct coupling is averaged to zero by rapid molecular tumbling, and scalar hyperfine coupling of much smaller magnitude is seen instead. This scalar coupling between atoms is transmitted essentially by s type atomic orbitals which have non-zero electron density at or near the nucleus, and this has been used to measure s character in molecular orbitals. The effect falls off rapidly with distance between nuclei and is not normally detectable over more than four bonds.

Relaxation processes in n.m.r. have been much studied. Two relaxation processes may be identified: the loss of energy to the surroundings corresponding to a return of spins in the excited state to the ground state, and characterised by a time T_1, and the exchange of energy between different spins characterised by T_2. T_2 determines (together with instrumental features) the width of the n.m.r. signal. The presence of quadrupolar nuclei coupled to the n.m.r. nucleus (or a quadrupolar nucleus itself studied by n.m.r.) can greatly reduce the relaxation times, giving a signal that is too broad to be observed. This is less of a problem for nuclei with high magnetic moments and low quadrupole coupling constants (resulting from a relatively low quadrupole moment such as ^{2}D, ^{11}B, and ^{14}N or, more usually, from a symmetric environment) or those with high quadrupole moments but low magnetic moments (such as ^{193}Ir, ^{197}Au) where the spin-spin coupling is feeble.

Paramagnetic ions or compounds can also greatly reduce relaxation times and broaden spectra – the effect is however dependent on the relaxation time of the magnetic field due to the electrons – if this is much faster than the characteristic n.m.r. measurement time ($\sim 10^{-7}$ sec) the time average of the magnetic field is zero and little effect is seen. If the magnetic field relaxes slowly, the relaxation times of the nuclear spins are reduced and the n.m.r. signal may not be seen. The full details of this phenomenon are quite complicated. The high local magnetic field due to a paramagnetic ion can be used profitably to spread out otherwise overlapping n.m.r. signals. This is the use of n.m.r. shift reagents. The most common are europium complexes which will readily accept an oxygen or nitrogen donor atom of a substrate – the local magnetic field changes the chemical shifts of nuclei in the substrate without broadening the resonance lines. The displacement of the chemical shift is proportional to any covalency in the lanthanide-donor bond (causing a delocalisation of spin density) and a term arising from a direct dipolar interaction between electronic and nuclear magnetic moments. This second, pseudo-contact term, has been suggested as a means of determining structures of ligands bound to the metal ion as the first (contact) term is very small for lanthanide ions, and the pseudo-contact shift is a function of the position of the shifted nucleus relative to the metal ion. The calculations are however far from straightforward.

The final application of n.m.r. is as a technique for following chemical reactions. If two non-equivalent n.m.r. nuclei are exchanging with each other in a time comparable with the n.m.r. observation time ($\sim 10^{-7}$ sec), the effects will be seen in the n.m.r. spectrum. If the exchange is much slower than 10^{-7} sec the two signals will be broadened slightly; if the exchange is much faster, one single signal will be seen, its position a weighted average of the two initial resonances. If the exchange rate is lowered, the single signal will broaden and separate into two poorly resolved humps, which on slowing further, will give the two initial signals. This rate change is generally accomplished by lowering the sample temperature. Thus SF_4 shows two distinct fluorine signals in solution at low temperatures, which collapse and merge slowly to give one signal at room temperature as a result of scrambling of the two different fluorine sites. N.m.r. has been used to study fluxional molecules (such as PF_5, $Fe(CO)_5$, Chap. 3). The analysis of n.m.r. spectra for kinetic data is quite complicated and it should be remembered that many effects can change the line shape especially when paramagnetic species are present.

From a practical point of view, n.m.r. suffers from being a rather insensitive technique requiring quite concentrated solutions. The advent of Fourier transform and double resonance techniques has considerably improved the sensitivity and applicability of n.m.r., but with a concomitant increase in price. ^{13}C n.m.r. without isotopic enrichment of samples is now almost routine and has been much used in organometallic chemistry. Apart from the proton, the most popular other nuclei for study are still ^{19}F and ^{31}P. Virtually all n.m.r. spectra are taken in solution although there is increasing interest in solid state and liquid-crystal spectra.

G. N.Q.R. Spectroscopy

Nuclear quadrupole resonance spectroscopy is the Cinderella of physical techniques in inorganic chemistry. The principle of the method derives from the fact that nuclei with a nuclear spin greater than or equal to 1 possess a nuclear electric quadrupole moment which, in an asymmetric electric field, will have two or more possible orientations in the field with differing energies (Fig. 8.1 a). N.q.r. spectroscopy measures the absorption of energy due to transitions between these levels. Since the nuclear quadrupole moment is a constant, the energy of the transition (the quadrupole coupling constant) is a measure of the asymmetry of the electric field experienced by the nucleus and may be related to the electron distribution near the nucleus. This may readily be related to the occupation of atomic orbitals, and thus gives a measure of the electron distribution in a molecule.

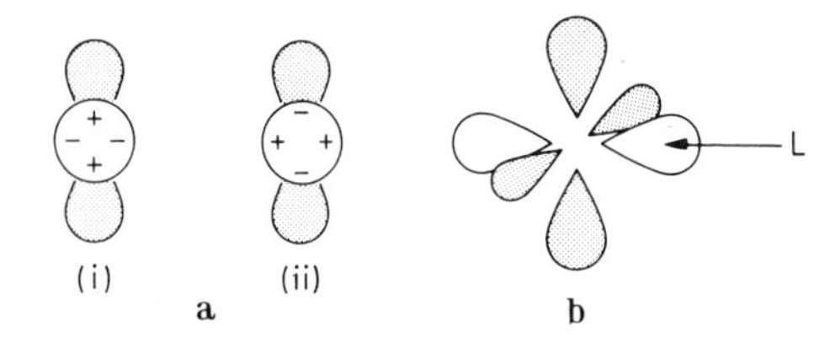

Fig. 8.1. (a) Two possible orientations of a quadrupole in a asymmetric electric field due to a p orbital. Orientation (i) has the lower energy, **(b)** increasing donation by L will lower the asymmetry around the chlorine atom

As an example we may consider chlorine, which has two stable isotopes both with quadrupole moments. A chloride ion will have a symmetrical occupation of its valence shell. If the chlorine is covalently bonded (as, for example, in Cl_2), the bond electron pair is delocalised over the chlorine and the 'ligand', and there will be a lower electron density along the bond axis than perpendicular to it, giving an asymmetric electric field (see Fig. 8.1 b). As L becomes more electronegative it will pull more electron density towards itself and increase the asymmetry around the chloride nucleus, increasing the quadrupole coupling constant. If we assume that a chlorine ion will have a zero coupling constant (although there may be a small 'lattice' contribution from asymmetrically disposed neighbouring ions) and that molecular chlorine has a pure, non-polar covalent bond, then we may linearly interpolate other values to calculate the bond covalency (as was done in Table 5.5). Some authors use a slightly more refined system allowing for participation by the chlorine s orbitals in bonding. The results generally agree with chemical intuition, although the n.q.r. coupling constant is very sensitive to chemical environment, and small differences (for chlorine, up to 1 MHz) may be due to solid state packing effects, hydrogen bonding, molecular vibrations, etc.

The analysis of the data for singly bonded chlorine is quite straightforward, but becomes slightly more complicated for species with coordination number greater than 1, as tensor sums over all the ligands are necessary. It is however easy to distinguish geometrical isomers, or establish an order of ligand strength, as measured by the electron density donated. The fact that n.q.r. data are so simply related to actual electron density makes them of interest in assessing the accuracy of M.O. calculations.

The disadvantages of n.q.r. are practical: the technique is insensitive and large samples (1–2 g) are required. Relaxation times are of paramount importance in n.q.r. spectroscopy and it is necessary to study highly crystalline materials to avoid the Heisenberg broadening of the signal as a result of rapid relaxation. Paramagnetic materials often have very fast nuclear relaxation times and are not easily studied by n.q.r. The sensitivity of n.q.r. to chemical environment is itself a problem since very large ranges of energy must be searched for the signal (and bonding theory is not sufficiently advanced to predict where the signal will appear).

Even when all factors seem favorable, it may be impossible to find the expected n.q.r. signal.

These difficulties clearly limit the use of n.q.r. as a routine method, but there is already a substantial amount of n.q.r. data published, and a large number of chemically interesting elements have nuclei amenable to n.q.r. studies (for example, Cl, Br, I, V, Mn, Co, Cu, B and Sb).

H. Mössbauer Spectroscopy

Mössbauer spectroscopy is another nuclear technique, the chemical information being obtained from the chemical perturbations of the energy of a nuclear excitation caused by the absorption of a γ–ray. Under certain conditions and for certain nuclear isotopes it is possible to measure the energy of a nuclear transition with a resolution of 1 part in 10^{12}. This very high resolution enables the effects of chemical environment to be distinguished with ease. The technique derives its name from the discovery of the resonant absorption of γ–rays by R.L. Mössbauer in 1957, but some workers prefer to describe it as nuclear gamma resonance (n.g.r.) spectroscopy by analogy with n.m.r. and n.q.r. Three effects of chemical interest may be related fairly simply to the electronic environment: the isomer (or chemical shift), the quadrupole splitting and the magnetic splitting.

The quadrupole and magnetic splittings are merely new manifestations of n.q.r. and n.m.r. respectively – if the nucleus is in a magnetic field or an asymmetric electric field, then, if either or both of the nuclear states possesses a magnetic dipole or electric quadrupole moment (as is invariably the case), the nuclear levels will be split by the interaction and several absorption lines will be seen in the Mössbauer spectrum. The splittings of the lines may readily be related to the splittings of the levels. In the case of quadrupole splittings, the interpretation is identical to n.q.r. spectra. The magnetic splittings are rather different from the splittings encountered as a result of an applied magnetic field, as they arise from the magnetic fields present in the sample. As the 'observation time' of Mössbauer spectroscopy is relatively long ($\sim 10^{-9}$s), fields due to sample paramagnetism generally average to zero and no magnetic splitting is seen. Ferro-, ferri- and anti-ferromagnetic compounds have large magnetic fields which relax slowly and give a large magnetic splitting. This gives a useful method of identifying magnetic exchange phenomena, since methods based on the change of bulk magnetism may be unable to distinguish, for example, exchange effects and spin-orbit coupling.

The chemical or isomer shift is a phenomenon unique to Mössbauer spectroscopy: it consists of a slight shift in the energy of the nuclear transition resulting from the variation of electron density at the nucleus. Whether the Mössbauer nuclear transition is increased or decreased in energy as the electron density increases depends on the nucleus in question, but for a given isotope it is relatively easy to establish, and the chemical shift may be taken as a measure of the electron density at the nucleus. This density varies from compound to compound as a result of two effects: the changing occupation of an s atomic orbital in the valence shell, and the 'screening' of the density at the nucleus by electrons in p, d or f type orbitals which themselves have zero density at the nucleus.

The effects are best illustrated by examples – the high spin ferric ion ($3d^5$) has a higher electron density at the nucleus than high spin ferrous ion ($3d^6$) as a result of the decreased screening by the 3d electrons. Iodine shows an increase in electron density at the nucleus as it passes from oxidation state –I ($5p^6$) to +V ($5p^0$), but shows a sharp fall in the +VII oxidation state as a result of the withdrawal of electron density from the 5s orbitals that are now involved in bonding. Further examples of the use of the isomer shift were discussed in Chap. 5 (page 196). Consideration of the isomer shift and quadrupole splitting may enable identification of the spectroscopic oxidation state.

From a practical point of view Mössbauer spectroscopy is limited (as are all nuclear methods) to those elements possessing suitable isotopes. All spectra must be taken in the solid state (although frozen solutions and matrix-isolated species may be studied), and for certain elements it is necessary to work at low temperatures. The sensitivity depends on the isotope studied, but Mössbauer spectra are not plagued by relaxation problems, and spectra of impure or poorly crystallised materials may easily be obtained – this gives some advantages over n.q.r. spectroscopy where a choice is possible. Mössbauer spectroscopists generally measure energies by the Doppler shift needed to produce resonant absorption between the sample and the reference source – date are usually quoted in mm/s.

The two most popular Mössbauer resonances occur in iron (^{57}Fe) and tin (^{119}Sn), and Mössbauer spectra of iron and tin compounds are almost routine. Although the range of compounds studied is rather more limited, a considerable amount of chemically interesting data has accumulated for compounds of ruthenium, antimony, tellurium, xenon, tantalum, tungsten, iridium, gold, europium and some other rare earths, and neptunium.

I. Electron Spin Resonance (E.S.R.) Spectroscopy

The principle of E.S.R. (or electron paramagnetic resonance, E.P.R.) spectroscopy is quite similar to n.m.r. spectroscopy: a system with unpaired spins will have a magnetic moment, and this magnetic moment will have various possible orientations in an applied magnetic field. The system will absorb energy corresponding to transitions between orientations of different energy (Fig. 8.2 a). This energy is given by the equation $E = h\nu = g\beta H_r$ where H_r is the resonant magnetic field, β the Bohr magneton and g the electronic g factor, equal to 2.0023 for a free electron. The g factor is however sensitive to the chemical environment and may thus be used to gain chemical information. The method of observation of the spectrum is similar to continuous wave n.m.r.: for a

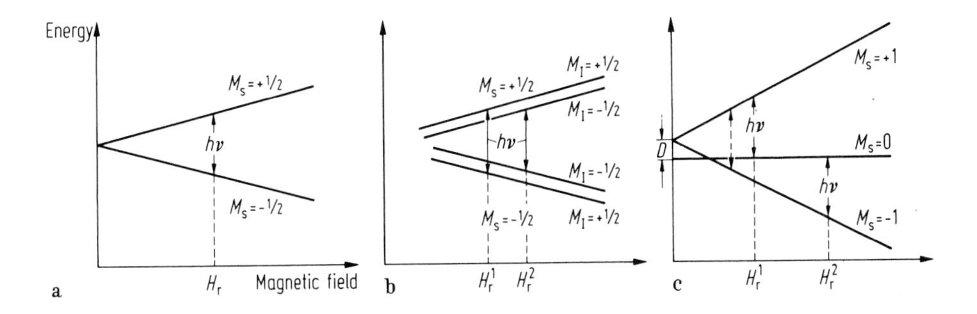

Fig. 8.2 a – c. The behaviour of the M_S levels in a magnetic field, and the observation of e.s.r. **(a)** Simple splitting of a doublet. Resonance seen at one field strength only. **(b)** Splitting of the doublet levels by hyperfine coupling with one I = 1/2 nucleus. Two resonances seen, **(c)** a triplet state with zero field splitting D. Two resonances seen. The dotted transition is the forbidden $\Delta M_S = 2$ transition

fixed energy of electromagnetic radiation, the magnetic field is varied until absorption is observed. As the electronic magnetic moment is much greater than nuclear magnetic moments, the electromagnetic energy is higher, and the magnetic field lower than for n.m.r. Two regions of energy are particularly popular – The X band with frequencies of 9 GHz (λ = 3 cm) and fields of the order of 3300 gauss, and the Q band with frequencies of 35 GHz and fields around 13000 gauss. The difficulties arising from the use of microwave energies are compensated by a much greater sensitivity in comparison with n.m.r.

The energy levels corresponding to the various M_s values of the system are also split by the hyperfine interaction with the nuclear spins present in the system, and this leads to a splitting of the e.s.r. absorption lines (Fig. 8.2 b). Since the electrons may be quite close to the nucleus, the splittings can sometimes be very large – in these cases the spectra become quite difficult to interpret. We will now consider the g value and the hyperfine splitting in some detail.

The g value

As well as a magnetic moment due to the electronic spin, the electron may possess orbital angular momentum and, as discussed in Chap. 4, this will also give a magnetic moment which will couple with the spin magnetic moment. This change in total magnetic moment will produce a considerable change in the g value. In most systems studied by E.S.R. the g value is, however, fairly close to the free electron value of 2.0023 as the orbital angular momentum is 'quenched' by the chemical environment. The slight deviations from the free electron value arise from the spin-orbit interaction of the electron spin magnetic moment with the orbital magnetic moment induced by the magnetic field – the origins of the temperature independent paramagnetism discussed in Chap. 4. For systems of cubic or high symmetry, or where molecular motion is so fast that a time average symmetric environment is present, the g value is isotropic. For systems of lower symmetry the g value is anisotropic and the g value observed depends on the orientation of the applied magnetic field with respect to the molecular axes. Thus, in single crystals, the g values is often highly sensitive to the sample orientation; in powders or frozen solutions the spectrum observed is the superposition of the spectra of all the different possible orientations.

Hyperfine coupling

Hyperfine coupling is also affected by the symmetry of the environment. There are two principal mechanisms of interaction: the direct nuclear spin: electron spin dipole interaction (the anisotropic hyperfine coupling), and the isotropic contact coupling between the nuclear magnetic moment and unpaired spin density in s orbitals centred on the nucleus. In cubic or spherical symmetry, or in systems where rapid rotation or tumbling gives a time averaged high symmetry, the anisotropic coupling is zero. In rigid systems of low symmetry, the anisotropic coupling is large and strongly dependent on the orientation of the molecular axes with respect to the field. The spectra are now complicated, but careful analysis can give much useful information.

In rapidly tumbling or cubic systems, only the isotropic hyperfine coupling is observed. In certain systems it is possible to use simple relationships such as that due to McConnell, putting the spin density proportional to the isotropic coupling, and thereby 'mapping' the molecular orbital containing the unpaired electron by its hyperfine coupling with magnetic nuclei.

The example of the coupling of the unpaired d electron in the t_{2g} M.O. of $[IrCl_6]^{2-}$ with chlorine nuclear spin was mentioned on page 77.

Electric quadrupole coupling with the nucleus may also be observed (notably for copper (II) compounds) but is fortunately rather rare.

Species with more than one unpaired spin

These systems, including triplet radicals and some transition metal ions, are more complicated. The different M_s values are not degenerate in the absence of a magnetic field as a result of dipole-dipole interactions between the spins, or spin orbit coupling effects; these effects give rise to a zero field splitting as shown in Fig. 8.2 c for a triplet. The $M_s = \pm 1$ level is at higher energy than $M_s = 0$ and the transitions corresponding to $\Delta M_s = \pm 1$ occur at different energies. Under certain conditions the normally forbidden $M_s = -1 \rightarrow M_s = +1$ transition may be seen as well. The zero field splitting can be quite large, and may be greater than the microwave energy used, in which case it may be impossible to observe an E.S.R. spectrum. Thus for high spin Mn^{2+} ($S = {}^5/2$) the $M_s = \pm {}^3/2, \pm {}^5/2$ levels might lie at high energy and transitions would be observed between $M_s = \pm {}^1/2$, split by the hyperfine coupling with the ^{55}Mn nucleus ($I = {}^5/2$). The zero field splitting is anisotropic, and vanishes in cubic symmetry.

Relaxation effects are important in E.S.R. spectra in a similar way to N.M.R., and chemical exchange effects may be observed. For species with more than one unpaired electron spin-lattice relaxation times are often very fast and the absorption lines may be broadened to a point where their observation is impossible, although this difficulty may sometimes be overcome by cooling the sample. This is a very common problem with transition metal ions. Electron exchange effects (such as those causing anti-ferromagnetism) have a considerable effect on the spectrum, and it is necessary to study magnetically dilute systems – thus the $[IrCl_6]^{2-}$ ion was studied at low concentration in the diamagnetic salt $Na_2PtCl_6 \cdot 6H_2O$.

E.S.R. is a very sensitive technique, and with a good sample one may obtain a spectrum with only 10^{11} spins present (about 10^{-12} moles). This has encouraged the use of E.S.R. for studies of paramagnetic impurities or defects in solids, and enables the study of otherwise unstable radicals formed in low concentrations in irradiated solids – for example ClO_2 may be formed and trapped by X irradiation of $NaClO_3$, as may CO_2^- from irradiated sodium formate. The sensitivity of E.S.R. has made it popular in the study of biological systems, and in many studies an artificial 'spin label' has been introduced by attaching a paramagnetic functional group (usually a nitroxide, $\rangle$N–O) and studying its E.S.R. spectrum.

We may now see why E.S.R. spectra may vary so much in complexity. A species with a single unpaired electron, tumbling rapidly in magnetically dilute solution will show only an average g value, and any isotropic coupling present – this spectrum will be easy to interpret. By contrast, a species with several unpaired spins, trapped in an anisotropic environment will give a much more complex spectrum, which may only be seen under certain specific conditions. If the relaxation time allows observation of the signal, E.S.R. spectroscopy may be used to study gas, liquid or solid samples, and the spectra are unaffected by the presence of large excesses of diamagnetic materials such as solvents.

J. Mass Spectroscopy

Mass spectroscopy involves the vaporisation of a compound, its ionisation (generally by collision with a beam of electrons), and the analysis of the positive ions produced to yield mass to charge ratios. For unipositive ions the ionic mass is thus determined. The ionisation generally produces excited gaseous ions which may undergo elimination reactions giving a series of peaks in the mass spectrum corresponding to the ionic fragmentation products.

The principal application of mass spectroscopy in inorganic chemistry is as an analytical tool. Since, in most cases, an ion corresponding to the singly ionised molecule (or molecular ion) is observed, molecular weight measurement is straightforward. Analysis of the fragmentation pattern may show which groups are lost easily, and give some idea of the structure of the ions. Analysis of the energy necessary to produce ionisation of the molecule followed by the appearance of a given fragmentation product (the appearance potential) can, under certain conditions, give useful thermodynamic data.

Mass spectroscopy requires the sample to be in the vapour phase, but this is a less severe limitation than it might at first appear, since the vapour pressure necessary is only 10^{-5}–10^{-6} Torr. Many involatile solids have vapour pressures of this order at quite low temperatures (< 250 °C), although heating solid samples to produce a vapour may also induce pyrolysis, and this possibility should aways be considered. In general, however, gases, liquids, and many solids (in particular molecular solids) will give mass spectra. An attraction of the method as an analytical tool is its high sensitivity – less than a milligram of compound is necessary to give a spectrum, and used as an analytical method, very small quantities of an element may be detected.

K. Conclusions

None of the quantities measured by the methods discussed above are related explicitly to chemical quantities. In each case approximations or assumptions must be made to obtain chemical information, and, as the precision of each method is improved, the errors due to these approximations may become increasingly important. As an example let us consider the determination of bond length, the distance between the nuclei of two chemically bonded atoms.

Neutron diffraction will be closest to a direct measure, but involves certain practical problems, and assumes that there will be no magnetic scattering of the neutrons. X–ray diffraction is more convenient practically, but the scattering is now due to the electrons around the nuclei. The distribution of electrons around the nuclei will not be completely spherical if they are involved in chemical bonding, and in lighter elements the distortion of the valence shell may be quite important. Although the assumption that the scattering electrons are spherically distributed about the nucleus is generally satisfactory, the deviations due to chemical bonding may be detected experimentally.

Measurement of nuclear spin coupling constants can also lead to bond lengths, but the direct dipole-dipole coupling interaction is also dependent on orientation, and there are other contributions to the measured coupling constant.

If the compound studied is not a solid, it must be studied in the gas phase, using either rotational spectroscopy, or electron diffraction. Electron diffraction again presents the problem of scattering by the electrons rather than the nuclei. Another problem, ignored in early work, is the phase change in the scattered electrons due to anomalous dispersion; this is particularly important when heavy atoms are present (this effect can also give rise to errors in X–ray diffraction). Rotational spectroscopy will yield the three moments of inertia of a molecule; with sufficient data from isotopically labelled molecules, bond lengths and angles may be calculated for fairly small molecules.

If care is taken, satisfactory agreement is found between the results of different methods, but the definition of a bond length is itself an approximation: all molecules are non-rigid inasmuch as they possess vibrational modes which involve the movement of nuclei. As there is a zero-point vibrational energy, all nuclei in a molecule must continually be in motion, and we can only define a time average of the internuclear distance. The internuclear distance that is measured is dependent on the *time of measurement.* If the measurement is made in a time much shorter than the 10^{-13} seconds required for a molecular vibration, the nuclei will be seen as 'frozen' at a particular configuration during the vibration (just as the Franck-Condon principle requires the nuclei to be frozen during an electronic transition). Experimentally, we do not measure the impact of only one photon on one single molecule of the compound, but the results of the impact of many millions of photons on the many millions of molecules in a crystal, or in a gaseous sample. The internuclear distance measured corresponds to the envelope of all the internuclear distances instantaneously observed, and the internuclear distances appear to be 'smeared out' as a result of the vibrational motion. The consequences of this for the particular case of X–ray diffraction are discussed below.

If the time of measurement is slow compared with the nuclear motion, one no longer observes the envelope of the instantaneous positions, but a well defined distance corresponding to the time average of the internuclear distances. This effect has already been mentioned in connection with n.m.r. studies of fluxionality.

If a 'sigma-bonded' or monohapto C_5H_5 group (see Fig. 3.21) is undergoing 'ring-whizzing' (1,5 shifts) more rapidly than the n.m.r. observation time, only one sharp proton n.m.r. signal will be seen. If the 1,5 shift is slower than the observation time, signals corresponding to five inequivalent protons will be seen. In any chemical system where there is continual change (e.g. due to vibrational motion or dynamic chemical equilibrium), it is necessary to compare the time scale of our measuring technique with that of the changes undergone by the system. This problem was first treated by Muetterties in his work on fluxionality. He established a list of 'times of measurement' such as that given in Table 8.1. It will be noticed that diffraction methods, electron and electronic spectroscopy are all faster than molecular vibrations, and will therefore show the sum of different signals corresponding to all the nuclear distances found during the vibrational cycle. We will consider the effects of this on X–ray diffraction.

As a result of molecular vibration, a particular atom in a crystal will be 'seen' in several slightly different sites, depending on the position of the nuclei at the instant of scattering. This may be allowed for, but the calculation involves the introduction of new parameters in the treatment of the diffraction data. All modern crystal structures show ellipsoids of vibration for atoms in the structure, the principal axes of the

Table 8.1. Typical times of measurement

Method	Time of measurement (sec)
Electronic spectroscopy	10^{-14}
X-ray diffraction	10^{-18}
Neutron diffraction	10^{-18}
Electron diffraction	10^{-20}
Electron and X-ray spectroscopy	10^{-18}
Vibrational spectroscopy	10^{-13}
Microwave spectroscopy	10^{-9}
Nuclear magnetic resonance[a]	$10^{-1} - 10^{-9}$
Nuclear quadrupole resonance[a]	$10^{-1} - 10^{-8}$
Mössbauer spectroscopy	10^{-8}
Electron spin resonance[a]	$10^{-4} - 10^{-9}$

[a] Values very variable, according to the system under investigation.

ellipsoids corresponding to the root mean spare amplitude of vibration along these axes; the ellipsoids are generally quite anisotropic.

At room temperature, a typical root mean square amplitude would be 0.1 Å; since accuracies of 0.01 Å or better are frequently quoted for bond lengths, it can be seen that vibrational effects are important. If there is a probability of rotation in the crystal, quite considerable errors may be introduced: consider a bent metal nitrosyl with the angle M–N–O close to 120°: if there is a rapid rotation of the NO group about the M–N axis (Fig. 8.3), the oxygen atom will be seen at all positions on the circle; the mean N–O bond length is now false, and it is difficult to distinguish a rotating, bent NO group from a vibrating, linear NO group. Any attempt to refine the treatment of the molecular motion will be hampered by the swamping of scattering from the nitrogen and oxygen atoms by the heavy metal atom scattering.

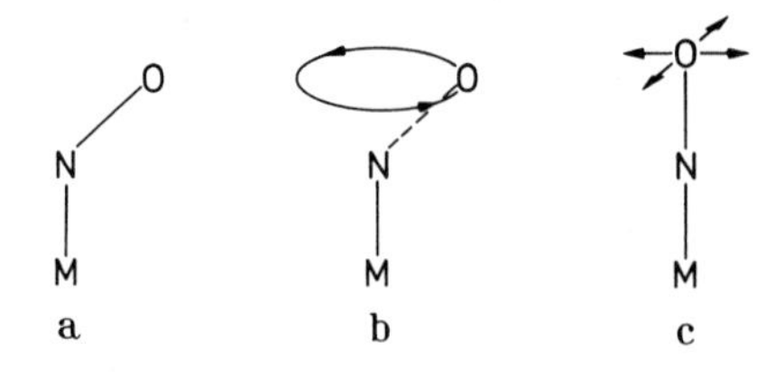

Fig. 8.3. **(a)** A bent M–N–O linkage, **(b)** rapid rotation causes the oxygen atom to appear symmetrically disposed about the M–N axis, (c) M–N–O bending of a linear M–N–O linkage gives the same average oxygen positions

This final section has been intended to show some of the problems which may be found in the use of the techniques described above, and to introduce the idea of the time of measurement. In capable hands, all techniques can yield much useful chemical information, and in many cases the results are quite unambiguous. The reader should however be aware of the possibilities of confusion, and should endeavour to avoid deducing too much from small differences in results obtained by unfamiliar techniques.

Bibliography

Atherton, N.M.: Electron spin resonance, theory and applications. Chichester: Ellis Horwood 1973

Atkins, P.W., Symons, M.C.R.: The structure of inorganic radicals. Amsterdam, New York: Elsevier 1967

Bacon, G.E.: Neutron diffraction in chemistry. London Butterworths 1977; Adv. Inorg. Chem. and Radiochem. *8,* 225 (1966)

Bancroft, G.M.: Mössbauer spectroscopy, an introduction. London, New York: McGraw-Hill 1973

Bancroft, G.M., Platt, R.H.: Introductory review of Mössbauer spectroscopy. Adv. Chem. and Radiochem. *15,* 59 (1972)

Baker, A.D., Brundle, C.R., Thompson, M.. Electron spectroscopy. Chem. Soc. Revs. *1,* 355 (1972)

Beattie, I.R.: Vibrational Spectroscopy. Chem. Soc. Revs. *4,* 107 (1975)

Brundle, C.R., Baker, A.D. (eds.): Electron spectroscopy: Theory, Techniques, and Applications, Vol. 1. London, New York, San Francisco: Academic Press 1977

Cockerill, A.F., Davies, G.L.O., Harden, R.C., Rackham, D.M.: Lanthanide shift reagents. Chem. Revs. *73,* 553 (1973)

Coppens, P.: Measurement of electron densities in solids by X–ray diffraction, in MTP Reviews of science, Physical Chemistry, Series 2, Vol. 11, J.M. Robertson (ed.). London, Boston: Butterworths 1975

Cotton, F.A.: Chemical applications of group theory, 2nd Ed. London, New York, Sydney: Wiley 1971

Drago, R.S.: Physical methods in inorganic chemistry. New York: Reinhold 1965

Farrar, T.C., Becker, E.D.: Pulse and Fourier transform NMR. London, New York, San Francisco: Academic Press 1971

Furlani, C., Canletti, C.: He(I) photoelectron spectra of d-metal compounds. Structure and Bonding *35,* 119 (1978)

Gans, P.: Vibrating Molecules. London: Chapman and Hall 1971

Gibb, T.C.: Principles of Mössbauer Spectroscopy. London: Chapman and Hall 1975

Gibb, T.C., Greenwood, N.N.: Mössbauer Spectroscopy. London: Chapman and Hall 1971. A more advanced text.

Glusker, J.P., Trueblood, K.N.: Crystal structure analysis. New York: Oxford University Press 1972

Goodman, B.A., Raynor, J.B.: ESR of transition metal complexes. Adv. Inorg. Chem. and Radiochem. *13,* 135 (1970)

Gordy, W., Cook, R.L.: Microwave molecular spectra. London, New York, Sydney: Wiley 1970

Harris, R.K.: N.m.r. and the periodic table. Chem. Soc. Revs. *5,* 1 (1976)

Hill, H.A.O., Day, P. (eds.): Physical methods in advanced inorganic chemistry. London, New York: Wiley 1968

Ibers, J.A.: Problem crystal structures, in Critical evaluation of chemical and physical structural information, page 186. Washington, D.C.: National Academy of Science 1974

Jaffe, H.H., Orchin, M.: Symmetry, orbitals, and spectra. London, New York, Sydney. Wiley 1971

Karnicky, J.F., Pings, C.J.: X–ray diffraction by liquids. Adv. in Chem. Phys. *34,* 157 (1976)

Kuchitsu, K.: Electron diffraction, in MTP Reviews of science, Physical chemistry, Series 1, Vol. 2, G. Allen (ed.). London, Boston: Butterworths 1972

Litzow, M.R., Spalding, T.R.: Mass spectrometry of inorganic and organometallic compounds. Amsterdam, New York: Elsevier 1973

Lloyd, D.R., DeKock, R.L.: Ultraviolet photoelectron spectroscopy. Adv. Inorg. Chem. and Radiochem. *16,* 66 (1974)

Long, D.A.: Raman spectroscopy. London: McGraw-Hill 1977

Lucken, E.A.C.: Nuclear quadrupole coupling constants. London, New York: Academic Press 1969

Lynden-Bell, R.M., Harris, R.K.: Nuclear magnetic resonance spectroscopy. London: Nelson 1969

Mayo, B.C.: Lanthanide shift reagents. Chem. Soc. Revs. *2,* 49 (1973)

Miller, J.M., Wilson, G.L.: Mass spectrometry in inorganic chemistry. Adv. Inorg. Chem. and Radiochem. *18,* 229 (1976)

Muetterties, E.L.: Timescales and non-rigidity. Inorg. Chem. *4,* 769 (1965), Acc. Chem. Res. *3,* 266 (1970)

Nakamoto, K.: Infrared spectra of inorganic and coordination compounds. 2nd Ed. London, New York, Sydney: J. Wiley 1970
Nieboer, E.: Biological uses of lanthanide shift reagents. Structure and Bonding *22,* 1 (1975)
Pople, J.A., Schneider, W.G., Bernstein, H.J.: High resolution nuclear magnetic resonance spectroscopy New York, Toronto, London: McGraw-Hill 1959
Sherwood, P.M.A.: Vibrational spectroscopy of solids. Cambridge: Cambridge University Press 1972
Smith, J.A.S.: Nuclear quadrupole resonance spectroscopy. J. Chem. Ed. *48,* 39, A77, A147, A243 (1971)
Spiro, T.G.: Resonance Raman spectroscopy. Acc. Chem. Res. *7,* 339 (1974)
Strommen, P.P., Nakamoto, K.: Resonance Raman spectroscopy. J. Chem. Ed. *54,* 474 (1977)
Sugden, T.M., Kenney, C.N.: Microwave spectroscopy of gases. London: Van Nostrand 1965
Turner, D.W.: Molecular photoelectron spectroscopy. London, New York: Academic Press 1970
Urch, D.S.: X–ray emission spectroscopy. Quart. Revs. *25,* 343 (1971)
Weissberger, A., Rossiter, B.W. (eds.): Techniques of chemistry, Vol. I, Physical methods of chemistry. London, New York: Wiley 1972. A multi-volume encyclopaedia of physical methods.
Wertz, J.E., Bolton, J.R.: Electron spin resonance. London, New York: McGraw-Hill 1972
Wheatley, P.J.: The determination of molecular structure. London: Oxford University Press 1968

Problems

1. The resonance Raman spectrum of $[MoO_3S]^{2-}$ shows a spectacular enhancement of the Mo–S vibration. Why?

2. In strong solutions of acid or base, the proton n.m.r. signal is shifted, but only one signal is observed. Why?

3. Discuss why U.V. and visible spectral bands are wide (a few percent of the energy of absorption) whilst n.m.r. and n.q.r. absorptions are very narrow.

4. Free radicals with the unpaired electron localised in an s type atomic orbital may often be distinguished easily from those with the electron localised in a p type orbital. Which measurement will permit this distinction?

5. Two X–ray crystal structure determinations of the structure of $[Ir(diphos)_2O_2]PF_6$ (Sec. 6.C.a) give rather different values for the O–O distance. What are the the problems that might be encountered in the solution of this structure?

6. What credence would you give to the localisation of the following vibrations:
   (i) S–O in $(CH_3)_2SO$; SOF_2 ; SO_4^{2-}
   (ii) M–Cl in Ph_3PAuCl ; $(C_5H_5)Fe(CO)_2Cl$; Al_2Cl_6 ?

7. Suggest possible methods for studying the following problems:
   (i) The molecular structure of a compound which polymerises on solidification
   (ii) Possible complex formation by phosphine ligands in solution
   (iii) Possible complex formation by nitrate ion in aqueous solution
   (iv) Surface corrosion of a metal
   (v) The distribution of iron in the various possible cation sites of a silicate mineral
   (vi) The structure of $U(BH_4)_4$
   (vii) Mixed valence behaviour in $(NIr_3(SO_4)_6(H_2O)_3)^{4-}$ and $(OIr_3(SO_4)_9)^{10-}$.

Appendix I

Tables for Group Theory

Our most frequent use of group theory is to investigate possible overlap and mixing of orbitals, and the possibilites of electronic transitions between orbitals. For this we need to know the I.R.s for which the orbitals under investigation are basis functions: we may either look this up in tables or determine the I.R.s by references to *character tables*. This appendix contains tables for the most common symmetries encountered in this book, but we will first give a simple illustration of the character table method.

Fig. AI.1. The two-fold axis of rotational symmetry is the z axis. The mirror planes are the xz and yz planes

Consider the two ligand σ orbitals in Fig. AI.1: the molecule ML_2 is bent and has symmetry C_{2v}. The four symmetry operations of the symmetry group are the identity operation E, a rotation of 180° about the z axis (C_2), and reflections in the xz and yz planes (σ_v(xz) and σ_v'(yz)). We may represent the effects of these operations on the orbitals ϕ_1 and ϕ_2 by matrices as described in Chap. 1:

Operation	Effect	Matrix	Trace
E	None	$\begin{pmatrix} 1 & 0 \\ 0 & 1 \end{pmatrix}$	2
C_2	$\phi_2 \rightarrow \phi_1, \phi_1 \rightarrow \phi_2$	$\begin{pmatrix} 0 & 1 \\ 1 & 0 \end{pmatrix}$	0
σ_v(xz)	$\phi_2 \rightarrow \phi_1, \phi_1 \rightarrow \phi_2$	$\begin{pmatrix} 0 & 1 \\ 1 & 0 \end{pmatrix}$	0
σ_v'(yz)	None	$\begin{pmatrix} 1 & 0 \\ 0 & 1 \end{pmatrix}$	2

It is found that the *trace* of the matrices representing the symmetry operations is sufficient to differentiate the I.R.s. The trace of a matrix is the sum of the elements on the leading diagonal (northwest-southeast) of the matrix; in group theoretical usage, the trace is often referred to as the *character* of the symmetry operation for a given representation. The I.R.s of the group are displayed in *character tables:*

I.R.	E	C_2	$\sigma_v(xz)$	$\sigma_v'(yz)$	
A_1	1	1	1	1	z, x^2, y^2, z^2
A_2	1	1	–1	–1	xy
B_1	1	–1	1	–1	x, xz
B_2	1	–1	–1	1	y, yz

The character for each symmetry operation is given for each I.R., and, at the extreme right of the table, functions which are basis functions for that I.R. are listed as well: thus, x and xz transform as the I.R. B_1. If the characters for a given I.R. are represented as vectors (e.g. (1 1 –1 –1) for A_2), it will be seen that the vectors corresponding to different I.R.s are orthogonal.

The representation for which ϕ_1 and ϕ_2 are basis functions may now be seen to have the characters:

E	C_2	$\sigma_v(xz)$	$\sigma_v'(yz)$
2	0	0	2

As there are no two-dimensional I.R.s for C_{2v} symmetry, this representation must be reducible: inspection shows that it is given by $A_1 + B_2$. The two basis functions for the I.R.s are $(\phi_1 + \phi_2)$ and $(\phi_1 - \phi_2)$ respectively.

If we wish to investigate overlap of the ligand σ orbitals with the orbitals of M, we may use the right hand side of the character table: the s orbital of M is totally symmetric and will transform as A_1; the p_z orbital will transform as the function z that is as A_1; the p_x and p_y will transform as $B_1(x)$ and $B_2(y)$ respectively. As x^2, y^2, and z^2 all transform as A_1, the orbitals $d_{x^2-y^2}$ and d_{z^2} will also transform as A_1.

In the following tables the symmetry properties of the more important ligand and central atom orbitals are indicated; where there are different ligand sites (e.g. for trigonal bipyramidal complexes) these are distinguished. σ and π distinctions are made with respect to the M–L bond axis.

Symmetry C_{2v}

A two -fold axis of symmetry, with two perpendicular mirror planes (see also above)
Example: bent AB_2, H_2O

Central atom

s		a_1
p	x	b_1
	y	b_2
	z	a_1
d	x^2-y^2, z^2	a_1
	xy	a_2
	xz	b_1
	yz	b_2

Ligands

σ	$a_1 + b_2$	
π	$a_1 + b_2$	(in yz plane)
π	$a_2 + b_1$	(in xy plane)

Symmetry C_{3v}

One three-fold axis of symmetry, with three planes of symmetry parallel to the axis
Examples: pyramidal ML_3, NH_3

Central atom

s		a_1
p	x, y	e
	z	a_1
d	z^2	a_1
	xz, yz	e
	x^2-y^2, xy	e

Ligands

σ	$a_1 + e$	
π	$a_2 + e$	(in xy plane)
π	$a_1 + e$	(out of xy plane)

Symmetry D_{3h}

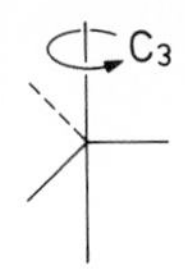

C_{3v} with a horizontal plane of symmetry added.
Examples: planar ML_3, BF_3, trigonal bipyramidal ML_5, PF_5

Central atom				*Ligands*	
s		a_1'		equatorial	
p	x, y	e'	σ	$a_1' + e'$	
			π	$a_2'' + e''$	(out of plane)
	z	a_2''	π	$a_2' + e'$	(in plane)
d	z^2	a_1'		axial	
	xz, yz	e''	σ	$a_1' + a_2''$	
	x^2-y^2, z^2	e'	π	$e' + e''$	

Symmetry D_{4h}

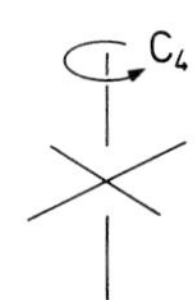

One four-fold axis of symmetry, with one mirror plane perpendicular and two parallel to the axis.
Examples: square planar ML_4; *trans*-octahedral ML_4L_2'

Central atom				*Ligands*	
s		a_{1g}		equatorial	
p	z	a_{2u}	σ	$a_{1g} + b_{1g} + e_u$	
			π	$a_{2u} + b_{2u} + e_g$	(out of plane)
	x, y	e_u	π	$a_{2g} + b_{2g} + e_u$	(in plane)
d	z^2	a_{1g}		axial	
	x^2-y^2	b_{1g}	σ	$a_{1g} + a_{2u}$	
	xy	b_{2g}	π	$e_g + e_u$	
	xz, yz	e_g			

Symmetry $D_{\infty h}$

—L—M—L—

C_∞

Linear, centrosymmetric
(one infinite axis of rotation)
Examples: MgF_2; CO_2

Central atom			*Ligands*	
s		σ_g	σ	$\sigma_u + \sigma_g$
p	z	σ_u	π	$\pi_g + \pi_u$
	x, y	π_u		
d	z^2	σ_g		
	xz, yz	π_g		
	xy, x^2-y^2	δ_g		

Symmetry T_d

Tetrahedral symmetry
Examples: ML_4, CH_4

Central atom			*Ligands*	
s		a_1	σ	$a_1 + t_2$
p		t_2	π	$e + t_1 + t_2$
d	x^2-y^2, z^2	e		
	xy, xz, yz	t_2		

Symmetry O_h

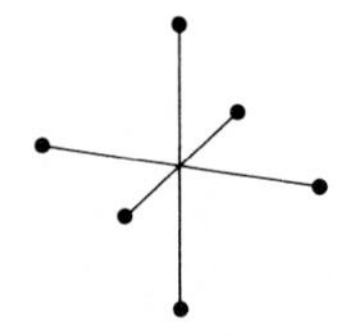

Octahedral symmetry
Examples: ML_6, SF_6

Central atom			*Ligands*	
s		a_{1g}	σ	$a_{1g} + e_g + t_{1u}$
p		t_{1u}	π	$t_{1g} + t_{1u} + t_{2g} + t_{2u}$
d	x^2-y^2, z^2	e_g		
	xy, xz, yz	t_{2g}		
f		$a_{1u} + t_{1u} + t_{2u}$		

Where there is only one element of symmetry, it is generally possible to classify orbitals by inspection. Thus:

Symmetry C_s	(one plane of symmetry)	
	Symmetric	A′
	Antisymmetric	A″
Symmetry C_2	(one two-fold axis)	
	Symmetric	A
	Antisymmetric	B
Symmetry C_i	(centre of symmetry)	
	Symmetric	A_g
	Antisymmetric	A_u

The dipole moment operator $\hat{\mu}$ always transform as the coordinates x, y, z, (i.e. the same way as the central atom p orbitals).

Direct Products

Only a few useful general rules and specific examples are given. For a comprehensive set of tables, see Tables for Group Theory, P.W. Atkins, M.S. Child, and C.S.G. Phillips, Oxford University Press, 1970.

In general:

symmetric · symmetric = antisymmetric · antisymmetric = symmetric
symmetric · antisymmetric = antisymmetric · symmetric = antisymmetric

Thus: $g \oplus g = u \oplus u = g$

$g \oplus u = u \oplus g = u$

and $A \oplus A = B \oplus B = A$

$A \oplus B = B \oplus A = B$

and similarly for symmetry with respect to a plane (′or″)

Particular examples:

For C_{3v} and D_{3h} $A_1 \oplus A_2 = A_2$; $A_2 \oplus E = E$; $E \oplus E = A_1 + A_2 + E$

For D_{4h} $B_1 \oplus A_2 = B_2$; $B_1 \oplus B_2 = A_2$; $E \oplus E = A_1 + A_2 + B_1 + B_2$;

$E \oplus A_2 = E \oplus B_1 = E \oplus B_2 = E$

For O_h and T_d $A_2 \oplus E = E$; $A_2 \oplus T_1 = T_2$; $A_2 \oplus T_2 = T_1$;

$E \oplus E = A_1 + A_2 + E$; $E \oplus T_1 = E \oplus T_2 = T_1 + T_2$

$T_1 \oplus T_2 = A_2 + E + T_1 + T_2$; $T_2 \oplus T_2 = T_1 \oplus T_1 = A_1 + E + T_1 + T_2$

Subscripts and superscripts may be added according to the rules given above:

for D_{3h}, $A_2' \oplus E'' = E''$; for O_h, $E_u \oplus T_{1u} = T_{1g} + T_{2g}$.

Problems

1. The symmetry operations of the group C_{3v} are the identity operator E, two rotations of 120° ($2C_3$) and three reflections in planes parallel to the axis of rotation (cf. C_{2v}). Taking the three ligand σ orbitals of a pyramidal ML_3 molecule as basis functions, write down the matrices corresponding to the identity, a rotation, and a reflection. Determine the character of each operation for this representation, and, using the character table below, identify the constituent I.R.s

C_{3v}	E	$2C_3$	$3\sigma_v$
A_1	1	1	1
A_2	1	1	–1
E	2	–1	0

(Note that although the matrices for the two rotations are different, their characters are identical. A similar result is true for the reflection).

2. Is overlap between ligand σ orbitals and central atom d_{xz} or d_{yz} orbitals possible for a C_{2v} ML_2 molecule?

3. Draw M.O. diagrams for (i) PF_5 and (ii) $Fe(CO)_5$

4. Calculate the direct products $\Gamma_x \oplus \Gamma_y$, $\Gamma_x \oplus \Gamma_z$, and $\Gamma_y \oplus \Gamma_z$ for C_{2v} symmetry. Predict the I.R.s for which xy, xz, and yz are basis functions, and compare your result with the table.

5. Ligand to metal charge transfer is often thought to involve excitation from occupied ligand π orbitals ($t_{1g} + t_{1u} + t_{2g} + t_{2u}$ for an octahedral complex) to empty metal d orbitals ($e_g + t_{2g}$). For which ligand π orbitals is the transfer allowed by the electric dipole mechanism?

6. What symmetry group would you use to discuss the structures of:
$[ICl_4]^-$; $[(Ph_3P)_2NiCl_2]$; $[Co(NH_3)_5Br]^{2+}$; $CO_3{}^{2-}$; $SO_3{}^{2-}$; $NO_2{}^-$

Appendix II

Abbreviations, Symbols, and Energy Units

The wide variety of abbreviations and units currently used by practising chemists is not such as to inspire great confidence in the universality of scientific method. I have tried to follow the IUPAC conventions (e.g. by replacing the kilocalorie by the kiloJoule), but have preferred the Ångstrom to the picometre for bond lengths.

Abbreviations

acac	acetylacetonate anion $(CH_3COCHCOCH_3)^-$
A.O.	atomic orbital
b.c.c.	body centred cubic
bipy	2,2′-bipyridyl
c.c.p.	cubic close packing
C.I.	configuration interaction
C.N.	coordination number
dien	diethylenetriamine (Table 5.3)
diphos	1,2 bis(diphenyphosphino)ethane $(Ph_2PCH_2CH_2PPh_2)$
E.A.	electron affinity
EDTA	ethylenediaminetetra-acetate anion (Fig. 6.21)
en	ethylene diamine (Fig. 6.21)
ESR, EPR	electron spin or electron paramagnetic resonance
Et	ethyl
f.c.c.	face centred cubic
$\mathscr{H}$	a Hamiltonian operator
h.c.p.	hexagonal close packing
I.E.	ionisation energy
I.R.	irreducible representation
i.r.	infra-red
J, K	electron repulsion integrals (Sec. I.C.a)
J	total electronic angular momentum quantum number
L	total electronic orbital angular momentum quantum number
L	any ligand or simple Lewis base
LCAO	linear combination of atomic orbitals

M	any metal or Lewis acid
M	quantum number giving the total electronic orbital angular momentum along the z axis
M_s	quantum number giving the total electronic spin angular momentum along the z axis
Me	methyl
M.O.	molecular orbital
n.m.r.	nuclear magnetic resonance
n.q.r.	nuclear quadrupole resonance
Ph	phenyl
phen	1.10 phenanthroline
py	pyridine
R	an aliphatic radical (e.g. C_3H_7)
r	spherical polar coordinate
S	total electronic spin angular momentum quantum number
S_{ij}	overlap integral between two orbitals i and j
U(MX)	lattice energy of the ionic compound MX
U.P.S.	ultraviolet photoelectron spectroscopy
U.V.	ultraviolet
V.B.	valence bond
VSEPR	valence shell electron pair repulsion
Xα	Slater Xα approximation (Sect. 2.A)
X.P.S.	X–ray photoelectron spectroscopy
Z	atomic number
α	Coulomb integral $\mathscr{H}_{ii}$ of an orbital i (Sect. 2.A)
β	resonance intergral $\mathscr{H}_{ij}$ between two orbitals i and j (see also Sect. 2.A)
Γ_j	the I.R. corresponding to the function or functions j
ΔH^o_{diss}	standard enthalpy of dissociation
ΔH^o_f	standard enthalpy of formation
ϵ	molar extinction coefficient
ζ	one electron spin-orbit coupling constant
η	see *hapto* nomenclature, page 92
θ, ϕ	spherical polar coordinates
λ	wavelength *or* many electron spin-orbit coupling constant ($\zeta/2S$)
$\hat{\mu}$	electric dipole moment operator
μ	denotes a bridging ligand (page 92)
ν	wavenumber or frequency
τ	spatial coordinates
χ	magnetic susceptibility
$\hbar$	Planck's constant/2π
$\Gamma_i \oplus \Gamma_j$	direct product of Γ_i and Γ_j
$\langle a \mid \mathscr{O} \mid b \rangle$	bracket notation (see page 5)

Scripts

Operators	are denoted by	script letters
Quantum numbers	"	italics
Vectors or tensors	"	bold face type
Electronic states	"	capital letters
One electron orbitals	"	lower case letters
Oxidation states	"	Roman numerals

Energy conversion table

		eV	kJ	cm^{-1}	Hz
1 eV	=	1	96.48	8066	2.42×10^{14}
1 kJ	=	1.04×10^{-2}	1	83.59	2.51×10^{12}
1 cm^{-1}	=	1.24×10^{-4}	1.20×10^{-2}	1	2.998×10^{10}
1 Hz	=	4.14×10^{-15}	3.99×10^{-13}	3.34×10^{-11}	1

1 calorie = 4.184 J

1 Å = 10^{-10} m = 100 pm = 0.1 nm

h = 6.626×10^{-34} Js

N = 6.022×10^{23} (Avogadro's number)

k = 1.381×10^{-23} J K^{-1} (Boltzmann's constant)

c = 2.998×10^{8} m s^{-1}

Subject Index

Inorganic Chemistry Concepts

Editors:
M. Becke, C.K. Jørgensen, M.F. Lappert, S.J. Lippard, J.L. Margrave, K. Niedenzu, R.W. Parry, H. Yamatera

The ensuing advances in inorganic chemistry will be the subject of this new series. The focus will be on fields of endeavor which as yet remain open and are therefore topical. The series is edited by an international panel of experts.

Volume 1
R. Reisfeld, C.K. Jørgensen

Lasers and Excited States of Rare Earths

1977. 9 figures, 26 tables. VIII, 226 pages
ISBN 3-540-08324-3

Contents:
Analogies and Differences Between Monatomic Entities and Condensed Matter. – Rare-Earth Lasers. – Chemical Bonding and Lanthanide Spectra. – Energy Transfer. – Applications and Suggestions.

Volume 2
R.L. Carlin, A.J. van Duyneveldt

Magnetic Properties of Transition Metal Compounds

1977. 149 figures, 7 tables. XV, 264 pages
ISBN 3-540-08584-X

Contents:
Paramagnetism: The Curie Law. – Thermodynamics and Relaxation. – Paramagnetism: Zero-Field Splittings. – Dimers and Clusters. – Long-Range Order. – Short-Range Order. – Special Topics: Spin-Flop, Metamagnetism, Ferrimagnetism and Canting. – Selected Examples.

Volume 3
P. Gütlich, R. Link, A. Trautwein

Mössbauer Spectroscopy and Transition Metal Chemistry

1978. 160 figures, 19 tables, 1 folding plate. X, 280 pages
ISBN 3-540-08671-4

Contents:
Basic Physical Concepts. – Hyperfine Interactions. – Experimental Mathematical Evaluation of Mössbauer Spectra. – Interpretation of Mössbauer Parameters of Iron Compounds. – Mössbauer-Active Transition Metals Other Than Iron. – Some Special Applications.

Volume 4
Y. Saito

Inorganic Molecular Dissymmetry

1979. 107 figures, 28 tables. Approx. 180 pages
ISBN 3-540-09176-9

Contents:
Introduction. – X-Ray Diffraction. – Conformational Analysis. – Structure and Isomerism of Optically Active Complexes. – Electron-Density Distribution in Transition Metal Complexes. – Circular Dichroism.

Lecture Notes in Chemistry

Publication of Lecture Notes is intended as a service to the international chemical community, in that a commercial publisher, Springer-Verlag, can offer a wider distribution to documents which would otherwise have a restricted readership. Once published and copyrighted, they can be documented in the scientific literature.

Volume 1
G. H. Wagnière

Introduction to Elementary Molecular Orbital Theory and to Semiempirical Methods

1976. 33 figures. V, 109 pages
ISBN 3-540-07865-7

Volume 2
E. Clementi

Determination of Liquid Water Structure
Coordination Numbers for Ions and Solvation for Biological Molecules

1976. 32 figures, 18 tables. VI, 107 pages
ISBN 3-540-07870-3

Volume 3
S. R. Niketic, K. Rasmussen

The Consistent Force Field

A Documentation

1977. 24 figures, 26 tables. IX, 212 pages
ISBN 3-540-08344-8

Volume 4
A. Graovac, I. Gutman, N. Trinajstić

Topological Approach to the Chemistry of Conjugated Molecules

1977, 34 figures, 2 tables. IX, 123 pages
ISBN 3-540-08431-2

Volume 5
R. Carbó, J. M. Riera

A General SCF Theory

1978. 25 figures, 19 tables. XII, 210 pages
ISBN 3-540-08535-1

Volume 6
I. Hargittai

Sulphone Molecular Structures

Conformation and Geometry from Electron Diffraction and Microwave Spectroscopy; Structural Variations
1978. 40 figures. VII, 175 pages
ISBN 3-540-08654-4

Volume 7

Ion Cyclotron Resonance Spectrometry

Editors: H. Hartmann, K.-P. Wanczek
1978. 66 figures, 32 tables. V, 326 pages
ISBN 3-540-08760-5

Volume 8
E. E. Nikitin, L. Zülicke

Selected Topics of the Theory of Chemical Elementary Processes

1978. 6 tables, 41 figures.
IX, 175 pages
ISBN 3-540-08768-0

Volume 9
A. Julg

Crystals as Giant Molecules

1978. 8 figures, 21 tables. VII, 135 pages
ISBN 3-540-08946-2